THE
TRILLION-DOLLAR
CONSPIRACY

ALSO BY JIM MARRS

The Rise of the Fourth Reich: The Secret Societies That Threaten to Take Over America

The Terror Conspiracy: Deception, 9/11, and the Loss of Liberty

Inside Job: Unmasking the 9/11 Conspiracies

Rule by Secrecy: The Hidden History That Connects the Trilateral Commission, the Freemasons, and the Great Pyramids

Alien Agenda: Investigating the Extraterrestrial Presence Among Us

Crossfire: The Plot That Killed Kennedy

THE TRILLION-DOLLAR CONSPIRACY

HOW THE NEW WORLD ORDER,

MAN-MADE DISEASES,

AND ZOMBIE BANKS

ARE DESTROYING AMERICA

JIM MARRS

wm

WILLIAM MORROW

An Imprint of HarperCollins*Publishers*

HarperCollins books may be purchased for educational, business, or sales promotional use. For information please write: Special Markets Department, HarperCollins Publishers, 10 East 53rd Street, New York, NY 10022.

FIRST EDITION

Designed by Joy O'Meara

Library of Congress Cataloging-in-Publication Data

Marrs, Jim.
　　The trillion-dollar conspiracy : how the new world order, man-made diseases, and zombie banks are destroying America / Jim Marrs. — 1st ed.
　　　　p. cm.
　　Includes index.
　　ISBN: 978-0-06-197068-9
　　1. United States—Economic policy—2009- 2. United States—Economic conditions—2009- 3. Finance—Government policy—United States. 4. Banks and banking—United States. I. Title.
　　HC106.84.M37 2010
　　330.973—dc22

2010007358

---

10 11 12 13 14　ov/RRD　10 9 8 7 6 5 4 3 2 1

ACKNOWLEDGMENTS

The author would like to gratefully acknowledge the many persons who provided encouragement and support in the writing of this book. In particular, sincere thanks go to Maritha Gan, Thomas Ruffner, and Nick Redfern, as well as to the conscientious editing of Danny Goldstein and Henry Ferris of HarperCollins Publishers, and, of course, my forbearing wife, Carol. Much honor and gratitude also goes to all those in the world who have devoted so much time and energy in the pursuit of truth.

During times of universal deceit, telling the truth becomes a revolutionary act.
—GEORGE ORWELL

CONTENTS

PART IV—HOW TO FREE ZOMBIES: THE THREE BOXES OF FREEDOM

THE SOAP BOX

THE BALLOT BOX

THE AMMO BOX

INTRODUCTION

Single acts of tyranny may be ascribed to the accidental opinion of the day; but a series of oppressions, begun at a distinguished period, and pursued unalterably through every change of ministers too plainly proves a deliberate, systematic plan of reducing us to slavery.

—THOMAS JEFFERSON

TODAY JEFFERSON'S WORDS MIGHT read, "An occasional act of tyranny may be excused as a momentary lapse of judgment by officials, but a continuous series of such acts pursued through both Democratic and Republican administrations clearly proves there is a deliberate and systematic plan to reduce once-free Americans to slavery."

To be a zombie is to exist under the most onerous bonds of slavery—bonds that allow for no thought to one's action. Zombies are controlled both mentally and physically by some outside force, whether through a virus causing them to seek blood, or voodoo magic. A zombie is neither dead nor alive and usually under the control of someone else, as in the old Hollywood films. Zombies stumble about, largely unaware of the world around them, intent on purposes that others have created for them through alchemy or electromagnetism. In old horror movies, actors Bela Lugosi and John Carradine controlled zombies, causing them to commit acts that ran against human nature.

And now zombies are everywhere. They're highly popular grists for movies, books, comics, and computer games. Is it just a coincidence that zombies are so popular, or is it possible that Americans like zombies because they offer us a reflection of how we perceive ourselves—as barely alive and living mindlessly?

Many individuals today stumble through their daily chores without caring or knowing why they do so. These people might be numbed by drugs or the incessant bombardment by the broadcast media, but whatever the case, many Americans seem like zombies in so many ways.

The term "zombie" is being applied to more and more aspects of life in modern America. Adderall is one of the most popular of the faddish new psychiatric drugs to combat attention-deficit/hyperactivity disorder (ADHD). This drug consists of equal amounts of the stimulants amphetamine and dextroamphetamine. In fact, ADHDTreatment.org describes a side effect of Adderall as "zombie" demeanor.

The word "zombie" has so pervaded our society that it has worked its way into the scientific community's lexicon. In mid-May 2009, researchers at the University of Texas and Texas A&M's AgriLife Extension Service in east Texas reported that they had found a way to control the state's fire ant infestation. They discovered that the tiny phorid fly, a native of the South American region where the fire ants in Texas originated, could "dive-bomb" the ants and lay eggs on them. The eggs would hatch inside the ants and eat away their brains, turning them into what scientists called "zombie ants." The ants would wander aimlessly for about two weeks until their heads fell off.

Not only does the word describe how we view our own existence, but it has—and can still be—applied to the dissolution of the pillars of our society, most notably our banks. During the recent financial fiasco, the banks whose liabilities exceeded their assets were called "zombie banks." As author Bill Sardi, a regular contributor to LewRockwell.com, explained, "Zombie banks are defined as a financial institution with an economic net worth that is less than zero, but which continues to operate because its ability to repay its debts is shored up by implicit or explicit government credit support." In a sense, they were dead but still going through the motions of life.

America is now confronted with an economic situation that is being compared to the Great Depression, and the only solutions seem to lie in aggregating debt, deflating the value of the dollar, and moving riches around. Not only that, but the scale of our economic problems has vastly increased. On October 1, 2008, the national debt was $10 trillion, but during 2009 it climbed to nearly $12 trillion, the single largest increase in a year. If every American man, woman, and child were to liquidate every asset he or she owns, the total could not equal this debt.

The term "trillion" is bandied about lightly by the mass media. Yet most people cannot truly conceive of the significance of such a number. A trillion square miles would encompass 3.7 million states the size of Texas (which covers approximately 270,000 square miles). A trillion dollars— on the other hand—could be made of one-dollar bills stretching all the way to our sun and back. If banking institutions that operate at a deficit are called "zombie banks," then couldn't we call a country whose debts exceed its assets a zombie nation? And perhaps this epithet also could be applied to its citizens?

In 2008, many saw the nation turn from National Socialism to Marxist Socialism when the totalitarian Bush administration turned over power to the Obama administration—an administration that favors socialist Medicare policies and redistributions of wealth. America's zombies now face a further loss of individual freedoms due to corrupt politics, corporate malfeasance, and legislation that continues to curtail individual freedom. As readers of *Rule by Secrecy* and *The Rise of the Fourth Reich* will understand, the global financiers—the global plutocrats of Wall Street, London, and Switzerland—have manipulated Western history for at least the past century, first by creating the Federal Reserve in America by deceitful political machinations, then communism in Russia by funding the Bolsheviks rather than the White Russians, and followed by financing National Socialism (Nazis) in Germany. Now these global financiers have taken control of the United States and are changing it in such ways that we now live in a society unimaginable to citizens of just two decades ago.

For instance, in the 1950s, Ronald Reagan publicly warned against socialized medicine. Today, the argument is just how socialized it will be. The information highway is a traffic jam of hype, misinformation, disinformation, distractions, and propaganda. New attempts to transform technology, health care, education, and the political system are being reported every day.

More and more Americans have been forced to focus on the dark side of contemporary life. Now, we live under the tyranny of a New World Order—a world that has been reordered by a small group of wealthy financiers and industrialists centered within secret societies such as the Council on Foreign Relations, the Trilateral Commission, and the Bilderberg group. Also, one must consider how this new world has become more and more a surveillance society and police state existing under a financially unstable infrastructure and fed by corporations that hold monopolies on food, water, and drugs. Are we living under a fascistic government? Possibly. The *American Heritage Dictionary of the English Language* defines fascism as "a philosophy or system of government that advocates or exercises a dictatorship of the extreme right, typically through the merging of state and business leadership together with an ideology of belligerent nationalism." Today, of course, the dictatorship would be of the extreme left, but nevertheless would include the power of both the government and the corporations that have their hands in public affairs.

Many people today believe the United States is going to hell in a handbasket—that the United States is no longer a vibrant republic based on constitutional law but rather a brain-dead and decaying empire being taken over by an entrenched financial elite who seek a worldwide socialist order to dominate. This belief grows among Americans as they read the daily headlines and listen to the electronic mass media. People are seeking true change, not mere political rhetoric, but they feel befuddled as they can't understand who precisely has hijacked their country.

Some Americans are acting out. In "tea parties" and in a massive demonstration in Washington on September 12, 2009, tens of thousands of Americans displayed their dissatisfaction with where the nation is going.

What does one call a country that seems to be merely mimicking the robust republic it used to be, whose population has been dumbed down by controversial educational programs, drugged out by an ever-growing pharmaceutical industry, and frightened into submission by constant threats of terrorism and economic collapse?

Would this not be a zombie nation? A nation that goes through the motions in commerce, politics, health, and education but without a spark of life, verve, or enthusiasm?

This is the true horror story.

What happened? How did the nation get this way? Was it simply a case of inattention by the electorate, the natural evolution of a society grown prosperous and complacent, overreaching greed and lust for profit by corporate leaders? Or could it have been a conspiracy?

PART I

A ZOMBIE NATION

I see in the near future a crisis approaching that unnerves me and causes me to tremble for the safety of my country. As a result of the war, corporations have been enthroned and an era of corruption in high places will follow, and the money power of the country will endeavor to prolong its reign by working upon the prejudices of the people until all wealth is aggregated in a few hands, and the Republic is destroyed.

—ATTRIBUTED TO ABRAHAM LINCOLN

ECONOMIC DECLINE

TIMES ARE TOUGH FOR AMERICA.

Thanks to what Treasury Secretary Timothy Geithner called the failure of America's financial system, by the start of 2010 more than $5 trillion of household wealth had evaporated. About one in every eight mortgages was in default or foreclosure. It is predicted that there will be ten million foreclosures on homes through 2012. One in every eight adults and one in four children now subsist on government food stamps.

All of these problems were exacerbated by high rates of unemployment. According to an Associated Press report, one in every five Americans is unemployed or underemployed, with the number expected to rise in 2010, causing the second-highest unemployment figure since World War II.

Dissension and dissatisfaction are widespread, and they're linked to the poor economy. If the economy were the hands of a zombie, those hands would be bound by debt.

Charles K. Rowley is a professor of economics at George Mason University and general director of the Locke Institute in Fairfax, Virginia. He is widely considered to be a major voice in political and economic thought. In an article for the United Kingdom's *Daily Telegraph*, Rowley wrote: "The US economy suffers from a growing culture of indebtedness that has increasingly contaminated the federal government since 2001 and has spilled over dramatically into private household behavior." He also raised a popular question, asking, "If excessive government indebtedness is a major source of the problem, why increase the government debt? Why encourage households to go yet further into debt?" Ominously, Rowley predicted "it is not impossible that the US will experience the kind of economic collapse from first- to third-world status experienced by Argentina under the national socialist governance of Juan Peron." In other words, if the U.S. government cannot find ways of living within its means, as most families are forced to do, the nation may fall into third-world status, complete with scarcities of food and water, consumer goods, and socialized government control.

One of the barely noticed aspects of the financial crisis is the substantial drop in tax revenues, even as the Obama administration and Congress spend more to stimulate the economy. According to CNN, through the end of August 2009, the federal government collected 25 percent less tax revenue than for the same eight-month period in 2008. The Congressional Budget Office predicted tax receipts would fall to 14 percent of the gross domestic product, a sharp decline from the historical average of 18.3 percent. Additionally, individual income tax revenues fell 20 percent while corporate income taxes dropped a whopping 56 percent. Predictions for 2010 were not much better.

And the loss of governmental revenue has filtered down to local governments. Increasing unemployment has caused thirty-two state unemployment insurance trust funds to fall below the recommended federal level, indicating these states will require massive federal loans to continue assistance for the jobless. Officials in Vigo County, Indiana, announced in mid-2009 that they could no longer afford to bury a dead person if that dead person had no savings, insurance, or family money set aside for a funeral. In Atlanta, citizens' groups have tried to stop city plans to demolish its remaining public housing units. More than twenty counties in Michigan have reverted paved roads to gravel in an effort to save money, according to the County Road Association of Michigan.

The *Wall Street Journal* has reported that 90 percent of all U.S. businesses are family owned or controlled. The financial crisis has forced many to close their doors. In fact, the Bureau of Labor Statistics estimated between the fourth quarter of 2007 and the fourth quarter of 2008, some four million firms with nineteen or fewer employees went out of business.

The American public in 2009 managed to actually increase their savings, but runaway deficit spending by the government undermined their efforts. Peter Schiff, the author of *Crash Proof,* explained, "The simple truth is that government debt is our debt. So if a family manages, at some cost to their lifestyle, to squirrel away an extra $1,000 in saving this year, but the government adds $20,000 in new debt per household (each family's approximate share of the $1.8 trillion fiscal 2009 deficit), that family ends up owing $19,000 more than they did at the beginning of the year!"

SOCIALISM AND LOSS OF INDIVIDUALITY

SOCIALISM IS A KEY word in understanding what has happened to America. Most dictionaries define "socialism" as the collective ownership and administration of the means of production and distribution of goods and services. Invariably, a centralized authority is needed to administer these means.

The communist leader Vladimir Ilyich Lenin foresaw a worker's paradise where "Each person will be voluntarily engaged in work according to his capacities, and each will freely take according to his needs." But, as Lenin noted, before a person could freely take from the State, that person must become subordinate to the State.

"All our lives we fought against exalting the individual," said Lenin. Espousing the same agenda of the early-day Western globalists who funded the Bolsheviks during the Russian Revolution of 1917, Lenin proclaimed, "The aim of socialism is not only to abolish the present division of mankind into small states and all-national isolation, not only to bring the nations closer to each other, but also to merge them." He also may have foreseen the methods being used to bring down the American Republic when he said, "The surest way to destroy a nation is to debauch its currency" and "Give me four years to teach the children and the seed I have sown will never be uprooted."

As former assistant secretary of the Treasury Paul Craig Roberts stated in a treatise on the first principles of freedom, "A person born before the turn of the [20th] century was born a private individual. He was born into a world in which his existence was attested by his mere physical presence, without documents, forms, permits, licenses, orders, lists of currency carried in and out, identity cards, draft cards, ration cards, exit stamps, customs declarations, questionnaires, tax forms, reports in multuplicate [*sic*], social security number, or other authentications of his being, birth, nationality, status, beliefs, creed, right to be, enter, leave, move about, work, trade, purchase, dwell. . . . Many people take private individuals for granted, and they will find what I am saying farfetched. But private individuals do not exist in the Soviet Union or in China where the claims of the state are total and even art and literature must be subservient to the interests of the state. . . ."

Roberts presented an example of how bureaucracy has begun to erode the liberties of American citizens: "[In the 1970s] US District Judge Wilbur Owens instructed the Board of Regents of the University System of Georgia to use involuntary transfers of faculty members between system institutions to achieve racial balance among the faculties. As long as the involuntary transfers of teachers was intra-city and confined to elementary and high school teachers, my liberal colleagues saw it as social progress. But once they faced inter-city involuntary transfers, they called it fascism. It is true that until the liberal progress of the 1960s, government direction of labor in this century was unique to the Hitler and Stalin regimes. As is often the case, people realize the consequences of statist ideas only when their own private individualities are touched."

But the fleecing of America did not merely start in the 1970s. It's been going on for many more decades. Consider a 1934 editorial cartoon published in the *Chicago Tribune,* entitled "Planned Economy or Planned Destruction?" In the drawing there are men identified as "Young Pinkies from Columbia and Harvard," who are shoveling money from a cart. Beneath the cart sits a disheveled Leon Trotsky writing, "Plan of action for U.S.—Spend! Spend! Spend! Under the guise of recovery—Bust the Government—Blame the capitalists for the failure—Junk the Constitution and declare a dictatorship." This cartoon might well have been drawn by a conservative cartoonist of today.

A few older citizens may recall the words of Norman Mattoon Thomas, a pacifist who ran for president six times between 1928 and 1948 under the Socialist Party of America banner, "The American people will never knowingly adopt Socialism," he said. "But under the name of 'liberalism' they will adopt every fragment of the Socialist program, until one day America will be a Socialist nation, without knowing how it happened."

In a 1948 interview, Thomas said he was retiring from American politics because both the Democratic and Republican parties had adopted every plank of the Socialists' platform and there was no longer a need for the alternative Socialist Party.

If Thomas was possibly correct in 1948, he is undoubtedly correct now. Many people see what once was termed "creeping socialism" in the United States now full-blown policy in Washington. This perception was reflected on the February 16, 2009, cover of *Newsweek* that declared, "We Are All Socialists Now." Many Americans cringed at the nationalization of the banking and auto industries. They feared more would follow.

TEA PARTIES

BEGINNING IN APRIL 2009, protests against "out-of-control" government spending, the wars in Iraq and Afghanistan, and the squabble over health care spread nationwide in citizen meetings termed "tea parties." The name came from the original Boston Tea Party of 1773, when American colonists tossed shipments of tea into Boston Harbor in protest of the British government's "taxation without representation." Many modern wits have pointed out, "If the colonists thought taxation without representation was bad, they should see taxation WITH representation."

In 2009, the spirit of protest spilled over into several town hall meetings, where members of Congress, off for the summer recess, were shouted at and, in some cases, chased from the hall by constituents angered by what they saw as President Obama's socialist health-care plan and general government malfeasance. This groundswell of public protest continued into 2010, with even more tea parties and demonstrations of anger over perceived socialist giveaway programs, the health-care crisis, corporate bailouts, and the destruction of the U.S. economy, all of which will be discussed later.

NEW WORLD ORDER

MANY CONCERNED CITIZENS TURNED to alternative radio talk shows and Internet blogs to learn more about a plan by globalists to control the world, one that President George H. W. Bush called the "New World Order." It's a term that Adolf Hitler once used. Self-styled globalists are those people who believe themselves above petty nationalism. These men and women deal with the planet Earth as their sphere of influence. Many view the United States as a not-so-profitable division of their multinational corporations. Globalists adhere to the old Illuminati philosophy of "The end justifies the means," although most would disdain any connection to that elder secret society or to the Nazis who carried this philosophy to its political extremes.

In the book *Shadow Elite,* Janine Wedel described globalists as "flexians," members of a transnational elite, the "mover and shaker who serves at one and the same time as business consultant, think-tanker, TV pundit, and government adviser [and] glides in and around the organizations that enlist his services. It is not just his time that is divided. His loyalties, too, are often flexible."

Despite the scoffs of "flexians" within the corporate mass media and bought-off politicians, a New World Order does exist and it often makes far-reaching plans. President T. Woodrow Wilson wrote that the bulk of money sent to Russia from the United States at the time of the Russian Revolution went to the Bolsheviks, the forerunners of the Communists. These funds came from the Rockefellers and other Wall Street capitalists such as Jacob Schiff, Elihu Root, J. P. Morgan, and the Harriman family (W. Averell Harriman became U.S. ambassador to the Soviet Union during World War II). These men and others also provided initial funding for the Council on Foreign Relations.

When these same globalists became fearful of worldwide communism (they needed separate national or economic blocs to play off against each other for the tensions necessary for maximum profit and control), they supported National Socialism in Germany. German army intelligence agent Adolf Hitler was funded to provide a bulwark against the Communist tide by enlarging his National Socialist German Workers Party (Nazis), in turn sowing the seeds of World War II. Three prominent Americans who were instrumental in funding the Nazis were National City Bank (now Citicorp) chairman John J. McCloy; Schroeder Bank attorneys Allen Dulles and his brother, John Foster Dulles; and Prescott Bush, a director of Union Banking Corporation and the Hamburg America shipping line. It is interesting to note that, following World War II, McCloy became the high commissioner of occupied Germany; John Foster Dulles became President Eisenhower's secretary of state; Allen Dulles became the longest-serving CIA director; and Bush, as a senator from Connecticut, was instrumental in forming the CIA. It might also be noted that both McCloy and Allen Dulles sat on the largely discredited Warren Commission assigned by President Lyndon B. Johnson to investigate the assassination of President John F. Kennedy. After World War II, the globalist agenda was advanced by the creation of the United Nations. An earlier attempt to create a transnational organization, the League of Nations, failed because the U.S. Senate thought that ratification would end American sovereignty.

Nick Rockefeller, a participant in the World Economic Forum and a member of the Council on Foreign Relations, may have revealed the agenda of the New World Order in a casual comment. According to the late Hollywood producer Aaron Russo, Rockefeller told him, "The end goal is to get everybody chipped, to control the whole society, to have the bankers and the elite people control the world."

Catherine Austin Fitts, assistant secretary of housing during the George H. W. Bush presidency, wrote in early 2009: "In the fall of 2001 I attended a private investment conference in London to give a paper, 'The Myth of the Rule of Law or How the Money Works: The Destruction of Hamilton Securities Group.' The presentation documented my experience with a Washington–Wall Street partnership that had engineered a fraudulent housing and debt bubble; illegally shifted vast amounts of capital out of the US; used 'privatization' as a form of piracy—a pretext to move government assets to private investors at below-market prices and then shift private liabilities back to government at no cost to the private liability holder. Other presenters at the conference included distinguished reporters covering privatization in Eastern Europe and Russia. As the portraits of British ancestors stared down upon us, we listened to story after story of global privatization throughout the 1990s in the Americas, Europe, and Asia."

Fitts reiterated Rockefeller's statement about a New World Order ruled by a global elite. She noted, "As the pieces fit together, we shared a horrifying epiphany: the banks, corporations and investors acting in each global region were the exact same players. They were a relatively small group that reappeared again and again in Russia, Eastern Europe, and Asia accompanied by the same well-known accounting firms and law firms. Clearly, there was a global financial *coup d'etat* underway."

Walter Cronkite, the legendary anchor of CBS News, often referred to as "the most trusted man in America," also stated his belief that the country was ruled by a small elite. Shortly before his death in July 2009, Cronkite was asked if there was a ruling class in America. "I am afraid there is," he replied. "I don't think it serves the democracy well, but that is true, I think there is. The ruling class is the rich who really command our industry, our commerce, our finance. And those people are able to so manipulate our democracy that they really control the democracy, I feel."

With the bulk of the public both manipulated and distracted by political parties and the corporate mass media, no one seems capable of discerning, much less opposing, this New World Order of elitists with corporate, family, and class connections and common interests.

Until the real rulers of America are identified and confronted, no amount of hand-wringing, letter writing, or demonstrating can have any meaningful effect.

DISSENSION IN THE RANKS

THE FINANCIAL CALAMITY OF 2008 exposed the New World Order to be in slight disarray even before it was firmly established. Though the Obama administration is rife with men and women well connected to the centers of wealth and power, as will be seen, control over both the economic and social conditions in the United States appeared to be getting out of their hands. There was even dissension in the ranks at the University of Chicago, which many consider to be the center of globalist thinking. The university's 1995 Nobel Memorial Prize winner in Economic Sciences, Robert E. Lucas, claimed the Obama administration's stimulus plans are "schlock economics," while his colleague, Professor of Finance John H. Cochrane, stated they were based on discredited "fairy tales." Their cry was reminiscent of the term "voodoo economics," used by George H. W. Bush against Ronald Reagan's free-enterprise plans during the Republican presidential primaries in 1980.

Paul Krugman, a *New York Times* op-ed columnist and winner of the 2008 Nobel Memorial Prize in Economic Sciences, wrote, "As I see it, the economics profession went astray because economists, as a group, mistook beauty, clad in impressive-looking mathematics, for truth. Until the Great Depression, most economists clung to a vision of capitalism as a perfect or nearly perfect system. That vision wasn't sustainable in the face of mass unemployment, but as memories of the Depression faded, economists fell back in love with the old, idealized vision of an economy in which rational individuals interact in perfect markets, this time gussied up with fancy equations . . . the central cause of the profession's failure was the desire for an all-encompassing, intellectually elegant approach that also gave economists a chance to show off their mathematical prowess.

"Unfortunately, this romanticized and sanitized vision of the economy led most economists to ignore all the things that can go wrong. They turned a blind eye to the limitations of human rationality that often lead to bubbles and busts; to the problems of institutions that run amok; to the imperfections of markets—especially financial markets—that can cause the economy's operating system to undergo sudden, unpredictable crashes;

and to the dangers created when regulators don't believe in regulation."

Conspiracy theorists have long been ridiculed for their claims that the Great Depression was manufactured by globalist bankers. Krugman added much weight to that argument with a narrative involving a statement by the current chairman of the Fed's board of governors, Ben Bernanke: "At a 90th birthday celebration for Milton Friedman, Ben Bernanke declared of the Great Depression: 'You're right. We did it. We're very sorry. But thanks to you, it won't happen again.' The clear message was that all you need to avoid depressions is a smarter Fed."

So we see that a plan is in play to debase the U.S. economy and impose a socialist system—whether Obama's Marxist Socialism or Bush's National Socialism apparently makes no difference to those wealthy or powerful enough to control the central bureaucracy of the state.

These globalists, who have manipulated world history for decades, if not centuries, are working a plan to turn the once-free and prosperous Republic of the United States into a socialist state populated by dumbed-down and destitute zombies by draining dry the nation's money supply.

It is truly a trillion-dollar conspiracy.

PART II

HOW TO CREATE ZOMBIES

All Socialism involves slavery.

—HERBERT SPENCER, British author, economist, and philosopher, 1884

FREE PEOPLE CAN TRAVEL anywhere at any time they like. They can start a business or a new profession, or even take a vacation for as long as they wish. One sure way to create a slave is to ensure a person is indebted. After all, anyone who cannot do any of the things a free person can do because he or she has a mortgage, bills of all sorts, and the need for a monthly paycheck should be considered a slave of sorts—a debt slave.

POLITICAL HACKING

Government is not reason; it is not eloquent; it is force. Like fire, it is a dangerous servant and a fearful master.

—GEORGE WASHINGTON

We shall consider politics the representative head of a zombie nation. Politics is a necessary partner in any widespread and high-level conspiracy. There is an inseparable blend of political and financial control in modern America. This powerful combination can be found within the Federal Reserve System, in the corridors of Washington and Wall Street, and even in corporate news stories dealing with both politics and finance.

Americans do not need an economics degree to figure out that the nation is past bankruptcy. Using the most conservative estimates, there is more than $70 trillion of American debt compared with about $13 trillion in gross domestic production. This does not include the $300 trillion or more in toxic derivative debt.

FOREIGN TRADE AND BONDS

INTERNATIONAL TRADE DEFICITS HAVE been draining the nation's reserves by $30 billion to $150 billion each year and have been for the past twenty years. Furthermore, our industrial, mining, and agricultural institutions have not only been weakened, but in many ways decimated by the movement toward globalization. No new steel foundries have been built in the United States since World War II.

The issue of debt is fundamental to understanding the machinations that formed the current economic crisis. By 2008, industry, banking, government, households, and individuals were smothered in debt. Eliminating debt will result in a society that looks far different from the one we have experienced in the past. The *New York Times* noted in a May 9, 2009, front-page report, "[T]he forces that enabled and even egged on consumers to save less and spend more—easy credit and skyrocketing asset values—could be permanently altered by the financial crisis that spun the economy into recession."

The "forces" mentioned in the *Times* article means bloated salaries, one of the few remaining options to corporations for cutting expenses and balancing the budget.

What is seen then is the culmination of a restructuring process that has taken place for more than two decades. Whereas the living standard has increased in many former dictatorships such as Russia and China, it has decreased in the United States thanks to these "forces," controlled by the New World Order plutocrats.

Given the consistent transfer of money between nations, is it possible that the economic meltdown was not accidental? Some people claim the so-called bailout is nothing but the largest transfer of wealth in Western history, a panicked effort to shore up the U.S. dollar. Additionally, not only was the U.S. dollar in danger, but its bonds were too. Dollar-based Grand Net bonds' net inflow dropped from an early 2007 high of about $950 billion to a 2009 low of nearly $200 billion, indicating a lack of faith in U.S. money. "The foreign creditors are moving away from the United States, plain and simple," wrote statistical analyst Jim Willie.

Willie went on to say, "The US dollar stewards are NOT [original emphasis] demonstrating control, discipline, or even anything remotely resembling honesty or integrity. . . . If not for the US Fed buying most of the US Treasury [bonds] issued, the long-term interest rates would be rising rapidly and with alarm [hyperinflation]. . . . They put the US dollar at grave risk. The Weimar territory lies directly ahead! . . . The Chinese financial market is actually leading the US market on directional turns. Sadly and tragically, the US dollar is stuck in mud, running out of time, awaiting a meat cleaver by foreign creditors."

Both China, the world's largest holder of foreign-currency reserves, and Russia wield that cleaver; and both have called for a new global currency to replace the dollar as the dominant place to store reserves.

One little-known and also one of the most unsettling aspects of the 2008 financial tsunami was the 2009 report that China's State-owned Assets Supervision and Administration Commission (SASAC) might support large enterprises in defaulting on the derivatives contracts that they purchased in 2008 from international banks. The Chinese business had purchased the contracts to protect themselves from rising commodity prices, and if they default on these contracts, it would deal a serious blow to investment banks hoping to sell more derivative hedges in China, which is the world's fastest-expanding major economy and top commodities consumer.

Another side to the problem is simply that any money China spends on bonds and derivatives is money they cannot loan to us. "[I]f China really wanted to spur domestic consumption, the best way to do so would be to stop buying our debt. Even better, they could sell Treasuries they already own and distribute the proceeds to their citizens to spend," wrote Peter Schiff, author and president of Euro Pacific Capital. "However, the Obama administration is heavily lobbying the Chinese to get them to step up to the plate and buy record amounts of new Treasury debt. Obama cannot have it both ways. He cannot claim he wants the Chinese to spend more, but then beg the Chinese government to take money away from Chinese consumers and loan it to the United States Treasury. In the end, Obama will get precisely what he publicly claims to desire but privately dreads. The Chinese government will come to its senses and stop buying

Treasuries. This will cause the U.S. dollar to collapse, but it will also allow Chinese citizens to fully enjoy the fruits of their labor."

Yet, as the Chinese people begin to buy more of their own products, it will mean fewer products available for export to America. And, as they spend more money on goods and services, there will be less money to loan to America. This could only lead to a deeper economic crisis.

The situation the United States finds itself in today is in many ways worse than that of the 1930s. More banks have failed than during the Great Depression, and unemployment is reaching levels of that time. But unlike the individuals of the 1930s—many of whom had come from an agricultural background and knew how to fend for themselves—the people in modern America can only look to government for their basic necessities. Could this push to government-regulated socialism be the real agenda behind the contrived financial meltdown of recent years?

The difference between today and the Great Depression is primarily about the worth of money. The 1930s experienced a monetary depression. Money retained its value because it was simply hard to come by and prices were depressed to reflect its scarcity. Today, America is experiencing an inflationary depression. Prices continue to rise because of an inflated money supply. The more money that's in circulation, the less it is worth.

LIARS' LOANS

WILLIAM K. BLACK, a professor of economics and law at the University of Missouri School of Law in Kansas City, suggested that more than simple greed and incompetence brought about the economic crisis of 2008. In the 1980s, Black lead the prosecution against miscreants in the savings and loan scandal. According to Black, the mortgage debacle was centered on the creation of triple-A-rated bonds that did not use verified incomes, assets, or employment. These were known as "liars' loans." Black pointed out that the liars' loans were deceitful and fraudulent, and the banks involved knew it.

"Fraud is deceit. And the essence of fraud is, 'I create trust in you, and then I betray that trust, and get you to give me something of value.' And as a result, there's no more effective acid against trust than fraud, especially fraud by top elites, and that's what we have," Black told PBS commentator Bill Moyers in April 2009. "The Bush Administration essentially got rid of regulation, so if nobody was looking, you were able to do this with impunity and that's exactly what happened. Where would you look? You'd look at the specialty lenders. The lenders that did almost all of their work in the sub-prime and what's called Alt-A [risky Alternative A-paper loans], liars' loans. . . . They knew that they were frauds."

Black said liars' loans were accomplished by failing to check the information provided by those seeking the loan. He said that often loan applicants were even told they could get a better deal if they inflated their income, job history, and assets. "We know that they said that to borrowers," said Black.

He pointed out that IndyMac, the Federal Savings Bank that failed on July 11, 2008, specialized in liars' loans—in 2006 it sold $80 billion worth of them—thus producing more losses than the entire savings and loan debacle of the 1980s.

And it was all based on fraud. Black explained, "Liars' loans . . . were known to be extraordinarily bad. And now it was getting triple-A ratings. Now a triple-A rating is supposed to mean there is zero credit risk. So you take something that not only has crushing risk . . . and you create this fiction that it has zero risk. That itself . . . is a fraudulent exercise. And again, there was nobody looking during the Bush years. . . . When they finally did look, after the markets had completely collapsed, they found . . . the appearance of fraud in nearly every file. . . ."

Black and others have compared the bad loans to the Ponzi scheme charged against Wall Street investment consultant Bernie Madoff. "Everybody was buying a pig in the poke with a pretty pink ribbon, and the pink ribbon said, 'Triple-A,' " said Black.

Although there is no specific law against liars' loans, Black argued that the bankers involved knew they had been made under false representation and that they would never be repaid. The loans were based on deceit, which lies at the heart of the legal definition of criminal fraud. Why was no one prosecuted for these acts of fraud? According to Black, federal investigators did not begin to scrutinize the major lenders until the market had actually collapsed, despite early warnings.

"The FBI publicly warned, in September 2004, that there was an epidemic of mortgage fraud, that if it was allowed to continue it would produce a crisis at least as large as the Savings and Loan debacle," said Black.

But the investigation didn't happen. Due to the war on terrorism, the Bush Justice Department transferred five hundred white-collar specialists in the FBI to national terrorism and refused to replace them. Today, Black noted, "There are one-fifth as many FBI agents [detailed to investigating mortgage fraud] as worked the Savings and Loan crisis."

GRAMM AND DEREGULATION

ONE OF THE PROTECTIONS against "banksters" (a derogatory term combining "bankers" with "gangsters") was the Glass-Steagall Act, which went into effect in 1934 following government hearings revealing how big banks of that day had looted customers for the benefit of a small group of insiders. The act separated normal banking activities (checking and savings accounts and commercial loans) from speculative investment banking (hedge funds, derivatives, and Wall Street investments) in the eyes of the law and allowed for regulation of the latter type of activity.

According to former U.S. Commodity Futures Trading Commission (CFTC) chairperson Brooksley Born, beginning in the Clinton years, almost all such protective regulation was stripped away. In a 2003 interview with *Washington Lawyer,* she stated, "One major issue was the enormous growth of over-the-counter (OTC) derivatives. OTC derivatives had been legally permitted for the first time in 1993 by a regulatory exemption that Wendy [Lee] Gramm had adopted as virtually her last act as CFTC chair. This allowed the growth of a business that is now estimated at over a hundred trillion dollars annually in terms of the notional value of contracts worldwide. Alan Greenspan had said that the growth of this market was the most significant development in the financial markets of the 1990s. The market was virtually unregulated and many, many times as big as the trading on the futures exchanges. The commission had kept some nominal authority over this market, but there were no mechanisms for enforcing the rules. For example, anti-fraud rules were retained, but no reporting was required. The market was completely opaque. Neither the commission nor any other federal regulator knew what was going on in that market!"

While Mrs. Gramm was chairing the CFTC, from 1988 to 1993, that body exempted Enron from regulation in trading of energy derivatives. Gramm later resigned from the CFTC and took a seat on the Enron board of directors where she served on its Audit Committee. Enron, the giant energy corporation whose bankruptcy in late 2001 was the largest in U.S. history to that date, drained more than $10 billion from shareholders and resulted in new regulations and legislation to enhance the reliability of financial reporting for public companies. Due to the massive fraud involved, several Enron executives, including founder Kenneth Lay and President Jeffrey Skilling, were sentenced to prison terms. The accounting firm of Arthur Andersen was found guilty of shredding Enron documents and eventually dissolved, putting eighty-five thousand persons out of work.

It should be noted that Wendy Lee Gramm is the wife of former Texas Republican senator Phil Gramm, who was forced to resign as senior economic adviser in John McCain's 2008 presidential campaign after describing Americans protesting the economic losses due to malfeasance as "a nation of whiners." As a senator, Gramm was the chairman of the U.S. Senate Committee on Banking, Housing, and Urban Affairs during the Clinton administration, and he led efforts to pass banking deregulation laws such as the landmark Gramm-Leach-Bliley Act in 1999. The act removed Depression-era laws that prevented banks from engaging in insurance and brokerage activities and was passed by an overwhelming majority of the House and by the Senate unanimously and was signed into law by President Clinton. Supporters of the bill used an old trick that was used to pass the Federal Reserve Act of 1913. Like the Federal

Reserve Act, the Gramm-Leach-Blilely Act was introduced on the last day before the Christmas holiday and was never debated by either congressional body. This bill, fully initiated by and supported by Republicans and passed with the support of Democrats during a Democratic administration, clearly demonstrates the collusion of the two political parties when it comes to corporate business.

Many economists claim the Gramm-Leach-Bliley Act's undermining of the Glass-Steagall Act was a significant cause of the 2007 subprime mortgage crisis and the 2008 global economic crisis. Economist Paul Krugman has described Phil Gramm as "the high priest of deregulation" and named Gramm and Fed chairman Alan Greenspan as the top two culprits responsible for the economic crisis. Gramm's culpability was echoed by CNN, *Time,* and Britain's the *Guardian.*

Brooksley Born described how, during the Clinton years, her commission questioned the bailout of large OTC derivatives dealers because they held $1.25 trillion worth of contracts yet held a mere $4 billion in supporting capital, which meant the dealers had far overextended themselves, leaving the market vulnerable to the very meltdown that occurred in 2008–09: "I became enormously concerned about OTC derivatives and thought the market was a nightmare waiting to happen," recalled Born. "I was particularly concerned that there was no transparency. No federal regulator knew what kind of position firms like Long-Term Capital Management and Enron had in the derivatives markets." Warren Buffett later called OTC derivatives the financial weapons of mass destruction.

Born said the Fed and Congress rebuffed the CFTC's efforts to reinstate some public protection over the financial field. "It wasn't a regulatory effort. We were just asking questions! The concept release didn't propose any rules. Alan Greenspan, Arthur Levitt, and Robert Rubin all said that these questions should not be asked and urged Congress to pass a bill that would forbid the commission from taking any regulatory steps on over-the-counter derivatives. There were no hearings on that bill, but during a congressional conference committee meeting on an appropriations bill, an amendment was added preventing the commission from taking any action on over-the-counter derivatives for six months. This occurred within a month after Long-Term Capital Management's collapse!"

Professor William Black pointed to the experience with AIG (American International Group) as an example of how the lack of regulation led to obscene profits and market manipulation. The taxpayer-backed bailout of AIG in late 2008 ended up totaling more than $180 billion, a cost equaling the entire savings and loan scandal of the 1980s.

In September 2008, AIG's credit ratings were downgraded and the Fed issued $85 billion in credit to keep the international insurance giant afloat. But the Fed also took a stock warrant for nearly 80 percent of AIG's equity. The government eventually increased AIG's credit to as much as $182.5 billion. Public outrage ensued from news reports that AIG had retained millions of dollars in bailout money, some of it going for executive bonuses and lavish junkets. AIG bondholders and counterparties were paid at one hundred cents on the dollar by taxpayers, yet the taxpayers had no claim to future profits. In other words, the benefits of the bailout went to the AIG banks while the taxpayers suffered the costs.

"AIG made bad loans but with guarantees and charged big fees up front," Black explained. "So, they booked a lot of income. Paid enormous bonuses. . . . And they got very, very rich. But, of course, then they had guaranteed this toxic waste. . . . [T]hose liars' loans are going to have enormous losses. And so, you have to pay the guarantee on those enormous losses. And you go bankrupt. Except that you don't in the modern world, because you've come to the United States, and the taxpayers play the fool. Under Secretary [of the Treasury Timothy] Geithner and Under Secretary [Henry] Paulson before him . . . took $5 billion . . . in U.S. taxpayer money and sent it to a huge Swiss Bank called UBS [through AIG]. [UBS] was defrauding the taxpayers of America. And we were bringing a criminal case against them. We eventually get them to pay a $780 million fine, but wait, we gave them $5 billion. So, the taxpayers of America paid the fine of a Swiss bank. And why are we bailing out somebody who is defrauding us?"

Some suggested that UBS was given $5 billion because AIG was the largest contributor to Obama's campaign and held much of the toxic derivative paper of Goldman Sachs, the major globalist investment firm once headed by Paulson. Though many Americans saw the AIG deal as simply a massive theft that debased our economy, no one in upper management—other than former figurehead and NASDAQ chairman Bernard L. "Bernie" Madoff—was ever charged with a crime.

According to TARP (Troubled Asset Relief Program) inspector Neil Barofsky, even by mid-October 2009, AIG executives still hadn't repaid half of the $45 million they promised to return. But by March 2009, the public became enraged when it learned that AIG had paid at least $165 million in executive bonuses from the $180 billion in taxpayer loans to keep the company afloat. AIG chief executive officer Edward M. Liddy told a House committee hearing that he had asked employees to voluntarily give back at least half of their bonuses, although he admitted he had no authority to force them to do so.

In December 2008, the U.S. government also took hold of the financing arm of one of the nation's largest manufacturers—General Motors. William Black and others have criticized the government takeover of General Motors (GM) as mere nationalization and have questioned why the president of GM was fired while the bankers who created the economic mess were not. "There are two reasons," Black said. "One, [government officials are] much closer to the bankers. These are people from the banking industry. And they have a lot more sympathy. In fact, they're outright hostile to autoworkers, as you can see. They want to bash all of their contracts. But when they get to banking, they say, 'contracts, sacred.' But the other element of your question is we don't want to change the bankers, because if we do, if we put honest people in, who didn't cause the problem, their first job would be to find the scope of the problem. And that would destroy the cover-up.

"Geithner is . . . covering up. Just like Paulson did before him. Geithner is publicly saying that it's going to take $2 trillion—a trillion is a thousand billion—$2 trillion taxpayer dollars to deal with this problem. But they're allowing all the banks to report that they're not only solvent, but fully capitalized. Both statements can't be true. It can't be that they need $2 trillion, because they have massive losses, and that they're fine. These are all people who have failed. Paulson failed, Geithner failed. They were all promoted because they failed. . . ."

Geithner denied any failure, claiming he was never supposed to regulate the banking business. During congressional testimony in March 2009, Geithner, who was the president of the New York Fed during much of the credit boom, indicated he had little interest in scrutinizing other banks' activities. "I've never been a regulator, for better or for worse," stated Geithner with surprising candor, adding, "And I think you're right to say that we have to be very skeptical that regulation can solve all of these problems. We have parts of our system that are overwhelmed by regulation."

"Overwhelmed by regulation!" lamented journalist Bill Moyers over Geithner's comments. "It wasn't the absence of regulation that was the problem, it was despite the presence of regulation you've got huge risks that build up." Black agreed, saying, "Well, he may be right that he never regulated, but his job was to regulate. That was his mission statement. As president of the Federal Reserve Bank of New York, [he was] responsible for regulating most of the largest bank holding companies in America. And he's completely wrong that we had too much regulation in some of these areas. I mean, he gives no details, obviously. But that's just plain wrong."

As 2009 drew onward, more financial institutions fell by the wayside, even as the media pumped out heartening stories of an economic rebound and more stimulus activity. In the face of criminal charges, the Alabama bank Colonial BancGroup, Inc., was closed by regulators in August 2009, becoming the seventy-seventh failed bank since the start of the year. It was also the largest bank failure since the loss of Washington Mutual, Inc., in 2008. Colonial posted a $606 million second-quarter loss in 2009, primarily due to loans to developers and home builders in Florida, a state where the housing industry tanked quickly. The bank failed to meet capital requirements to qualify for TARP funds because it simply did not have enough financial reserves to be eligible for TARP support.

One problem, said Robert Auerbach, formerly an economist with the Financial Services Committee of the U.S. House of Representatives, is that central bank officials are often too close to the banks they are meant to keep in check. "The boards of directors of every Fed bank, including the New York Fed, have nine directors. Six of them are elected by the banks in the district," said Auerbach. "So you have the banks in New York electing the directors that are supposed to supervise them."

One proven means for keeping the true condition of some banks from the public eye during any reorganization is to retain the officers responsible for the problem in the first place. "[A]s long as I keep the old CEO who caused the problems, is he going to go vigorously around finding the problems? Finding the frauds?" asked Black in Moyers's interview. He added, "We adopted a law after the Savings and Loan crisis, called the Prompt Corrective Action Law. And it requires [bank officers] to close these institutions. And they're refusing to obey the law."

When asked if Geithner and others in the Obama administration have engaged in a cover-up along with the banks, Black responded, "Absolutely, because they are scared to death . . . of a collapse. They're afraid that if they admit the truth, that many of the large banks are insolvent. They think Americans are a bunch of cowards, and that we'll run screaming to the exits. And we won't rely on deposit insurance."

DOWNSIZING AMERICA

PEOPLE LIKE BLACK AND Moyers who are in prestigious positions fail to mention that the motive behind Geithner's and the banks' financial antics can be traced to secretive globalist organizations such as the Council on Foreign Relations. Moyers also usually fails to mention that he is a member of the CFR, having obviously passed its stringent globalist eligibility requirements. It is in examples such as this that one can see the guiding hand of the globalists in both the world of commerce and of journalism.

Another person close to secretive society members was Henry "Hank" Paulson, the George W. Bush Treasury secretary who oversaw the bailout of AIG. During both the Bush and Obama administrations, AIG was used to funnel taxpayer funds to certain banks like UBS and Goldman Sachs, where Paulson had previously been the CEO.

In 2006, when Bush named Paulson to head the Treasury, the CFR explained the president's agenda in an op-ed piece: "Bush essentially set five goals for the new Treasury secretary. Keep taxes low. Curb federal government spending to curb the budget deficit. Deal with international imbalances. Keep investment markets open. Support innovation and risk-taking in the private sector to boost US economic growth. . . . Paulson is the right man at the right time to take on issues like these."

Despite the fact that IndyMac had failed only days before, on July 20, 2008, Paulson reassured the public that "it's a safe banking system, a sound banking system. Our regulators are on top of it. This is a very manageable situation."

Paulson has been identified as a key figure in the economic debacle that began in 2008. *Time* magazine stated, "If there is a face to this financial debacle, it is now his."

Noting that Goldman Sachs got the lion's share of taxpayer bailout money—$12.9 billion—William Black declared, "Now, in most stages in American history, that would be a scandal of such proportions that he wouldn't be allowed in civilized society. . . . The tragedy of this crisis is it didn't need to happen at all."

Black, along with many other commentators, saw losses in workers' income, securities, pensions, and futures as the result of the misconduct of "a relatively few, very well-heeled people, in very well-decorated corporate suites . . . and their ideologies, which swept away regulation." *Forbes* magazine in 2006 estimated Paulson's personal wealth at $700 million.

Black and others acknowledged that the destruction of the U.S. financial system came about due to a lack of integrity on the part of several high government and banking officials as well as massive conflicts of interest and a loss of morality. But this is simply the view of those unwilling to address the true issue—conspiracy.

After studying three separate government reports predicting a coming "fiscal doomsday," the chairman of the investment counseling firm the Weiss Group Inc., Martin D. Weiss, had yet another word in mind. "When our leaders have no awareness of the disastrous consequences of their actions, they can claim ignorance and take no action. Or when our leaders have no hard evidence as to what might happen in the future, they can at least claim uncertainty. But when they have *full knowledge* of an impending disaster . . . they have *proof* of its inevitability in ANY scenario . . . and they so *declare* in their official reports . . . but STILL don't lift a finger to change course . . . then they have only one remaining claim: INSANITY!" he wrote (original emphasis). But it would be insane to actually believe that the nation's money masters are truly insane. The only alternative is conspiracy. The financial meltdown happened because it was engineered to happen.

The belief that the economic collapse was orchestrated even reached the mainstream media. In early 2009, Washington insider Dick Morris pointed out to Fox News commentator Sean Hannity how the International Monetary Fund (IMF) was attempting to bring the U.S. economy under international control by using the excuse that it would merely be coordinating "regulatory efforts." "The conspiracy theorists who have talked about the New World Order and the UN taking control, they are right. . . . It's happening!" he exclaimed.

No matter how clearly Dick Morris saw things, only a few in Congress seemed to be getting the message. Texas Republican representative Kay Granger got it. In an August 2009 letter to constituents, she wrote, "Something happened this week that has serious consequences for each and every one of us, but you probably didn't even know it happened. On Tuesday [August 25, 2009], the Office of Management and Budget (OMB) released their Midsession Review. . . . The Midsession Review showed that our country is going to be $2 trillion deeper in debt than the White House originally told us at the beginning of this year. That's nearly $6,700 more debt for every man, woman, and child in America. If this doesn't show that the policy of spend, spend, spend isn't working, I don't know what does."

The only answer that Washington seems to come up with to deal with all problems is to spend more money on central government programs. Is this merely ineptitude or is this proof of a hidden agenda, one designed to force the American republic into a tightly controlled socialist society?

DEBT SLAVES

Permit me to issue and control the money of a nation, and I care not who makes its laws.

—AN OFT-REPEATED PARAPHRASE OF AMSCHEL MAYER ROTHSCHILD'S 1838 QUOTE, "I care not what puppet is placed upon the throne of England to rule the Empire on which the sun never sets. The man who controls Britain's money supply controls the British Empire, and I control the British money supply."

ECONOMICS IS THE LIFEBLOOD of any nation. Many compared President Barack H. Obama's $787 billion economic stimulus package in 2009 to giving blood to a corpse. They feared the stimulus was simply throwing good money after bad, especially in light of health and data-gathering provisions that seemed out of place in financial legislation.

As the U.S. economy deteriorated, President Obama expanded the Bush administration's policies for bailing out banks and other financial institutions. President Obama explained that sending money directly to taxpayers might seem more appealing, but said it wouldn't be as effective in stimulating the economy, saying that "A dollar of capital in a bank can actually result in eight or ten dollars of loans to families and businesses, a multiplier effect that can ultimately lead to a faster pace of economic growth."

STIMULUS PACKAGE

OBAMA DID NOT COMMENT on criticism raised over the many improprieties connected to the economic crisis, nor did he comment on the argument that his "economic growth" actually was nothing other than an austerity budget based on war. Michel Chossudovsky, a professor of economics at the University of Ottawa and director of the Centre for Research on Globalization, noted that "[Obama's] austerity measures hit all major federal spending programs with the exception of Defense and the Middle East War, the Wall Street bank bailout, [and] Interest payments on a staggering public debt.

"At first sight, the budget proposal has all the appearances of an expansionary program, a demand oriented 'Second New Deal' geared towards creating employment, rebuilding shattered social programs and reviving the real economy. The realities are otherwise. Obama's promise is based on a *mammoth austerity program* [original emphasis]. The entire fiscal structure is shattered, turned upside down." Understandably, Chossudovsky concluded that the Obama plan "largely serves the interests of Wall Street, the defense contractors and the oil conglomerates." He warned that the Bush-Obama bank bailouts will lead America into a spiraling public debt crisis. "The economic and social dislocations are potentially devastating," he added.

What this means is that the American taxpayer has been made the lender of last resort for the two government-sponsored private enterprises—the Federal National Mortgage Association (Fannie Mae) and the Federal Home Loan Mortgage Corporation (Freddie Mac), whose combined debt of $5.4 trillion has been effectively transferred to the nation's balance sheet. In addition to personal debt, every American now has a financial responsibility for Fannie Mae and Freddie Mac, as well as other financial institutions.

What is even more maddening was the use of some bailout funds to create extravagant "golden parachute" retirement and severance payments to financial executives who would have to leave their failing companies. These garnered unfavorable publicity in late 2008, as did the revelations of shady dealings between Wall Street and its regulators. Take, for instance, Charles Millard, former director of the Pension Benefit Guaranty Corp. (PBGC), an independent federal corporation that protects the pension plans of nearly forty-four million American workers and retirees. In May 2009, Millard was called to testify before the Senate Aging Committee over charges that he had cozy and improper contacts with Wall Street firms. Millard, citing his constitutional right to avoid self-incrimination, declined to answer questions. The PBGC, which insures corporate pensions, announced in late May 2009 that it had suffered a $33.5 billion deficit for the first half of the fiscal year, up considerably from a $10.7 billion deficit in 2008.

According to hearing testimony by PBGC inspector general Rebecca Anne Batts, Millard directly participated in granting more than $100 million in PBGC contracts to the international investment firms of Black-Rock Inc., JP Morgan, and Goldman Sachs, against the advice of senior corporate management. Telephone and e-mail records showed Millard had contacts with his prospective bidders prior to hiring them to manage real estate and private equity investments. Millard's experience illustrates both the incestuous relationship between persons in government who are supposed to be protecting the public and Wall Street. It is also noteworthy that Millard invoked the Fifth Amendment just like Mafia gangsters in the past. If there had been no wrongdoing, then why refuse to testify?

BEFORE THE CRASH

BEFORE THE MARKET CRASH in 2008, stress reached deep into certain strata of American life. Many retirees who once believed their money was safe saw principal losses of up to 80 or 90 percent of their investment.

Serious market slowdown began when investment banks across the globe refused to buy one another's credit—an unusual move—and when mortgage-purchasing companies Freddie Mac and Fannie Mae decided they could make more money by buying subprime mortgages. It was all part of the Bush administration's policy of conforming to the United Nations' Millennium Development Goals, which were unveiled in 2000. These goals addressed such issues as the eradication of extreme poverty and hunger, universal primary education, gender equality, health improvement, and ensuring environmental sustainability. It was laudable goals such as these that led to government pressure on lending institutions to issue subprime mortgages. The result? Hundreds of thousands of unsold homes.

Although it's well known that the economic mess began with the banks, mortgage lenders, and real estate companies, the current housing and mortgage mess actually was the result of maneuvering by both Democrats and Republican politicians, a fact that adds considerable weight to the argument that both major parties are controlled by the same globalists seeking to install a worldwide socialist system.

During the 1990s, Bill Clinton's Democratic administration was pressuring Fannie Mae, the nation's largest underwriter of home mortgages, to expand mortgage loans to low- and moderate-income borrowers. After all, granting low-income families the chance for home ownership sounded good on paper.

"Fannie Mae has expanded home ownership for millions of families in the 1990s by reducing down payment requirements," Franklin D. Raines, chairman and CEO of Fannie Mae, told the *New York Times* in 1999. The newspaper noted that at least one study seemed to indicate racial prejudice in this lending as it reported that 18 percent of such subprime loans went to black borrowers as compared to 5 percent for all other groups. With great prescience, *Times* writer Steven A. Holmes noted in 1999, "In moving, even tentatively, into this new area of lending, Fannie Mae is taking on significantly more risk, which may not pose any difficulties during flush economic times. But the government-subsidized corporation may run into trouble in an economic downturn, prompting a government rescue similar to that of the savings and loan industry in the 1980s."

While Fannie Mae was lowering loan qualifications its stockholders were pressuring for greater profits, creating a recipe for financial disaster. And, as usual, both the political and financial machinations involved crossed party lines but not the agenda of the globalists.

Larry Summers—a Treasury secretary under Clinton, Obama's head of the National Economic Council, and a member of the Council on Foreign Relations—is an advocate of cutting both corporate and capital gains taxes and convinced Clinton to sign into law several Republican bills that allowed banks to expand their powers. One of these bills repealed the 1933 Glass-Steagall Act, which prevented the merger of commercial banks, insurance companies, and brokerage firms such as Goldman Sachs and Merrill Lynch. Additionally, Summers supported the Commodity Futures Modernization Act just before the 2000 election, which denied the governmental Commodity Futures Trading Corporation the ability to conduct oversight on the trading of financial derivatives. In the wake of Obama's stimulus package in April 2009, Summers was criticized for collecting $2.7 million in speaking fees from Wall Street companies that had received government bailout money.

Summers was paving the way for the abuse of America's financial system. Meanwhile, his protégé, Under Secretary for International Affairs Timothy Geithner, was making political gains. In 2002, during the first George W. Bush administration, Geithner left the Treasury Department to join the Council on Foreign Relations as a senior Fellow in the International Economics Department. Also a protégé of Henry Kissinger, Geithner had previously served as president of the New York Federal Reserve Bank. By 2009, Geithner was Obama's Treasury secretary. Again, here we see two men (Summers and Geithner) connected to the same secretive globalist society—the Council on Foreign Relations (CFR)—freely moving between both Democratic and Republican administrations. The CFR is secretive because it does not publicly announce its agenda or decisions, nor does it allow anyone to join without an invitation, and then only after careful vetting of the candidate's propensity to favor globalization.

Princeton-educated economics researcher F. William Engdahl wrote that Treasury Secretary Geithner's "dirty little secret" was that during the credit crisis, he only tried to save the five largest banks—banks that held "96 percent of all US bank derivative positions in terms of nominal value, and an eye-popping 81 percent of the total net credit risk exposure in event of default." A derivative is a financial instrument whose worth is derived from another resource, whether property, goods, or services, called the underlying asset. Derivatives have been used in complex financial dealings to hedge against loss by allowing speculators to sell or trade the derivative and to gamble on gaining great profit by acquiring derivatives in the hope that the underlying asset will maintain or increase its value. In declining order, the five banks that had the most derivatives are JP Morgan Chase, Bank of America, Citibank, Goldman Sachs, and the recently merged Wells Fargo–Wachovia. The leadership of these five banks is full of CFR members.

BANK STRESS TESTS

IN EARLY MAY 2009, after months of foot-dragging, federal regulators finally released the results of their bank "stress tests," which test whether or not a certain bank can repay its debts and survive harsh economies. From the five banks listed above, only JP Morgan Chase passed the test. This means it was not required to raise more capital to prevent further losses.

The Charlotte-based Bank of America tested the worst on the stress tests. Government regulators informed the bank that it needed almost $34 billion in additional capital, which accounted for almost half of its total deficit. This news worsened problems for the banking giant, already under criticism for receiving more than $45 billion in government aid and for acquiring the investment bank Merrill Lynch.

Bank of America wasn't the only one with problems. Among others, Wells Fargo needed to raise $13.7 billion, GMAC Financial Services (formerly known as General Motors Acceptance Corporation) needed $11.5 billion, and Citigroup needed $5.5 billion. All told, the nation's large banks needed $74.6 billion to build a capital cushion, according to federal regulators.

Federal Reserve chairman Ben Bernanke was publicly upbeat about the tests, describing them as a "fair and comprehensive effort." "[Markets] can be reassured that banks will be strong and be able to lend even if the economy is worse than currently expected," he told CNBC. However, banks that failed the government's stress test would be required to quickly come up with a plan to raise additional resources. One such plan was for the federal government to convert preferred shares bought by the U.S. Treasury into common stock. Douglas Elliott, a former JP Morgan Chase investment banker now with the Brookings Institution, told the Associated Press, "Essentially what we'll be doing is swapping a kind of loan for actual ownership of a part of the bank. So it increases the taxpayers' risk but also increases the potential return."

Increased taxpayer risk? This does not seem such a good idea in shaky financial times. "Continuing to pour taxpayer money into these five banks without changing their operating system is tantamount to treating an alcoholic with unlimited free booze," said F. William Engdahl. "The government bailout of AIG, at more than $180 billion [as of April 2009], has primarily gone to pay off AIG's credit default swap obligations to counterpart gamblers Goldman Sachs, Citibank, JP Morgan Chase and Bank of America, the banks who believe they are 'too big to fail'. In effect, these institutions today believe they are so large that they can dictate the policy of the federal government. Some have called it a bankers' coup d'etat. It is definitely not healthy."

So the big banks pocket the money and the poor, strapped taxpayers are left with the bill, not to mention ownership of banks that continued to be troubled financially well into 2010.

By mid-2009, Americans were driving less and spending less and the economy was deflating. Even though products became cheaper in the face of inflation, people stopped buying what they couldn't afford. The housing market, which is a key indicator of economic strength, continued to lag far behind projections. Housing start-ups were doing particularly poorly. In April 2009, the U.S. Department of Housing and Development announced that non-government-backed housing starts, even after seasonal adjustments, were 54 percent lower (458,000) than the April 2008 rate of

1,001,000. Privately backed housing starts are any homes being built that are not being financed by the government. These have long been a prime indicator of the national economy.

There was also blame tossed at the unequal distribution of money. Chuck Collins, director of the Program on Inequality and the Common Good for the Institute for Policy Studies, said, "In our view, extreme inequalities contributed to the economic collapse. . . . This matters because wealth is power—the power to shape the culture, to distort elections, and shape government policy. A plutocracy is a 'rule by wealth'—and more and more the priorities of the society are shaped by the interests of organized wealth."

IMPROPRIETIES AND DEATH

APPARENTLY THE STRESS CREATED by the gargantuan amounts of money involved in the economic squeeze can be hazardous to your health as well as your wealth. Stress may have contributed to the untimely deaths of at least five high-profile financial officers who died in the months following financial collapse in October 2008.

In January 2009, German billionaire Adolf Merckle apparently threw himself under a train after losing money shorting Volkswagen stock. Patrick Rocca, an Irish property speculator who was close to both President Bill Clinton and British prime minister Tony Blair, was found shot in the head following the crash of the real estate market. Chicago real estate mogul Steven Good was found fatally shot in his car. Financial adviser Rene-Thierry Magon de la Villehuchet reportedly committed suicide in his Manhattan office just before Christmas 2008 after losing both his and his clients' money in the Bernie Madoff scandal.

One particularly troubling death was that of Freddie Mac acting chief financial officer David Kellermann, who was found, the apparent victim of suicide, in his Vienna, Virginia, home on April 22, 2009. In 2008, the U.S. Treasury Department had to pump $45 billion into the government-sponsored mortgage firm to shore up $50 billion in losses. Questions immediately arose over reports about Kellermann's role in the massive losses at Freddie Mac and about the nature of his death. One police spokesman told All Headline News that Kellermann died from a gunshot wound. Strangely enough, however, another police officer initially said he had hanged himself.

There was more controversy when reporters found that Kellermann was deeply involved in the Securities and Exchange Commission's and the U.S. Justice Department's investigations into questionable bookkeeping practices within Freddie Mac. "Kellermann figured in several recent controversies at Freddie Mac," reported the *Washington Post* in April 2009. "He and a group of company attorneys tussled with regulators in early March as the firm prepared to file its quarterly earnings report with the Securities and Exchange Commission. [Kellermann's] group insisted that Freddie Mac inform shareholders of the cost to the company in helping carry out the Obama administration's housing recovery plan. The regulators urged the company not to do so."

"This isn't the story of a guy who was trying to cover something up. It's the story of a guy who was trying to do the right thing," commented one housing industry veteran, who asked for anonymity, apparently suspecting the possibility of danger in telling the truth in such matters.

More than one conspiracy-minded researcher believed that something more than suicide was at work in Kellermann's death and that there may have been other deaths connected to an effort to silence insiders who might have knowledge of the situation that someone does not want made public.

In a statement from his political action committee, perennial office seeker and conspiracy advocate Lyndon LaRouche said, "There is no evident motive for suicide in this case, but there is a motive for suppressing making Kellermann's views known. The guy is killed, probably murdered. He deserves justice. His right to justice is overriding. The question is what else did David Kellermann know which influential circles did not want him to reveal?"

THE RICH GET RICHER

IT HAS LONG BEEN said that the rich get richer while the poor get poorer. Many researchers equate the term "plutocracy"—rule by the wealthy—with the New World Order.

Although the belief that an organized plutocracy controls the world has long been derided as merely a "conspiracy theory," G. William Domhoff, a professor in psychology and sociology at the University of California, Santa Cruz, has the statistics to prove its existence. Domhoff's first book, *Who Rules America?*, was a controversial 1960s bestseller that argued that the United States is dominated by an elite political and economic ownership class.

Using updated figures, Domhoff stated in a posting: "In the United States, wealth is highly concentrated in a relatively few hands. As of 2007, the top 1 percent of households (the upper class) owned 34.3 percent of all privately held wealth, and the next 19 percent (the managerial, professional, and small business stratum) had 50.3 percent, which means that just 20 percent of the people owned a remarkable 85 percent, leaving only 15 percent of the wealth for the bottom 80 percent (wage and salary workers). In terms of financial wealth (total net worth minus the value of one's home), the top 1 percent of households had an even greater share: 42.2 percent."

Domhoff defined "total assets" as the gross value of owner-occupied housing plus other real estate owned by the household, cash and savings deposits, money market accounts, stocks and bonds, retirement plans, and other financial securities. He defined "total liabilities" as mortgage debt; consumer debt, including auto loans; and any other debt.

According to Domhoff, wealth distribution has been extremely concentrated throughout American history. During the nineteenth century, the top 1 percent of wealth owners owned 40 to 50 percent of assets in large port cities like Boston, New York, and Charleston. He said this disparity remained stable during the twentieth century, "although there were small declines in the aftermath of the New Deal and World War II, when most people were working and could save a little money. There were progressive income tax rates, too, which took some money from the rich to help with government services.

"Then there was a further decline, or flattening, in the 1970s, but this time in good part due to a fall in stock prices, meaning that the rich lost some of the value in their stocks," wrote Domhoff. "By the late 1980s, however, the wealth distribution was almost as concentrated as it had been in 1929, when the top 1 percent had 44.2 percent of all wealth. It has continued to edge up since that time, with a slight decline from 1998 to 2004, before the economy crashed in the late 2000s and little people got pushed down again."

Domhoff recorded that as of 2007, "income inequality in the United States was at an all-time high for the past 95 years, with the top 0.01 percent ... receiving 6 percent of all U.S. wages, which is double what it was for that tiny slice in 2000; the top 10% received 49.7%, the highest since 1917."

The numbers are even more shocking when viewed on a global scale. Using numbers from the World Institute for Development Economics Research, Domhoff concluded the top 10 percent of the world's adults control about 85 percent of global household wealth. "That compares with a figure of 69.8 percent for the top 10 percent for the United States. The only industrialized democracy with a higher concentration of wealth in the top 10 percent than the United States is Switzerland at 71.3 percent," he noted. At the same time, the U.S. government's income is declining. According to the White House, 2008 individual income tax receipts were estimated at $1.168 trillion. Yet when tax receipts were tallied, the total was $155 billion less than that at $1.043 trillion.

Domhoff's work presents a strong argument that wealth indeed equals power. Such power comes with the ability to donate to political parties, engage lobbyists, and provide grants to experts to think up new policies beneficial to the wealthy. Money also can hire public relations firms to improve one's image or make large donations to universities and cultural entities such as museums, music halls, and art galleries. Wealth in the form of stock ownership can be used to control whole corporations, which today have inordinate influence in society, media, and government.

And just as wealth can lead to power, so can power lead to wealth. Recent presidents such as Lyndon B. Johnson and Richard M. Nixon entered office without an extraordinary amount of money but left as millionaires. This is because those who control a government can use their positions to feather their own nests. Domhoff said this can be done by means of a favorable land deal for relatives at the local level or perhaps a huge federal government contract to a new corporation run by friends who will hire you when you leave government. "If we take a larger historical sweep and look cross-nationally, we are well aware that the leaders of conquering armies often grab enormous wealth, and that some religious leaders use their positions to acquire wealth," commented Domhoff.

PUBLIC DEBT, PRIVATE PROFIT

WHETHER RICH OR POOR, most Americans believe their finances are safe, thanks to a federal government corporation created in the Great Depression year of 1933.

About eight-four hundred American banks participate in the Federal Deposit Insurance Corporation (FDIC), an independent agency created by the Congress to maintain stability and public confidence in the nation's financial system by insuring deposits, supervising banks for safety and soundness, and managing receiverships. These banks allocate a small portion of their profits to collectively insure bank deposits in cases where a bank fails.

And fail they did in late 2008 and 2009. Between the two years, 111 banks failed and many more teetered on collapse, effectively depleting the FDIC reserve fund from $52.8 billion in 2008 to a mere $10.4 billion in the first quarter of 2009, its lowest point since the height of the savings and loan scandal in 1992.

But what is more disturbing is that this reserve fund, much like Social Security, is merely an illusion.

In 2008, the former chairman of the FDIC, William M. Isaac, wrote an article titled "The Mythical FDIC Fund," in which he revealed the FDIC's insolvency: "When I became Chairman of the FDIC in 1981, the FDIC's financial statement showed a balance at the U.S. Treasury of some $11 billion.... I decided it would be a real treat to see all of that money, so I placed a call to [then] Treasury Secretary Don Regan."

The conversation went like this:

ISAAC: Don, I'd like to come over to look at the money.

REGAN: What money?

ISAAC: You know ... the $11 billion the FDIC has in the vault at Treasury.

REGAN: Uh, well you see, Bill, ah, that's a bit of a problem.

ISAAC: I know you're busy. I don't need to do it right away.

REGAN: Well ... it's not a question of timing.... I don't know quite how to put this, but we don't have the money.

ISAAC: Right ... ha ha.

REGAN: No, really. The banks have been paying money to the FDIC, the FDIC has been turning the money over to the Treasury, and the Treasury has been spending it on missiles, school lunches, water projects, and the like. The money's gone.

ISAAC: But it says right here on this financial statement that we have over $11 billion at the Treasury.

REGAN: In a sense, you do. You see, we owe that money to the FDIC, and we pay interest on it.

ISAAC: I know this might sound pretty far-fetched, but what would happen if we should need a few billion to handle a bank failure?

REGAN: That's easy—we'd go right out and borrow it. You'd have the money in no time ... same day service most days.

ISAAC: Let me see if I've got this straight. The money the banks thought they were storing up for the past half century—sort of saving it for a rainy day—is gone. If a storm begins brewing and we need the money, Treasury will have to borrow it. Is that about it?

REGAN: Yep.

ISAAC: Just one more thing, while I've got you. Why do we bother pretending there's a fund?

REGAN: I'm sorry, Bill, but the President's on the other line. I'll have to get back to you on that.

There is no record that Regan ever got back to Isaac.

"Why do we bother pretending there's a fund?" asked Darryl Robert Schoon, economic commentator and author of *How to Survive the Crisis and Prosper in the Process*. "[T]he answer is obvious. Modern economics, i.e. central banking, is a shell game where bankers with the aid of governments have foisted a highly lucrative fraud on society; and, while the fraud of the FDIC fund is egregious, it is no more egregious than the fraud of the Fed or of the economy itself."

And the fraud does not stop with the FDIC. Schoon and others believe modern banking is essentially a Ponzi scheme on a global scale, in which bankers loan nonexistent money and receive repayment of the nonexistent funds plus compounding interest in return.

"In economies based on the fraudulent issuance of money as debt, there are only predators and victims. Bankers are the predators, society is the victim (businessmen are victims who often believe they're predators) and governments are the well-paid-off referees in the rigged game being played out in today's capital markets," Schoon wrote.

At the heart of this combination Ponzi scheme and shell game lies the privately owned Federal Reserve System. But you and I, dear reader, will get to that.

Chris Martenson, a businessman with a doctorate in neurotoxicology from Duke University and an MBA in finance from Cornell, wrote, "Our entire monetary system, and by extension our economy, is a Ponzi economy in the sense that it really only operates well when in expansion mode. Even a slight regression triggers massive panics and disruptions that seem wholly inconsistent with the relative change, unless one understands that expansion is more or less a *requirement* of our type of monetary and economic system. Without expansion, the system first labors and then destroys wealth far out of proportion to the decline itself. What fuels expansion in a debt-based money system? Why, new debt (or credit), of course! So one of the things we keep a very close eye on, as they do at the Federal Reserve, is the rate of debt creation."

Martenson and others believe a major theme in the current credit bubble collapse is the extent to which private credit has been crumbling while the Federal Reserve has been purchasing debt and the federal government has been increasing its borrowing. "In essence, *public* debt purchases and new borrowing has attempted to plug the gap left by a shortfall in *private* debt purchases and borrowing [original emphasis]. That's the scheme right now—the Federal Reserve is creating new money out of thin air to buy debt, while the US government is creating new debt at the most fantastic pace ever seen. The attempt here is to keep aggregate debt growing fast enough to prevent the system from completely seizing up," explained Martenson.

Martenson, who said he continually seeks to accept or reject his own hypotheses based on the evidence at hand, explained that the Federal Reserve has been monetizing far more U.S. government debt than has openly been revealed by allowing foreign central banks to swap their agency debt for Treasury debt. "This is not a sign of strength and reveals a pattern of trading temporary relief for future difficulties," Martenson wrote. "When the full scope of this program is more widely recognized, more pressure will fall upon the dollar, as more and more private investors shun the dollar and all dollar-denominated instruments as stores of value and wealth. This will further burden the efforts of the various central banks around the world as they endeavor to meet the vast borrowing desires of the US government. One possible result of the abandonment of these efforts is a wholesale flight out of the dollar and into other assets. To US residents, this will be experienced as rapidly rising import costs and increasing costs for all internationally-traded basic commodities, especially food items. For the rest of the world, the results will range from discomforting to disastrous, depending on their degree of dollar linkage. . . . The shell game that the Fed is currently playing does not change the basic equation: Money is being printed out of thin air so that it can be used to buy US government debt." It has been long understood that creating more money leads to inflation since the more currency in circulation, the less it's worth, especially paper money that has no intrinsic value.

As to the government buying private debt, a crude example of what has happened goes like this:

Tom has a mortgage on a very nice house. He has a good job and his credit is good. Dick lives in a run-down home badly in need of serious repairs and has been in and out of jobs so he has a low credit score. Yet, due to government pressure on the lending industry to provide housing to all, Dick has a mortgage on his home. Through a scheme called "bundling," Tom's mortgage and a few others like his are combined with Dick's mortgage and many others like his. By sleight of hand, this combined package of mortgages is given an A-1 rating and the package is sold to venture capital firms as a good investment. With these investment packages growing in number, the economy booms. But when the housing bubble breaks, the investment firms, many of the largest filled with globalists, turn to the government for relief with the argument that if they go bankrupt, the whole national economy will suffer. The government then pays these firms for their investment at full value, even though many of the houses are substandard (subprime) and not worth full value. The government pays with taxpayer money, then orders more money printed to cover the shortfall. The investment firms are also paid with the condition that their money comes in the form of government bonds, which means even more paper is spread around, causing further inflation and devaluation. It is robbery on a grand scale, with the strapped taxpayer taking the hit while the middle-men financiers continue to make a profit. To add insult to injury, many of these financiers are banks and investment houses outside the United States, which means U.S. taxpayers are paying back foreign investors for making bad investments.

HOW IT ALL BEGAN

OUR NATION'S ECONOMIC DECAY did not start with the Obama administration or even with the George W. Bush regime; rather, it began decades earlier in the early twentieth century with the founding of a privately owned banking syndicate known as the Federal Reserve System, a government-sanctioned cartel of private banks that was created in a conspiratorial manner and is under heavy criticism to this day, even being blamed for the current financial woes.

Joan Veon, a businesswoman and international reporter who has covered more than a hundred global conferences on financial and trade matters, wrote that the recent bailouts were simply the latest moves by the globalists to solidify their control over the United States. "The bailout of Freddie and Fannie provided us with the latest excitement in the diabolical saga of the raping, robbing, and pillaging of America. Interestingly enough, it took place 13 months after the beginning of the credit crunch . . . it was planned and managed destruction in order to accomplish the final transfer of America's financial sovereignty," she noted.

Former secretary of housing Catherine Austin Fitts agreed, stating that in the attempt to build a global American-run military empire, trillions of dollars have been shifted out of the United States by both legal and illegal means to reinvest in Asia and emerging markets through taxpayer bailout money coupled with Fed loans to foreign banks. In doing so, she said we have left economic sovereignty behind. "Finally, the expense and corruption of empire resulted in bailouts of $12–14 trillion, delivering a new financial war chest to the people leading the financial engineering [the globalists]. Now we have exploding unemployment, an exploding federal deficit, an Inspector General for the TARP [Troubled Asset Relief Program] bailout program predicting that the ultimate bailout cost could rise to $23.7 trillion . . . ," said Fitts.

With this lost money came lost jobs. Unemployment figures are usually a good gauge of the nation's economy. In mid-2009, unemployment was officially 9.4 percent. If for some reason this number seems low, one must note that that these numbers do not include "those who would like a job but have stopped looking—so-called discouraged workers—and those who are working fewer hours than they want," said Dennis Lockhart, president and CEO of the Federal Reserve Bank in Atlanta. With these numbers included, the unemployment rate would move from the official 9.4 percent to 16 percent. As 2010 progressed, so did the unemployment figures, which began to match the numbers of the Great Depression.

Yet unlike the 1930s, money was still available; and money is the life-blood of a zombie nation. The trappings of wealth and bankers' lifestyles are often admired by outsiders with a fervency bordering on religious, yet only those who live these lifestyles understand the inner workings of the money cult. And they work hard to keep these inner workings secret.

Consider the 1966 essay "Gold and Economic Freedom" by Alan Greenspan, who from 1987 to 2006 was chairman of the Fed. Greenspan wrote, "Deficit spending is simply a scheme for the 'hidden' confiscation of wealth. Gold stands in the way of this insidious process. It stands as a protector of property rights. If one grasps this, one has no difficulty in understanding the statists' antagonism toward the gold standard." In other words, spending paper money you don't have runs up debt that, with interest due, earns much more than the original debt, especially if it is not repaid promptly. This is the "hidden confiscation of wealth." Paper money can be devalued, but a gold piece will always retain some value and is therefore a good hedge against both inflation and devaluation, which is why the globalists seeking a strong central authority (statists) are generally opposed to a gold standard, because it robs them of the means of robbing the public through high interest rates, service charges, late payments, and monetary exchanges.

Following a talk by Greenspan at the Economic Club of New York in 1993, Dr. Lawrence Parks, the executive director of the Foundation for the Advancement of Monetary Education (FAME), approached the Fed chairman and asked if he still agreed with his 1966 conclusions on deficit spending and gold. "Absolutely," Greenspan responded. Parks then asked why Greenspan did not speak out about his knowledge and the response was, "Some of my colleagues at the institution I represent [the Fed] do not agree with me."

Whether Greenspan was fibbing or he was mistaken about his colleagues, the Fed actually shared Greenspan's opinion on gold—they just didn't want the public to know. The nonprofit Gold Anti-Trust Action Committee Inc. (GATA) was organized in 1999 to oppose the illegal collusion over the price and supply of gold and related financial securities. According to the committee, in 2009, the Federal Reserve System disclosed to Congress that it had made gold swap arrangements with foreign banks, but it does not want the public to know about them. This disclosure directly contradicted the Fed's earlier denials of making gold swaps to GATA back in 2001. A GATA news release also suggested that the Fed was indeed very much involved in the surreptitious international central bank manipulation of the gold price particularly and the currency markets generally,

Earlier in 2009, GATA sought information on current gold swaps, a practice denied by Alan Greenspan, then Fed chairman, back in 1995. But this question was rebuffed by the Fed, which claimed this information was exempt from Freedom of Information Act requests. GATA appealed to the Fed's board. But in a September 2009 letter to GATA's lawyer, Federal Reserve Board member Kevin M. Warsh upheld the denial of information by stating, "In connection with your appeal, I have confirmed that the information withheld under Exemption 4 consists of confidential commercial or financial information relating to the operations of the Federal Reserve Banks that was obtained within the meaning of Exemption 4. This includes information relating to swap arrangements with foreign banks on behalf of the Federal Reserve System and is not the type of information that is customarily disclosed to the public. This information was properly withheld from you."

GATA claimed the letter was not the first admission of the Fed making gold swaps but that "it comes at a sensitive time in the currency and gold markets." According to a GATA news release, "The U.S. dollar is showing unprecedented weakness, the gold price is showing unprecedented strength, Western European central banks appear to be withdrawing from gold sales and leasing, and the International Monetary Fund is being pressed to take the lead in the gold price suppression scheme by selling gold from its own supposed reserves in the guise of providing financial support for poor nations."

It is now expected that a lawsuit will be filed in federal court to appeal the Fed's denial of GATA's freedom-of-information request concerning gold swaps. Those people stocking up on gold for safekeeping might keep in mind that gold and silver—in fact, just about anything considered a financial asset—may be seized by federal authorities in wartime or any officially declared "emergency." Those who hoard gold against the possible devaluation or collapse of the dollar might remember that during the Great Depression, the hoarding and use of gold as a medium of exchange was outlawed.

According to the GATA website, government confiscation of gold has never been a serious or imminent threat, but in any "emergency," this could swiftly change. "While the U.S. Government in 1933 did demand the exchange of circulating government-issued coins for paper money (proceeding to devalue the paper money after the gold was surrendered), that gold then was a huge part of the country's money supply, and amid the national economic collapse at that time the government could make a plausible complaint against 'hoarding.' There are no such circumstances today, gold no longer being in general circulation as currency. . . . But of course lately the arrogance and imperiousness of the U.S. government have far exceeded even the paranoia of precious metals investors. Certainly capital controls may be imposed in the United States in the next currency crisis, and it's not far from capital controls to even more brutal interventions in the economy."

Such concern intensified with a 2005 letter to GATA in which the former chief counsel for the Treasury Department's Office of Foreign Assets Control, Sean M. Thornton, explained the scope of the government's power in making financial seizures. "It took GATA six months and a little prodding to get answers from the Treasury, but the Treasury's reply, when it came, was remarkably comprehensive and candid.

"The government's authority to interfere with the ownership of gold, silver, and mining shares arises . . . from the Trading with the Enemy Act, which became law in 1917 during World War I and applies during declared wars, and from 1977's International Emergency Economic Powers Act, which can be applied without declared wars.

"While the Trading with the Enemy Act authorizes the government to interfere with the ownership of gold and silver particularly, it also applies to all forms of currency and all securities. So the Treasury official stressed that it could be applied not just to shares of gold and silver mining companies but to the shares of all companies in which there is a foreign ownership interest. Further, there is no requirement in the law that the targets of the government's interference must have some connection to the declared enemies of the United States, or, really, some connection to foreign ownership. Anything that can be construed as a financial instrument, no matter how innocently it has been used, is subject to seizure under the Trading with the Enemy Act and the International Emergency Economic Powers Act."

USURY

"USURY" IS A TERM that has all but disappeared from our language. Once, "usury" was defined as any interest charged for a loan, but modern dictionaries softened this definition to merely "excessive" interest. The Texas Constitution once defined "usury" as any interest in excess of 6 percent. This ceiling was increased over the years until the whole concept was deleted.

Those who know the Bible recall that Jesus was crucified by those in power for chasing "money changers" out of the temple. Public anger today is being directed at the financial moguls of both Wall Street and Washington, D.C.

"Charging interest on pretended loans is usury, and that has become institutionalized under the Federal Reserve System," argued G. Edward Griffin, author of *The Creature from Jekyll Island*. This has been accomplished by masking the operations of the Fed in secrecy and arcane economic terms. "The . . . mechanism by which the Fed converts debt into money may seem complicated at first, but it is simple if one remembers that the process is not intended to be logical but to confuse and deceive," Griffin added.

Former *Washington Post* editor William Greider wrote, "The details of [the Fed's] actions were presumed to be too esoteric for ordinary citizens to understand." Some believe this ignorance may be a blessing. Henry Ford was quoted as saying, "It is well enough that the people of the nation do not understand our banking and monetary system for, if they did, I believe there would be a revolution before tomorrow morning."

"Most Americans have no real understanding of the operation of the international moneylenders," stated the late senator Barry Goldwater. "The bankers want it that way. We recognize in a hazy sort of way that the Rothschilds and the Warburgs of Europe and the houses of J. P. Morgan, Kuhn, Loeb and Company, Schiff, Lehman and Rockefeller possess and control vast wealth. How they acquire this vast financial power and employ it is a mystery to most of us. International bankers make money by extending credit to governments. The greater the debt of the political state, the larger the interest returned to the lenders. The national banks of Europe are actually owned and controlled by private interests." These same "private interests" now own and control the Federal Reserve System.

MONEY FOR FAITH AND DEBT

ACCORDING TO WILLIAM GREIDER, the Fed has assumed a cult-like power: "To modern minds, it seemed bizarre to think of the Federal Reserve as a religious institution. . . . Yet the conspiracy theorists, in their own demented way, were on to something real and significant. . . . [The Fed] did also function in the realm of religion. Its mysterious powers of money creation, inherited from priestly forebears, shielded a complex bundle of social and psychological meanings. With its own form of secret incantation, the Federal Reserve presided over awesome social ritual, transactions so powerful and frightening they seemed to lie beyond common understanding. . . .

"Above all, money was a function of faith. It required implicit and universal social consent that was indeed mysterious. To create money and use it, each one must believe and everyone must believe. Only then did worthless pieces of paper take on value."

Money today is increasingly mere electronic blips in a computer accessed by plastic cards at ATMs. There is nothing to back it up. As money is loaned at interest by great institutions, its worth decreases as more and more of it comes into existence. This is called inflation, which in some ways is a built-in tax on the use of money. And inflation can be manipulated upward or downward by those who control the flow of money, whether it be through paper or the electronic blips.

"The result of this whole system is massive debt at every level of society today," wrote author William Bramley. "The banks are in debt to the depositors, and the depositors' money is loaned out and creates indebtedness to the banks. Making this system even more akin to something out of a maniac's delirium is the fact that banks, like other lenders, often have the right to seize physical property if its paper money is not repaid."

THE FEDERAL RESERVE ANOMALY

IN AMERICA, THE BANKERS of the Federal Reserve System have the greatest control of the nation's money. Because the Fed is at the center of U.S. monetary policy control, it has become the central bank of the United States. By changing the supply of money in circulation, the Fed influences interest rates, which in turn affects millions of families' mortgage payments. It also can cause financial markets to boom or collapse and the economy to expand or contract into recession.

The Fed is "the crucial anomaly at the very core of representative democracy, an uncomfortable contradiction with the civic mythology of self-government," wrote William Greider. His 1987 book *Secrets of the*

Temple: How the Federal Reserve Runs the Country disparages "nativist conspiracy theories" yet presents an eloquent conspiracy argument for the Fed's control.

Consider that a paper bill is simply a promissory note to be traded at some point for something of value. It thus makes sense to perceive paper money as valuable as real goods or services. This viewpoint worked well before the invention of interest. The early goldsmiths in Europe who warehoused gold coins used their stockpiles as the basis for issuing paper money. Since it was highly unlikely that everyone would demand their gold back at the same time, the smiths became bankers, loaning out a portion of their stockpile at interest for profit. This practice—loaning the greater portion of wealth while retaining only a small fraction for emergencies—became known as fractional reserve, or fractional banking. This system worked well until everyone suddenly wanted their deposits back and started a "run" on the bank. Bank runs, or depositors demanding their money back all at one time, were a major cause of financial damage during the Great Depression of the 1930s. But runs are not just history. In early 2008, Northern Rock Bank, the fifth-largest bank in the United Kingdom, was nationalized by the government due to financial problems created by the subprime mortgage crisis and a run on its branch banks.

After the invention of fractional banking came the implementation of "fiat" money—intrinsically worthless paper money made valuable by law or decree of government. An early example of this system was recorded by Marco Polo during his visit to China in 1275. Polo noted the emperor forced his people to accept black pieces of paper with an official seal on them as legal money under pain of imprisonment or death. The emperor then used this fiat money to pay all his foreign debts.

"One is tempted to marvel at the [emperor's] audacious power and the subservience of his subjects who endured such an outrage," wrote G. Edward Griffin, "but our smugness rapidly vanishes when we consider the similarity of our own Federal Reserve Notes. They are adorned with signatures and seals; counterfeiters are severely punished; the government pays its expenses with them; the population is forced to accept them; they—and the 'invisible' checkbook money into which they can be converted—are made in such vast quantity that it must be equal in amount to all the treasures of the world. And yet they cost nothing to make. In truth, our present monetary system is an almost exact replica of that which supported the warlords of seven centuries ago."

Nowhere was the art of making money out of money more developed than in the ancient Khazar Empire, which evolved from nomadic raider-clans operating on the east-west caravan routes in the Caucasus Mountain region north of Iraq and between the Black Sea and Caspian Sea. By the tenth century, the Khazars had created a wealthy empire that stretched from north of the Black Sea to the Ural Mountains and west of the Caspian Sea to the Dnieper River.

The warlords of the Khazars thought that exchanging and loaning money would be more profitable and less hazardous than raiding caravans. There was one problem. The Khazar Empire was almost evenly divided among Christians, Muslims, and Jews. Both Christians and Muslims believed that charging interest on a loan, then called usury, was a sin. Only Jews could openly charge interest on loans. Whether they did it out of pragmatism or actual religiosity, the Khazar aristocrats professed a conversion to Judaism. According to the *Random House Encyclopedia,* "Some scholars believe they [the Khazars] are the progenitors of many Eastern European Jews." This would include the renowned Rothschild family, who financially ruled Europe for more than a century. Conspiracy researchers claim they still dominate the world financial order and have been the financial backers of the Rockefellers and other wealthy families. It might be noted that none of these converted Khazarians had any connection whatsoever to Palestine, yet these were among the Russian progenitors of the political movement known as Zionism.

The 1917 Balfour Declaration, a statement by British foreign secretary Alfred Balfour that guaranteed a Jewish home in Palestine and was later approved as a mandate by the League of Nations, is acknowledged as the foundation for the creation of the state of Israel. This letter originally was a reply to a leading Zionist, Baron Walter Rothschild, the first unconverted Jewish peer in England's House of Lords.

The money-management methods of the Rothschild banking dynasty have been emulated for decades by the globalist financiers, whether Jewish or otherwise. One key component of this management is secrecy. Utilizing bought-off politicians, who catch the public rage and scrutiny, major globalists are able to operate out of the public eye almost with impunity. Derek Wilson, who chronicled the Rothschild empire in his 1988 book *Rothschild: The Wealth and Power of a Dynasty,* wrote, "Even when, in later years, some of them [Rothschilds] entered parliament, they did not feature prominently in the assembly chambers of London, Paris or Berlin. Yet all the while they were helping to shape the major events of the day: by granting or withholding funds; by providing statesmen with an official diplomatic service; by influencing appointments to high office; and by an almost daily intercourse with the great decision makers."

The invention of the printing press, which allowed for the printing of paper money as well as the Bible, led to the Age of Enlightenment and the decline of the Roman Church. Money replaced religion as the new control mechanism of the wealthy elite. And despite the popular myth, the American colonial revolt against England occurred more over concern for its own currency than a small tax on tea. Benjamin Franklin wrote, " . . . the inability of the colonists to get the power to issue their own money permanently out of the hands of George III and the international bankers was the prime reason for the Revolutionary War." As previously discussed, wealth equals power. And the American revolutionists knew that to gain true freedom, they had to break the power of the Rothschild-dominated Bank of England, which had outlawed their money—colonial script.

Once America's freedom was secured, Founding Fathers Thomas Jefferson and Alexander Hamilton began arguing over whether or not to adopt a central bank. Hamilton believed in a strong central government with a central bank overseen by a wealthy elite. "No society could succeed which did not unite the interest and credit of rich individuals with those of the state," Hamilton wrote. Supporters of Hamilton's elitism formed America's first political party, the Federalists. Hamilton, once described as a "tool of the international bankers," argued that "A national debt, if it is not excessive, will be to us a national blessing. It will be a powerful cement to our nation. It will also create a necessity for keeping up taxation to a degree which, without being oppressive, will be a spur to industry."

America's first central bank, the Bank of North America, was created in 1781 by Continental congressman Robert Morris, who modeled the bank after the Bank of England. The bank was formed before the Constitution was drafted and was wrought with fraud and plagued by inflation caused by the creation of baseless "fiat" currency. The bank lasted for three years. Morris's former aide, Alexander Hamilton, became secretary of the Treasury and in 1791 headed the next attempt at a central bank by establishing the First Bank of the United States. He was strongly opposed by Jefferson and his followers. In 1811, the charter of the First Bank of the United States was not renewed.

Jefferson knew from British and European history that a central bank trading on interest could quickly become the master of a nation, noting to John Taylor in 1816 that " . . . the other nations of Europe have tried and trodden every path of force or folly in fruitless quest of the same object, yet we still expect to find in juggling tricks and banking dreams, that money can be made out of nothing. . . . [B]anking establishments are more dangerous than standing armies; and that the principle of spending money to be paid by posterity, under the name of funding, is but swindling futurity on a large scale." Jefferson added, "Already they have raised up a money aristocracy. . . . The issuing power should be taken from the banks and restored to the people to whom it properly belongs."

Jefferson believed that instituting a central bank would be unconstitutional. "I consider the foundation of the Constitution as laid on this ground [enshrined in the Tenth Amendment]: That 'all powers not delegated to the United States, by the Constitution, nor prohibited by it to the States, are reserved to the States or to the people.' To take a single step beyond the boundaries thus specially drawn around the powers of Congress, is to take possession of a boundless field of power, no longer susceptible of any definition. The incorporation of a bank, and the powers assumed by this bill, have not, in my opinion, been delegated to the United States, by the Constitution."

Despite Jefferson's lobbying, the financial chaos that resulted from the War of 1812 prompted Congress to issue a twenty-year charter to the Second Bank of the United States in 1816. Andrew Jackson, the first president from west of the Appalachian Mountains, denounced the central bank as unconstitutional and as "a curse to a republic; inasmuch as it is calculated to raise around the administration a moneyed aristocracy dangerous to the liberties of the country." This central bank ended in 1836, after President Jackson vetoed a congressional bill to extend its charter.

Much to bankers' dismay, Jackson fully eliminated the national debt by the end of his two terms as president. It was probably no coincidence that America's first assassination attempt was made on Jackson by a man named Richard Lawrence, a man who claimed to be in touch with "the powers in Europe," who had promised to intervene if any attempt was made to punish him. Lawrence was a painter, and many speculate that at the time the lead in his paints had caused him to become mentally unbalanced and fancy himself the rightful king of England. After stalking Jackson for several weeks, on January 30, 1835, a particularly humid day, he approached the president coming from a funeral. Stepping suddenly from behind a pillar, Lawrence pulled two pistols but both misfired, most likely due to damp powder. Lawrence was swiftly wrestled to the ground by onlookers, including Congressman Davy Crockett aided by Jackson. At his trial, Lawrence was prosecuted by Francis Scott Key, author of "The Star-Spangled Banner." The jury took only five minutes to find Lawrence insane and he spent the rest of his life in mental institutions, dying in 1861. Although many persons, including Jackson, believed Lawrence was part of a larger conspiracy, at the time there was no evidence to prove whether he was merely a lone-nut assassin or an early-day patsy somehow manipulated into attacking Jackson, an implacable enemy of the international bankers. However, it might be worth noting that in two successful presidential assassinations—those of Abraham Lincoln and John F. Kennedy—both men were attempting to thwart the international bankers—Lincoln by issuing his own money, greenbacks, and Kennedy in bypassing the Fed with U.S. notes in 1963.

"While most people understand what took place when the American

Revolution was fought, many are not aware of the permanent financial revolution that [was] being fought over the world's monetary system since 1694 when the Bank of England was created," explained international reporter Joan Veon. "At that time, a group of private individuals decided that they could make a great deal of money if they changed the laws of the land to shift control of the country's finances from the government to them. The Bank of England, which is England's 'central bank,' is a private corporation which earns a continuous stream of income when the British government borrows from them to run the country. England was the ingenious country that recognized they could run the world's finances if they established private corporations in all the countries of the world. The combined debt of all the world's country's [*sic*] would create an income stream of unbelievable amounts. In 1913, Congress passed the Federal Reserve Act creating our central bank. Most Americans don't know that this organization is a private corporation established to control America's monetary system through the banking industry."

Other attempts were made to resurrect a central bank in America but none succeeded until the creation of the Federal Reserve System at the hands of a well-documented conspiracy. "The situation we are confronted with did not happen in the last few years, but began in 1913 when a group of cunningly deceitful legislators passed the Federal Reserve Act on December 24 at 11:45 p.m., after those who were opposed went home for Christmas," Veon noted.

"[T]here was an occasion near the close of 1910, when I was as secretive, indeed, as furtive as any conspirator. . . . I do not feel it is any exaggeration to speak of our secret expedition to Jekyll Island as the occasion of the actual conception of what eventually became the Federal Reserve System . . . ," wrote Frank A. Vanderlip, one of the men who created the Fed. He went on to become president of New York's National City Bank, a forebear of today's Citibank.

What Vanderlip was referring to was a secretive trip on the night of November 22, 1910, by seven men who perhaps held as much as one-fourth of the world's wealth. Jekyll Island was J. P. Morgan's fashionable hunting retreat off the coast of Georgia, and the men went under secrecy so strict that they only used first names when addressing one another and brought in new servants who were unaware of their identities.

During their week on Jekyll Island, the men worked on a plan for a banking reform that the government deemed necessary after a series of financial panics in 1879, 1893, and 1907. In fact, Princeton University president and future U.S. president Woodrow Wilson proclaimed that the solution to the financial panics laid in the appointment of "a committee of six or seven public-spirited men like J. P. Morgan to handle the affairs of our country." Cries arose for a stable national system that could regulate banking and prevent crises and panics. Today, many researchers believe these panics were artificially created as a pretext for the "reforms."

The seven men were Vanderlip, who represented William Rockefeller and Jacob Schiff's investment firm of Kuhn, Loeb & Company; Assistant Secretary of the Treasury Abraham Piatt Andrew; senior partner of J. P. Morgan Company Henry P. Davison; First National Bank of New York (a Morgan-dominated institution) president Charles D. Norton; Morgan lieutenant Benjamin Strong; Kuhn, Loeb & Company partner Paul Moritz Warburg; and Rhode Island Republican senator Nelson W. Aldrich. Though Aldrich was not technically a banker, he was an associate of J. P. Morgan. He was also the father-in-law of John D. Rockefeller Jr. Paul Warburg, an original founder of the Council on Foreign Relations, was the brother of Max Warburg, chief of the M. M. Warburg Company banking consortium in Germany and the Netherlands. In just a few years, Max Warburg would aid Lenin in crossing wartime Germany to found communism in Russia.

It must also be noted that senator Aldrich was chairman of the National Monetary Commission, charged with stabilizing the U.S. monetary system. Aldrich and his commission toured Europe at taxpayer expense and consulted with the top central banks of England, France, and Germany, which were all dominated by the Rothschilds. After spending $300,000 of tax dollars, the commission subsequently released a thirty-eight-volume history of European banking, focusing on the German Reichsbank, whose principal stockholders were the Rothschilds and M. M. Warburg Company.

The National Monetary Commission's final report was prepared by the very men who had secretly journeyed to Morgan's Jekyll Island Hunt Club ostensibly to hunt ducks. These men concluded that having one central bank in the United States was insufficient. Rather, several would be needed, and they would have to operated under the auspices of what would look like an official agency of the U.S. government. They also agreed that no one was to utter the words "central" or "bank," a pact that held up well—the Fed was never publicly referred to as "the central bank" until well into the 1980s, when the term was no longer as loaded.

Speaking before the American Banker's Association, Aldrich stated, "The organization proposed is not a bank, but a cooperative union of all the banks of the country for definite purposes." Paul Warburg had conceived of constructing a cooperative banking union in which restrictions on the banker would be removed in a manner palatable to both the bankers and the public.

But too many people saw the Aldrich Plan as a transparent attempt to create a system by the bankers and for the bankers. "The Aldrich Plan is the Wall Street plan," warned Representative Charles A. Lindbergh, father of the famed aviator. When Aldrich proposed his plan as a bill, it never got out of committee.

Aldrich needed a new tactic. It came by way of the House Banking and Currency Committee chairman, Representative Carter Glass of Virginia, who attacked the Aldrich Plan by openly stating it lacked government control and created a banking monopoly. Glass drafted an alternative, the Federal Reserve Act. Jekyll Island planners Vanderlip and Aldrich spoke out venomously against Glass's bill, even though entire sections of the bill were identical to the Aldrich Plan. By putting on a front of banker opposition, Aldrich and Vanderlip ingeniously garnered public support for the Glass bill in the major newspapers.

Meanwhile, another tactic was being played out in the political arena—dethroning the president. President William Howard Taft was already on the record pledging to veto any legislation creating a central bank. A more compliant leader was needed by the bankers. This leader was Woodrow Wilson, the academic who had been retained as president of Princeton

University by his former classmates Cleveland H. Dodge and Cyrus Mc-
Cormick Jr., both directors of Rockefeller's National City Bank of New
York.

"For nearly 20 years before his nomination, Woodrow Wilson had
moved in the shadow of Wall Street," wrote author Ferdinand Lundberg.
Wilson, who had praised J. P. Morgan in 1907, had been made governor of
New Jersey. With the approval of the nation's bankers, Wilson's nomination for president was secured by Colonel Edward Mandell House, a close
associate of Warburg and Morgan. House would go on to become Wilson's constant companion and adviser. "The Schiffs, the Warburgs, the
Kahns, the Rockefellers and the Morgans [all] had faith in House," noted
Professor Charles Seymour, who edited House's papers.

But there was a problem. Early polling indicated that the Democrat
Wilson could not defeat the Republican Taft. In a political maneuver
that has been used successfully several times since, former president Theodore "Teddy" Roosevelt—also a Republican—was encouraged to run as a
third-party candidate. Large sums of money were provided to his Progressive Party by two major contributors closely connected to J. P. Morgan.

The maneuver worked as well with the 1912 campaign as it would with
the subsequent campaigns of George Wallace, John B. Anderson, Ross
Perot, Ralph Nader, and Chuck Baldwin. Roosevelt pulled enough votes
away from Taft for Wilson to be elected by a narrow margin.

Wilson signed the Federal Reserve Act on December 23, 1913, the
same day a House-Senate conference committee had passed it along and
the day before Christmas Eve. Congress was already home and the average
citizen's attention was focused on the holidays. "Congress was outflanked,
outfoxed and outclassed by a deceptive, but brilliant, psycho-political attack," commented G. Edward Griffin.

Today, the Federal Reserve System is composed of twelve Federal Reserve banks that operate under the New York Federal Reserve bank. Each
serves a different section of the country. These banks are administered by
a board of governors, which is appointed by the president and confirmed
by the Senate. The confirmation is usually a rubber-stamp procedure.

As previously noted, the current chairman of the Fed's board of governors is Ben Shalom Bernanke, who succeeded Alan Greenspan in 2006 and was renamed chairman by President Obama in August 2009. In 2008, Bernanke was photographed leaving the yearly meeting of that secretive globalist group known as the Bilderbergers (see Jim Marrs's *Rule by Secrecy* for the history of the Bilderbergs) in Chantilly, Virginia. Also on the board of governors is Daniel Tarullo, a Georgetown law professor who specializes in international economic regulation, banking law, and international law, and who has served as a senior fellow at the Council on Foreign Relations.

The youngest governor in the history of the board is Kevin Maxwell Warsh, a vice president of Morgan Stanley, who was age thirty-five at his appointment in February 2006. Warsh was trained as a lawyer, not as an economist.

Today, most people recognize that the Fed is a pivotal force in the world economy, but few understand who controls it and why. It is a private organization owned by its member banks, which are owned by private stockholders. And who are these stockholders?

"An examination of the major stockholders of the New York City banks shows clearly that a few families, related by blood, marriage, or business interests, still control the New York City banks which, in turn, hold the controlling stock of the Federal Reserve Bank of New York," wrote Eustace Mullins. In his 1983 book *The Secrets of the Federal Reserve,* Mullins presented charts connecting the Fed and its member banks to the families of the Rothschilds, Morgans, Rockefellers, Warburgs, and others.

It is interesting to note that those who sit at the very top of the corporate, academic, and labor power hierarchy are listed as 2009 directors of the Federal Reserve Bank of New York. This list includes James Dimon, chairman and CEO of JPMorgan Chase & Co.; Charles V. Wait, president, CEO, and chairman of the Adirondack Trust Company of Saratoga Springs, New York; Jeffrey R. Immelt, chairman and CEO of General Electric Company, Fairfield, Connecticut; Lee C. Bollinger, president of Columbia University; Kathryn S. Wylde, president and CEO of Partnership for New York City; and board chairman Denis M. Hughes, president of the New York State AFL-CIO.

Some suspicious researchers have speculated on why so many secret society members—Greenspan, Bernanke, Tarullo (all members of the CFR)—and attorneys are needed to supervise the U.S. monetary system. It might be that bankers need their legal expertise. According to early conspiracy researcher and author Gary Allen, "Using a central bank to create alternate periods of inflation and deflation, and thus whipsawing the public for vast profits, had been worked out by the international bankers [aided by legal and public opinion experts] to an exact science."

In 1913, Congressman Charles A. Lindbergh said that the Federal Reserve System "establishes the most gigantic trust on earth. . . . When the President signs this act, the invisible government by the money power . . . will be legitimized. The new law will create inflation whenever the trusts want inflation. From now on, depressions will be scientifically created," he warned.

"Most Americans understand that the Fed controls our money system, but they believe its [*sic*] part of our government, as would be expected of any organization holding that much power over the destiny of our country," explained Stephen Zarlenga, director of the American Monetary Institute in New York State. "Americans also erroneously believe the banking business consists of accepting deposits from clients and then re-loaning them to borrowers at a higher rate of interest.

"Though the number is definitely growing, most Americans have no idea that money (or more accurately interest-bearing bank credits [acting as a] purchasing media which serves as money) is created by the banking system when loans are made, through the fractional reserve provisions. This is understood by few novices, and often economists and even bankers fail to comprehend that they function as part of a money creation system, when they issue credits, and deposit them into their client's accounts when loans are extended. Therefore most Americans would be surprised to learn that almost all of what we use for money is not issued by our government, but by private banks. They have been 'allowed' to form erroneous assumptions about our money and banking system that are far from reality and that serves to shield from closer scrutiny [from questions such as] whether the Fed is truly operating in the public interest or advancing more private agendas, either on purpose or by default."

Bruce Wiseman, president of the Citizens Commission on Human Rights and former chairman of the history department at John F. Kennedy University, explained the Fed's operations: "When the Fed prints the money or clicks the mouse, they have no money themselves. They are just creating it out of thin air. They just print it, or send it digitally. And then they charge interest on the money they lent to the Treasury. A hundred-dollar bill costs $0.04 to print. But the interest is charged on the $100. Go ahead: read it again; the words won't change.

"The interest on the national debt last year [2008] was $451,154,049,950.63. That's $1.23 billion a day. These are the same people that are now running our banks, insurance companies and automobile manufacturers. Reason weeps. Sure, I oversimplified it. The Fed doesn't own all the debt and they do some other things. But these are the basics. That is how a central bank works."

Wiseman and many others believe the goal of the current financial crisis is to destroy the U.S. dollar as the currency of world finance and, in the resulting chaos, put in its place a globalist-run monetary authority that pledges such a crisis shall not happen again.

FINANCIAL STABILITY BOARD

THE GLOBAL FINANCIAL AUTHORITY Wiseman alluded to may be found in the Financial Stability Board (FSB), created in April 2009 during the G-20 London Summit. The acronym G-20 refers to the group of twenty finance ministers and central bank governors from nineteen nations and the European Union. The FSB includes representatives from all G-20 nations.

The FSB evolved from the Financial Stability Forum (FSF), which was established in 1999 as a group within the Bank for International Settlements to "promote international financial stability." It is clear now that the Forum's agenda of stability did not work out so well. Following the

G-20 London Summit, this group expanded from the discussion group forum (FSF) to a policy-making board (FSB) that can set standards, policies, and regulations and then pass them on to the respective nations. Today, the FSB is made up of the central bankers from Australia, Canada, France, Germany, Hong Kong, Italy, Japan, the Netherlands, Singapore, Switzerland, the United Kingdom, and the United States plus representatives of the World Bank, the European Union, the IMF, and the Organization for Economic Co-operation and Development (OECD). Europe, in other words, has six of the twelve national members.

It has been noted that the G-20 will enlarge the FSB to include all its member nations. However, observers see a definite pro-European bias. The United States will have one vote, equal to that of Italy.

The governor of Italy's central bank, Mario Draghi, chairs the FSB and is former executive director of the World Bank. Like former Treasury secretary Henry Paulson, Draghi is a former executive with Goldman Sachs. Both Paulson and Draghi left the global investment firm in 2006 when Paulson went to Washington to head the up the Treasury and Draghi went to Rome to oversee Italy's financial system and the FSF.

America's commitment to the FSB was made on April 2, 2009, when President Obama signed the G-20 communiqué in London and announced America's agreement to the new global economic union. "Henceforth, our SEC [Securities and Exchange Commission], Commodities Trading Commission, Federal Reserve Board and other regulators will have to march to the beat of drums pounded by the Financial Stability Board, a body of central bankers from each of the G-20 states and the European Union," warned Dick Morris, bestselling author and former adviser to President Clinton. "The Europeans have been trying to get their hands on our financial system for decades. It is essential to them that they rein in American free enterprise so that their socialist heaven will not be polluted by vices such as the profit motive. Now, with President Obama's approval, they have done it."

Morris also opined on the FSB's ability for "implementing . . . tough new principles on pay and compensation and to support sustainable compensation schemes and the corporate social responsibility of all firms. . . .

That means that the FSB will regulate how much executives are to be paid and will enforce its idea of corporate social responsibility at 'all firms.'" Bruce Wiseman interprets President Barack Obama's signing of the United States into the FSB as "essentially turn[ing] over financial control of the country, and the planet, to a handful of central bankers, who, besides dictating policy covering everything from your retirement income to shareholder rights, will additionally have access to your health and education records."

Although the Fed is technically owned through shares held by its twelve regional banks, these banks are entirely owned by the private member banks within their respective districts. And who controls these banks? Their investors, many of whom may not even be Americans. Stephen Zarlenga argues that there may not be reason for concern here, however. "Stories that the Federal Reserve is 'owned' by foreign bankers . . . are not accurate and these types of rumors have mainly served to discredit wholesome criticism of the banking system. . . . The control of the Federal Reserve System is more difficult to untangle and is not just a matter of counting shareholder votes. While foreign bankers might indirectly own shares of the regional Federal Reserve Banks through ownership of American banking companies, such ownership would be reported to the SEC if any entity held more than 5 percent of the American corporation."

But, according to Zarlenga, there is one significant caveat: "The strong, potentially undue foreign influence, for example through the Bank for International Settlements (BIS)." Bringing the BIS into the financial mix is cause for further concern.

INTO A BIS

IN A 2003 ARTICLE titled "Controlling the World's Monetary System: The Bank for International Settlements," Joan Veon noted that the BIS is "where all of the world's central banks meet to analyze the global economy and determine what course of action they will take next to put more money in their pockets, since they control the amount of money in circulation and how much interest they are going to charge governments and banks for borrowing from them. . . . When you understand that the BIS pulls the strings of the world's monetary system, you then understand that they have the ability to create a financial boom or bust in a country. If that country is not doing what the money lenders want, then all they have to do is sell its currency."

The BIS has even seemed to be cryptically signaling that it may try to exert more global financial control. In September 2009, a BIS report stated, "The global market for derivatives rebounded to $426 trillion in the second quarter [2009] as risk appetite returned, but the system remains unstable and prone to crises." Within days of this report, the former chief economist for the BIS, William White, warned that the world has not tackled the problems at the heart of the economic downturn and is likely to slip back into recession. He added, "The only thing that would really surprise me is a rapid and sustainable recovery from the position we're in."

Considering the growing power of the BIS over the U.S. economy and the bank's Nazi history, BIS developments should be of serious concern to all Americans. It deserves much closer scrutiny than that provided by the corporate mass media. For one, the public should be aware that the BIS is essentially a sovereign state. Its personnel have diplomatic immunity for their persons and papers. No taxes are levied on the bank or the personnel's salaries. The grounds on which BIS offices sit are sovereign, as are the buildings and offices. No government has legal jurisdiction over the bank, nor do any governments have oversight over its operations.

It should also be noted that the BIS was originally owned in part by the Fed, the Morgan-affiliated First National Bank of New York, the Bank of England, Germany's Reichsbank, the Bank of Italy, the Bank of France, and other major central banks. The BIS, considered a "central bankers' bank," was created in 1930 in Basel, Switzerland, ostensibly to handle German war reparations.

The BIS was also heavily manipulated by secret societies. According to Carroll Quigley, a historian and a mentor to former President Bill Clinton, it was part of a plan "to create a world system of financial control in private hands able to dominate the political system of each country and the economy of the world as a whole . . . to be controlled in a feudalist fashion by the central banks of the world acting in concert by secret agreements arrived at in frequent meetings and conferences."

The BIS continued to control the finances between Germany and the Allied nations throughout World War II. According to Quigley, the BIS was administered by a multinational staff and was considered the "apex of the system" of bankers who secretly exchanged information and planning during World War II. Even worse, by the start of the war, the BIS was under total Nazi control. According to the bank's charter, which was agreed to by the governments that formed the bank, the bank was immune from seizure, closure, or censure even if its owners were at war. "The [BIS] bank soon turned out to be . . . a money funnel for American and British funds to flow into Hitler's coffers and to help Hitler build up his war machine," stated author Charles Higham.

"Over the years, I have watched as the BIS has continued to push the envelope further in a borderless world," wrote Joan Veon. "Some of their growing powers have come directly from governments like ours that have transferred the regulatory power they used to have over the banking system to the central bank while the rest comes from the simple fact that they do indeed control the monetary system of the world."

Veon, who had occasion to visit BIS headquarters, believes the bank has gained more power in global finance than most people know. This power stems from "[the BIS's] very powerful committees which include: the Basel Committee on Banking Supervision which has been working on how to regulate not only international banks of the world, but eventually . . . every national bank as well; the Committee on the Global Financial System, which monitors financial markets around the world with the objective of identifying potential risks for financial stability; and the Committee on Payment and Settlement Systems, [which] looks to strengthen the infrastructure of financial markets with regard to rules on how to transfer monies and how to make payments between member banks.

"The *Wall Street Journal* reported on a [2003] meeting which included [economist] Dr. [Jacob] Frenkel, former U.S. Fed Chairman Paul Volcker and Nobel Laureate and Columbia Economics Professor Dr. Robert Mundell. . . . Their theme was 'Does the Global Economy Need a Global Currency?' The thesis was that if the euro can replace the franc, mark and lira, why can't a new world currency merge the dollar, euro and yen? I submit to you that this is the next agenda of the central bankers. When this change occurs, I can assure you, they will make money on a new global currency. Time will tell if we do."

The globalist bankers make money on each dollar they print because the American taxpayer is available to make up for any losses incurred.

G. Edward Griffin quoted Paul Warburg, one of the founders of the Fed and its first chairman, as admitting, "While technically and legally the Federal Reserve note is an obligation of the United States Government, in reality it is an obligation, the sole actual responsibility for which rests on the reserve banks. . . . The government could only be called upon to take them up [on their obligation] after the reserve banks had failed."

"The man who masterminded the Federal Reserve System is telling us that *Federal Reserve notes constitute privately issued money with the taxpayers standing by to cover the potential losses of those banks which issue it* [original emphasis]," Griffin explained. Again, we see a clear example of private profit but public debt—the reserve banks take the profit while the taxpayers take the losses.

Perhaps Jefferson and Lindbergh were right after all when they warned about private control over a central bank. With the creation of the Fed, the major bankers finally fulfilled a long-standing goal—taxpayer liability for the losses of private banks. Some have called it "corporate socialism," whereby liabilities are assumed by the public treasury but profits are for the private gain of the bank officers and investors.

A taxpayer bailout was made manifest in the fall of 2008. The money that was used to cover government overspending and private corporation bailouts comes from a national income tax, which was invented by the same men who were behind creating the Fed. Sounding eerily like today's politicians, Wilson proclaimed his government was "more concerned about human rights than about property rights." Using this rhetoric as a smoke screen, Wilson pushed through more "progressive" legislation than any previous American administration. He created the Internal Revenue Service of the Treasury Department to enforce a graduated income tax, the Federal Farm Loan Act, which created twelve banks for farmers, and the Federal Trade Commission to regulate business.

To many people at the time, all of this legislation appeared necessary. Some still would argue that perhaps it is better that knowledgeable bankers be in charge of our nation's money supply. After all, a 1963 Federal Reserve publication states, "The function of the Federal Reserve is to foster a flow of money and credit that will facilitate orderly economic growth, a stable dollar, and long-run balance in our international payments."

ATTEMPTS TO AUDIT THE FED

IF THE TRUE FUNCTIONS of the Fed are to protect the nation's money, then it has failed miserably. In 2009, its failure brought demands for an audit of the Fed and possibly for its abolishment. Because no governmental audit of the Fed has been allowed since its inception, there has been no way to examine the Fed's true operating expenses or activities.

As far back as 1975, consumer advocate Ralph Nader asked, "Since other departments of government, including the departments of Defense and Treasury and other agencies that regulate banks, have long been subject to the audit of the General Accounting Office (GAO)—the investigative arm of Congress—why has the Federal Reserve been excluded? The answer is found in the secretive mixture of big power and big money of the banking goliaths and their Federal Reserve servants that for decades has kept such matters away from both [the] public and Congress, in order to retain their unperturbed control."

No matter how obscure the functions of the Fed are to the average citizen, according to Nader, its decisions and policies "affect the level of inflation, unemployment, home buying, consumer credit and other prices consumers and workers must bear. It also adds up to how few or how many financial corporations will dominate the economy." Despite Nader's support, as well as the backing of savings and loans institutions, credit unions, and some small bankers, a bill to provide for annual congressional audit of the giant Federal Reserve System was never passed in the 1970s.

Nothing much has changed more than thirty years later. Explanations that come from the Internet of how the Fed operates almost always come from government or Fed sources. Nevertheless, efforts have continued to rein in the Fed. On the pro-business site Forbes.com, Texas representative and dark horse presidential campaign contender Ron Paul wrote in May 2009, "One of the fallacies of modern economics is the idea that a central bank is required in order to keep inflation low and promote economic growth. In reality, it is the central bank's monetary policy that causes inflation and depresses economic growth. Inflation is an increase in the supply of money, which in our day and age is directly caused or initiated by central banks."

After noting the crumbling economy, Paul observed, "The necessary first step to restoring economic stability in this country is to audit the Fed, to find out the multitude of sectors in which it has involved itself and, once the audit has been completed, to analyze the results and determine how the Fed should be reined in. Proposals to push the Fed back into the shadows, or to give it an even greater role as a guarantor of systemic stability, are as misguided as they are harmful."

On February 26, 2009, Ron Paul introduced bill H.R. 1207, stating: "Serious discussion of proposals to oversee the Federal Reserve is long overdue. I have been a longtime proponent of more effective oversight and auditing of the Fed. . . . Since its inception, the Federal Reserve has always operated in the shadows, without sufficient scrutiny or oversight of its operations. While the conventional excuse is that this is intended to reduce the Fed's susceptibility to political pressures, the reality is that the Fed acts as a foil for the government. Whenever you question the Fed about the strength of the dollar, they will refer you to the Treasury, and vice versa. The Federal Reserve has, on the one hand, many of the privileges of government agencies, while retaining benefits of private organizations, such as being insulated from Freedom of Information Act requests.

"The Federal Reserve can enter into agreements with foreign central banks and foreign governments, and the GAO [the government's General Accountability Office] is prohibited from auditing or even seeing these agreements. Why should a government-established agency, whose police force has federal law enforcement powers, and whose notes have legal tender status in this country, be allowed to enter into agreements with foreign powers and foreign banking institutions with no oversight? Particularly when hundreds of billions of dollars of currency swaps have been announced and implemented, the Fed's negotiations with the European Central Bank, the Bank of International Settlements, and other institutions should face increased scrutiny, most especially because of their significant effect on foreign policy. If the State Department were able to do this, it would be characterized as a rogue agency and brought to heel, and if a private individual did this he might face prosecution under the Logan Act, yet the Fed avoids both fates.

"More importantly, the Fed's funding facilities and its agreements with the Treasury should be reviewed. The Treasury's supplementary financing accounts that fund Fed facilities allow the Treasury to funnel money to Wall Street without GAO or Congressional oversight. Additional funding facilities, such as the Primary Dealer Credit Facility and the Term Securities Lending Facility, allow the Fed to keep financial asset prices artificially inflated and subsidize poorly performing financial firms. . . . The Federal Reserve Transparency Act would eliminate restrictions on GAO audits of the Federal Reserve and open Fed operations to enhanced scrutiny. . . . By opening all Fed operations to a GAO audit and calling for such an audit to be completed by the end of 2010, the Federal Reserve Transparency Act would achieve much-needed transparency of the Federal Reserve."

National polls indicated deep and widespread public support for Paul's proposed audit. A mid-2009 Gallup poll showed that only 30 percent of those surveyed thought the Fed was doing a good job. Additionally, a Rasmussen poll stated that 75 percent of respondents wanted Congress to audit the Fed. Taking these poll numbers into consideration, the passage of legislation to audit the Fed is a litmus test to see who wields more power in the United States—the people or the banking interests.

As of February 2010, Paul's attempt to pass legislation to audit the Fed had gained 319 cosponsors in the House and 32 sponsors in the Senate where it was known as the Federal Reserve Sunshine Act of 2009 (S. 604). In early 2009, H.R. 1207 was referred to the House Committee on Financial Services, chaired by Massachusetts Democrat Barney Frank. In a letter to a constituent, Frank wrote: "I agree with the general thrust of [Ron Paul's] bill. . . . There have already been some moves forward in increasing the transparency of the Federal Reserve, and I agree that there are further steps we can take. . . . I do believe that the Federal Reserve is exercising that power with some good effects recently, but it is not a power that should exist in a democratic society in the hands of an entirely unelected entity."

On July 6, 2009, South Carolina Republican senator Jim DeMint attempted to amend the Legislative Branch Appropriations Act by adding the entire text of Ron Paul's bill, but he was stopped by senior Nebraska Democratic senator Ben Nelson, who said the amendment violated Senate Rule 16, which prevents tacking legislation onto an appropriations bill. After DeMint pointed out that other GAO audits in the appropriations bill violated Rule 16, Vice President Joseph Biden, who also is president of the Senate, agreed but took no action and the bill passed without the amendment. After two readings, S. 604 was referred to the Committee on Banking, Housing, and Urban Affairs in March 2009. On November 19, 2009, the House Committee on Financial Services approved an amendment to the Financial Stability Improvement Act of 2009 (H.R. 3996) that included many provisions of Paul's bill, such as the removal of some GAO audit restrictions and review of Fed policies and agreements with foreign institutions. This amendment was opposed by Fed chairman Bernanke, Treasury Secretary Geithner, and other Obama administration officials. After further changes to the amendment, including a provision that provided for audits of the Fed's balance sheet but not its monetary policies, in December the Financial Stability Improvement Act was combined with several other financial bills to form the Wall

Street Reform and Consumer Protection Act of 2009—Financial Stability Improvement Act of 2009 (H.R. 4173), which was passed on December 11 in the House with a 223 to 202 vote. No Republicans voted for the bill, including Paul, who apparently saw this combining maneuver as an attempt to water down his original audit proposal.

Paul's vote apparently was especially addressed to those who continue to support a hands-off attitude of the Fed, such as Forbes columnist Thomas F. Cooley, the Paganelli-Bull professor of economics, and Richard R. West, dean of the NYU Stern School of Business, who writes a weekly column for *Forbes*.

In a spring 2009 *Forbes* column, Cooley argued that "it is important to have an independent central bank. . . . An independent central bank can focus on monetary policies for the long term—that is, policies targeting low and stable inflation and a monetary climate that promotes long-term economic growth. Political cycles, alas, are considerably shorter. Without independence, the political cycle would subject the central bank to political pressures that, in turn, would impart an inflationary bias to monetary policy . . . politicians in a democratic society are short-sighted because they are driven by the need to win their next election. This is borne out by empirical evidence. A politically insulated central bank is more likely to be concerned with long-run objectives."

Cooley quoted a Ron Paul statement that "auditing the Fed is only the first step towards exposing this antiquated insider-run creature to the powerful forces of free-market competition. Once there are viable alternatives to the monopolistic fiat dollar, the Federal Reserve will have to become honest and transparent if it wants to remain in business." In response to this, Cooley wrote, "Great! Obviously, monetary policy is so falling-off-a-log simple that your elected representatives can insert themselves via the demand for transparency into decisions of true complexity and subtlety. Why am I not feeling reassured?"

He added, "Anything that threatens the independence of the Fed threatens the long-term viability of monetary policy. It is really important that the expanded role of the Fed in the current crisis not threaten that viability." But does such viability include secrecy and arrogance?

FED ARROGANCE

THE ARROGANCE OF THE Fed today is such that its board members refuse to even reveal what they have done with this nation's wealth.

The amounts of wealth involved are staggering, both in losses, bailouts, and unaccounted-for funds. In mid-May 2009, Federal Reserve inspector general Elizabeth A. Coleman stunned a congressional panel by verifying that her office could not account for $9 trillion worth of off-balance-sheet transactions made by the Fed between September 2008 and May 2009. "We're actually conducting a fairly high-level review of the various lending facilities collectively," she said. She added that she could not provide any information on those investigations and that she had no authority to look into Fed practices but only to oversee the Federal Reserve's board of governors. Her inability to answer questions regarding the missing taxpayer funds prompted Florida Democratic representative Alan Grayson to state, "I am shocked to find out that nobody at the Federal Reserve, including the inspector general, is keeping track of this."

Even the Fed chairman apparently wasn't keeping an eye on the store. On July 21, 2009, Grayson confronted Fed chairman Ben Bernanke concerning the whereabouts of more than half a trillion dollars that the Fed had made as credit swaps with foreign banks. Bernanke's response: "I don't know."

Many Americans saw the Fed's economic recklessness as nothing less than an attempt by the financial rule makers to break the rules for themselves and their cronies in order to privatize profits and socialize losses. Americans were also concerned about the Federal Reserve System's great power over American monetary policy. Despite this concern, and despite his desire to see the Fed audited, Barney Frank, the chairman of the Financial Services Committee, and others in Congress have suggested that the Fed supervise the entire U.S. monetary system. A number of financial analysts disagreed. "I have intense concerns with the Fed as a regulator," said economist William K. Black. "Fed regulators have no power within the institution, and the institution is inherently hostile to vigorous regula-

tory action against the big banks." Conrad DeQuadros, a former economist at fallen investment giant Bear Stearns, agreed with Black's point,
writing, "There were obviously some significant lapses [at the Fed] . . . so
widening their regulatory authority isn't really what the system needs."
The Reuters news agency put the usual mild spin on the economists' criticisms by stating in an April 2009 article: "Yet given the institution's
opaqueness and its failure to prevent the current financial crisis, critics say
the country would not be well served if the central bank were anointed as
an all-powerful supra-regulator."

"Opaqueness" is an understatement.

In November 2008, the worldwide financial information network
Bloomberg filed suit against the Fed under the Freedom of Information
Act after the central bank refused to disclose details concerning eleven
Fed-created lending programs that paid out more than $2 trillion in U.S.
taxpayer money. Not only did Fed officials decline to say who received this
staggering amount of money, but they also would not detail what assets the
Fed had accepted as collateral. Bloomberg LP, majority owned by New York
mayor Michael Bloomberg, sued on behalf of its Bloomberg News unit.

The Fed responded to Bloomberg by reiterating its non-government
standing and claiming that, although it had found 231 pages of records on
the transactions, it was allowed to withhold such information as trade secrets and commercial information. Fed officials further argued the United
States is facing "an unprecedented crisis" in which "loss in confidence in
and between financial institutions can occur with lightning speed and
devastating effects."

The Bloomberg FOIA suit had argued that knowing what collateral
was received in exchange for public money is "central to understanding
and assessing the government's response to the most cataclysmic financial crisis in America since the Great Depression." However, in an e-mail
response to Bloomberg News, Jennifer J. Johnson, secretary for the Fed's
board of governors, wrote, "In its considered judgment and in view of current circumstances, it would be a dangerous step to release this otherwise
confidential information."

Various Internet wags have suggested the "don't delay us with questions or the whole economy will collapse" tactic has been used all too frequently to stall or prevent public scrutiny of financial wrongdoing. "If they told us what they held, we would know the potential losses that the government may take and that's what they don't want us to know," explained Carlos Mendez, a senior managing director at New York's global private investment house ICP Capital LLC.

In late August 2009, Manhattan chief U.S. district judge Loretta Preska rejected the Fed's argument of confidentiality and ordered the central bank to disclose details of the emergency loans. New Jersey Republican representative Scott Garrett wrote that Preska's decision was "strikingly good news. . . . This is what the American people have been asking for." But, because Judge Preska's decision is expected to be appealed by the Fed, there is now more reason for the central bank to be audited. Perhaps the reason more Americans are not upset by the financial improprieties today has something to do with what they ingest.

DEBILITATING FOOD AND WATER

[Nazi] German chemists worked out a very ingenious and far-reaching plan of mass control that was submitted to and adopted by the German General Staff. This plan was to control the population of any given area through mass medication of drinking water supplies . . . the real reason behind water fluoridation is not to benefit children's teeth. . . . The real purpose behind water fluoridation is to reduce the resistance of the masses to domination and control and loss of liberty.
—CHARLES ELIOT PERKINS,
U.S. chemist sent to reconstruct the I. G. Farben chemical empire after World War II

IT IS BOTH PARADOXICAL and tremendously ironic that the American public has more unlimited access to healthy food than any population in human history (long after World War II, a banana was considered a costly delicacy in England), yet Americans are on average unhealthy, obese, and overmedicated.

Many nutritionists believe the problem lies not only with the quantity of food consumed but the quality as well.

BAD FOOD AND SMART CHOICES

BY 2010, THE FOOD industry tried to bolster its responsibility with a new front-of-pack nutrition labeling program called Smart Choices. According to the food industry, the program was designed so that "shoppers [could] make smarter food and beverage choices within product categories in every supermarket aisle." The Smart Choices website said the program was "motivated by the need for a single, trusted and reliable front-of-pack nutrition labeling program that U.S. food manufacturers and retailers could voluntarily adopt to help guide consumers in making smarter food and beverage choices."

According to the program's website, "To qualify for the Smart Choices Program, a product must meet a comprehensive set of nutrition criteria based on the Dietary Guidelines for Americans and other sources of nutrition science and authoritative dietary guidance. The Smart Choices Program covers food and beverages in 19 distinct product categories, including cereals, meats, fruits, vegetables, dairy, and snacks, allowing shoppers to compare similar products,"

Critics, such as syndicated columnist and former Texas agricultural commissioner Jim Hightower, claim the program is nothing less than an industry scam, created and paid for by such outfits as Coca-Cola, ConAgra, General Mills, Kellogg's, Kraft, and PepsiCo.

"Under this handy consumer program, hundreds of approved food products in your supermarket are getting a bold, green checkmark printed right on the front of the package, along with the reassuring phrase, 'Smart Choices.' No need to read those tedious lists of ingredients on the back, for the simple green check mark is henceforth your guarantee of nutritional yumminess. For example, you'll find it on such items as Froot Loops and Fudgesicle bars," groused Hightower. "But even by industry standards, this is goofy. I mean—come on, Froot Loops? A serving of this stuff is 41 percent sugar. That's a heavier dose than if you fed cookies to your kids for breakfast. Wow, talk about setting a low bar for nutritional quality! Indeed, food manufacturers can slap a Smart Choice label on a product just by adding some vitamin C to it, even if the product also contains caffeine, saccharine, and chemical additives known to cause cancer and other diseases. That's not smart, it's stupid—and deceptive."

Deceptive, or just shrewd business? And do others do better or worse for eating nonnutritious food? A recent issue of the journal *Cancer Causes & Control* reported that a 1996–2003 study of Ohio's Amish community showed significantly lower incidences of cancer. The Amish, known for their horse-drawn wagons and simple diets, are far healthier than the rest of the American population.

An inadequate diet diminishes the ability of the body to fight disease and leads to lingering illness and even death. This plays well into the globalists' scheme to reduce the human population, as shall be seen. And they control the corporate food industry along with the mass media.

FALSE CLAIMS AND RECALLS

Sometimes even a manufacturer's standard marketing presentation leads to legal action. In early 2009, the Coca-Cola Company was notified of a class action lawsuit filed by the Center for Science in the Public Interest (CSPI) that claimed the company made deceptive and unsubstantiated claims on its VitaminWater line of beverages. "Coke markets VitaminWater as a healthful alternative to soda by labeling its several flavors with such health buzz words as 'defense,' 'rescue,' 'energy,' and 'endurance,'" stated a CSPI news release, which pointed out that the company makes a wide range of dramatic claims, including that its drinks variously reduce the risk of chronic disease, reduce the risk of eye disease, promote healthy joints, and support optimal immune function. However, CSPI nutritionists claim the 33 grams of sugar in each bottle of Vitamin-

Water do more to promote obesity, diabetes, and other health problems than the vitamins in the drinks do to perform the advertised benefits listed on the bottles.

CSPI also criticized MillerCoors, in the wake of a previous settlement with competitor Anheuser-Busch, over advertising for new beverages directed toward the youth market. CSPI described MillerCoors's Sparks as "an alcoholic energy drink that contained stimulant additives that are not approved for use in alcoholic drinks, including caffeine, taurine, ginseng, and guarana." Often called "alcospeed," Sparks contains more alcohol than beer, according to CSPI, which added, "No studies support the safety of consuming those stimulants and alcohol together, but new research does indicate young consumers of these type of drinks are more likely to binge drink, become injured, ride with an intoxicated driver, or be taken advantage of sexually than drinkers of conventional alcoholic drinks." Following a settlement with thirteen state attorneys general, MillerCoors agreed to remove stimulants from Sparks.

Many people still feel that the food they prepare from a supermarket or local grocery must be safe. After all, doesn't the federal government assure it's safe?

In 1993, more than five hundred people were sickened and four died in the Northwest from *E. coli* 0157:H7, then termed "hamburger disease" because it was found in undercooked beef. This particular pathogen, however, was found in other foods, including salami, lettuce, apple cider, and even raw milk, and it, as well as similar infectious bacteria, can survive and even multiply at refrigerator temperatures. A public outcry resulted, with the U.S. Department of Agriculture (USDA) issuing its "Pathogen Reduction: Hazard Analysis and Critical Control Points" (HACCP) rules in 1996. Under these regulations, the food industry was given the responsibility of ensuring the safety of its products. The government only had to verify this was being done.

In 2008–2009, a wide variety of food items were recalled for potential *Salmonella* contamination. These recalls included everything from snacks, cakes, candies, seafood, and dips to vegetables, fruits, eggs, meats, infant formula, and mouth rinse. An extensive listing of recalled products is available at http://www.recalls.org/food.html.

Eating on the run may help explain the rise in both cases and concern over tainted or unsafe food. In the United States, two out of three people ate their main meal away from home at least once a week in 1998. According to a 1997 study entitled "Impact of Changing Consumer Lifestyles on the Emergence/Reemergence of Food-borne Pathogens," a typical consumer more than eight years old ate food away from home at least four times per week. It also reported that half of each food dollar spent by Americans went to food prepared outside the home.

The nation's growing dependence on prepared food means that by the time consumers eat the food, it has been transported numerous times, cooked and cooled, and touched by many different people. Each step in processing could increase the risk of pathogens.

Although food once was grown and distributed locally in America, today large corporations produce food in centralized facilities and ship nationally and internationally, which means that a processing mistake will be felt nationwide or all over the world instead of just locally. Improper holding temperatures, inadequate cooking, contaminated equipment, food from unsafe sources, and poor personal hygiene by packagers can all lead to foodborne illnesses. According to Answers.com, in 1998 Sara Lee recalled thirty-five million pounds of hot dogs and lunch meat due to the presence of *Listeria*. "This is food contamination on a scale unprecedented a generation ago," stated the site. It's enough to make even a glutton think twice about the food he or she eats.

GROWING HORMONES

RECENTLY, GENETICALLY MODIFIED FOOD crops using growth hormones have come under increasing scrutiny for causing health irregularities. Monsanto first synthesized the hormone in large quantities in 1994 utilizing recombinant DNA technology. Cattle now are routinely given growth hormones to make them gain weight faster, thus reducing both the time and feed required prior to slaughtering. Regulation of these hormones is not possible because it is impossible to tell the difference between the added hormones and those made by the animal's own body.

Since the introduction of artificial growth hormones several reports have shown that boys are growing pubic hair and girls are developing breasts at younger ages than in the past. According to the official journal of the American Academy of Pediatrics, studies in the United States have shown an earlier onset of puberty in recent decades and there is evidence that the onset of puberty is changing, possibly related to environmental exposure to endocrine-disrupting chemicals that mimic estrogen in the body. It is hormone signals from the brain that trigger the onset of puberty.

Some experts argue that such premature puberty is merely the result of cosmetics and the desire for kids to emulate favorite celebrities. However, this does not explain why premature puberty has been noted in the United States and not in Europe, according to a May 2009 study cited in the *New York Times,* which added, "This discrepancy has led to speculation that the changes observed in the United States may really be due to differences in data collection methods among large-scale studies and changing ethnic demographics in that country." But such rationalization fails to mention growth hormones. Has the use of growth hormones in beef and milk-producing cattle escaped the consideration of these researchers? If the hormones will increase growth in the cows, it surely must promote accelerated growth in humans.

In a recent report based on a fifteen-year study of young girls in Denmark, researchers determined that the average age of breast development has begun a full year earlier compared with girls studied in the early 1990s. This may mean that as the use of growth hormones spreads, so does the accelerating maturation of youngsters.

This may not be just another conspiracy theory as, according to the *New York Times,* "Studies have documented that a number of chemicals, such as bisphenol-A used to make hard clear plastic containers, may act as endocrine disruptors and have estrogenic effects on the body."

Few large epidemiological studies have been conducted to determine whether early puberty is associated with growth-hormone-treated foods, and some that have, such as a study of recombinant bovine growth hormone (rbGH), were done by the manufacturer. So no clear connection has been established between chemicals having estrogenic effects and premature puberty. It is reminiscent of how the cigarette industry once fought health studies over the hazards of smoking.

Concerns over food safety can be dated back to 1902 when USDA chemists found that food preservatives contained harmful chemicals, a discovery that added to growing public concern. In 1906, the Pure Food and Drug Act and the Federal Meat Inspection Act were passed in an effort to assure the public that the government was attempting to protect them from impure foods and drugs. But then in 1933, Arthur Kallet and F. J. Schlink published *100,000,000 Guinea Pigs: Dangers in Everyday Foods, Drugs and Cosmetics,* a popular book attacking the 1906 Food and Drug Act and stating that the federal government was incapable of protecting the public from unsafe food and drugs due to incompetence and ineffective laws. The authors stated their book was "written in the interest of the consumer, who does not yet realize that he is being used as a guinea pig. . . ."

Noting the close connections between the government and the giant corporations that produce both the nation's food and drugs, they foresaw that "If the poison is such that it acts slowly and insidiously, perhaps over a long period of years . . . then we poor consumers must be test animals all our lives; and when, in the end, the experiment kills us a year or ten years sooner than otherwise we would have died, no conclusions can be drawn and a hundred million others are available for further tests."

THE RISE OF THE FDA

DUE TO THE POPULARITY of Kallet and Schlink's book, as well as federal whistle-blowers speaking out publicly, the mass media that tended to stand with the corporations was bypassed, leading to demands that action be taken to safeguard food and drug consumption. The result was passage of the federal Food, Drug, and Cosmetic Act (FDCA) of 1938. Today this act is considered the foundation of government food and drug regulation. It was meant to be enforced by the Food and Drug Administration, which was created in 1927 when existing federal offices were combined. The FDCA expanded the definition of contamination to include harmful bacteria or chemicals and allowed the FDA to inspect food manufacturing and processing facilities and monitor animal drugs, feeds, and veterinary devices. The act also required ingredients of nonstandard foods to be listed on labels, prohibited the sale of food prepared under unsanitary conditions, and authorized mandatory standards for foods, such as setting the allowable amount of rat feces in foodstuffs.

Some claimed such legislation was not enough. According to a citizen petition to the USDA, filed by the Physicians Committee for Responsible Medicine in 2001, "[T]he prevalence of food borne illness in this country caused by eating fecally contaminated meat and poultry remains staggeringly high, providing clear evidence that current inspection methods and regulations are insufficient and misdirected." The petition claimed that current inspection policies pertain only to that feces which is visible to the naked eye and does not protect consumers from unseen particulates.

Further promises of public protection came in August 1996 with passage of the Food Quality Protection Act (FQPA) that allowed the Environmental Protection Agency (EPA) to regulate pesticides used in the food production. However, the FQPA also eliminated the 1958 Delaney clause to the 1938 law that prohibited even tiny amounts of any cancer-causing substance added to food products. The Delaney clause, an amendment named after New York Democratic congressman James Delaney, had set a fixed risk standard of "zero cancer risk" for pesticide residue in food, whereas the FQPA softened this to a mere "reasonable certainty that no harm" would result from any type of exposure, including drinking water. Some saw the hand of the corporate globalists in this move to lessen public protection.

One method to protect corporate interests is to fill government posts with persons connected to both sides. One prime example of the revolving door between government regulation and corporate foodstuffs is Michael Taylor, who was named President Obama's new deputy commissioner for foods at the FDA in early 2010.

Fresh out of law school in 1976, Taylor began his career as an FDA staff attorney. He then moved to the law firm of King & Spaulding, which represented Monsanto as it was developing genetically engineered bovine growth hormone (BGH). Returning to the FDA in 1991 as deputy commissioner for policy, Taylor, while instituting tougher anticontamination measures for foods, supported the FDA decision to approve Monsanto's bovine growth hormone. He also was partly responsible for a controversial policy permitting milk from BGH-treated cows not to be labeled as such. Taylor then moved to the U.S. Agriculture Department in 1994 to oversee its food-safety program before returning to work for Monsanto as a vice president for public policy. After a time at George Washington University, in July 2009, Taylor became an adviser to the FDA commissioner.

GENETICALLY MODIFIED FOODS

ANOTHER PUBLIC CONCERN HAS been over nontraditional, genetically modified organisms (GMOs) in foods. Such organisms have had their genes altered by scientists in a laboratory to help the crop resist weeds, insects, and diseases; increase its nutrients; or lengthen its shelf life.

Beginning in 2006, more than twelve hundred lawsuits were filed against Bayer CropScience AG claiming damages caused by the firm's genetically modified (GM) rice seeds. Although the rice was not approved for human consumption, Bayer—along with Louisiana State University—had been testing the rice for resistance to the company's Liberty herbicide. Farmers in five states claimed the modified rice had escaped and contaminated commercial rice supplies in more than 30 percent of America's ricelands. When the USDA announced that trace amounts of the GM rice had been found in U.S. long-grain rice stocks, there was a 14 percent decline in rice futures, which meant lower prices paid for crops. Growers claimed this cost them $150 million.

"Bayer did not keep track of its genetically modified seed," argued attorneys for the rice growers. "This is a living, growing organism. That's why you have to be so careful."

But a major focal point of concern in the debate over GMOs is Monsanto. Headquartered in Creve Coeur, Missouri, this multinational agricultural biotechnology corporation is the world's leading producer of GM seeds as well as pesticides. In 2005, Monsanto was reaching into other areas of food. The company applied for two patents with the World Intellectual Property Organization (WIPO) in Geneva for exclusive ownership of GM pigs.

"If these patents are granted, Monsanto can legally prevent breeders and farmers from breeding pigs whose characteristics are described in the patent claims, or force them to pay royalties," warned Greenpeace researcher Christoph Then. "It's a first step toward the same kind of corporate control of an animal line that Monsanto is aggressively pursuing with various grain and vegetable lines."

Some semblance of sanity was brought to this issue on March 29, 2010, when U.S. district court judge Robert W. Sweet struck down two patents on human genes that had been linked to ovarian and breast cancer. This decision sent a chill through the multibillion-dollar corporations that today claim patent rights on about 20 percent of human genes. Judge Sweet's 152-page decision, involving gene patents of the Myriad Genetics company, stated the patents were "improperly granted" as they involved a "law of nature." He agreed with gene patent opponents, who argued that the idea that isolating a gene made it patentable was merely "a 'lawyer trick' that circumvents the prohibition on the direct patenting of the DNA in our bodies but which, in practice, reaches the same result."

Some researchers see Monsanto as attempting to dictate what farmers will grow and what consumers will eat. The agricultural giant produces patented seeds (termed "Terminator" seeds) designed to not reproduce, meaning farmers each year will have to buy more Monsanto seeds. Several recent court cases involved Monsanto attorneys suing farmers who illegally, or even unknowingly thanks to the winds, ended up with Monsanto's patented crops growing in their fields. Such activity has made Monsanto a prime target for antiglobalization and environmental activists.

Interest in modifying genetic material increased after a March 2009 report was released that stated that South African farmers lost millions of dollars when eighty-two thousand hectares of Monsanto GM corn failed to produce hardly any seeds. Although the manufacturer, Monsanto, offered compensation for the losses, Mariam Mayet, director of the Africa Centre for Biosecurity in Johannesburg, demanded an immediate ban on all GM foods and a government investigation.

But at least in this case only crops were lost. During 2008, an underreported epidemic took place in India, when thousands of desperate farmers were driven to suicide when they could not get out of debt. While Monsanto claimed that their weevil-resistant cotton would produce larger crops, they failed to mention they would require much more water, an ingredient in short supply. In 2003, more than seventeen thousand Indian farmers had committed suicide. The numbers have simply grown ever since, creating both mystery and controversy. Although the suicides were caused primarily by bankruptcy, many believe these bankruptcies in part came as a result of the promotion of Monsanto GM seeds.

Though the suicide epidemic seems complex to those studying it, there has been more and more scrutiny directed at the role of the World Trade Organization (WTO) and the biochemical firm Monsanto. Curiously, the suicides began around 1998, the same year the WTO allowed corporate giants like Monsanto into India's seed market. Nonrenewable genetically modified crops soon replaced the self-sustainable farming system that India had used for thousands of years. Farmers were obligated to purchase not only GM seed but also the chemical pesticides produced by Monsanto for those crops.

According to Jessica Long of Montreal's nonprofit Centre for Research on Globalization, "Seventy-five percent of cultivable Indian land exists in dry zones. Non-GM rice utilizes 3,000 liters of water in order to produce one kilo, while non-renewable hybrid rice requires 5,000 liters per kilo! . . . Continuous GM cotton crop failures resulted in the state of Andrha

Pradesh, the seed capital of India, prohibiting the sales of [*Bacillus thuringiensis,* a bacterium used as a pesticide] cotton varieties by Monsanto."

Due to the ongoing controversy over the use of GM seeds, in 2008 the Indian government forced Monsanto to reduce royalties received from its patented seeds.

"The economic disparity of Indian farmers only increases as they try to keep up with the lowest import prices. It is estimated that they are losing $26 billion annually," stated Long. "While 90 percent of farm loans come from money lenders, they are charged anywhere from 36–50 percent interest, placing them in a cyclical mode of poverty. Surely poverty alone cannot be responsible for such massive amounts of bloodshed! After all, poverty has always existed, so what is it about current conditions that have led to all this bloodshed? The fact is that mass suicides have transformed these farmers into agrarian martyrs for peasants everywhere."

Monsanto officials denied that their firm was behind the deaths, explaining on the company website: "The reality is that the tragic phenomena of farmer suicides in India began long before the introduction of Bollgard [Monsanto's herbicide] in 2002. Farmer suicide has numerous causes with most experts agreeing that indebtedness is one of the main factors. Farmers unable to repay loans and facing spiraling interest often see suicide as the only solution." Although bankruptcy was the obvious cause of most of India's suicides, many blamed Monsanto's genetically modified crops, which required more water than traditional crops, as well as Monsanto's herbicides for farmers' losses.

"By claiming global monopoly patent rights throughout the entire food chain, Monsanto seeks to make farmers and food producers, and ultimately consumers, entirely dependent and reliant on one single corporate entity for a basic human need. It's the same dependence that Russian peasants had on the Soviet Government following the Russian revolution. The same dependence that French peasants had on Feudal kings during the Middle Ages. But control of a significant proportion of the global food supply by a single corporation would be unprecedented in human history," warned Brian Thomas Fitzgerald of Greenpeace.

In January 2010, a study published in the *International Journal of*

Biological Sciences reported that researchers, after analyzing the effects of genetically modified foods on mammalian health, linked Monsanto's GM corn to kidney and liver damage in rats. Monsanto officials were quick to state that the research was "based on faulty analytical methods and reasoning and do not call into question the safety findings for these products." However, the study's author, Gilles-Eric Séralini, responded, "Our study contradicts Monsanto conclusions because Monsanto systematically neglects significant health effects in mammals that are different in males and females eating GMOs, or not proportional to the dose. This is a very serious mistake, dramatic for public health. This is the major conclusion revealed by our work, the only careful reanalysis of Monsanto's crude statistical data."

Awareness about GMOs in foods can be traced back as early as 2002. Although the FDA, EPA, and USDA all have stated that their research shows no long-term health risks from GMO foods, Dr. Stanley Ewen, a consultant histopathologist at Aberdeen Royal Infirmary and one of Scotland's leading experts in tissue diseases, warned in a report to a government health committee that eating GM food could cause cancer. In a report to a government health committee, Ewen expressed "great concern" about the use of the cauliflower mosaic virus as a "promoter" in GM foods that could increase the risk of stomach and colon cancers. Ewen wrote that the infectious virus is used like a tiny engine to drive implanted genes to express themselves and could encourage the growth of polyps in the stomach or colon. "The faster and bigger the polyps grow, the more likely they are to be malignant," he wrote, adding, "It is possible cows' milk will contain GM derivatives that can be directly ingested by humans as milk or cheese. Even a lightly cooked, thick fillet steak could contain active GM material."

Cancer was only one of some fifty harmful effects of GMO foods and growth hormones listed in a research article by nutritionist Nathan Batalion that included a warning from Harvard biology professor Dr. George Wald, a Nobel Laureate in Medicine.

"Our morality up to now has been to go ahead without restriction to learn all that we can about nature. Restructuring nature was not part of the bargain. This direction may be not only unwise, but dangerous. Potentially, it could breed new animal and plant diseases, new sources of cancer, novel epidemics," stated Wald.

Monsanto's growth hormone IGF-1 has been linked to increased risk of human colorectal and breast cancer in studies both in the United States and Canada. However, the FDA downplayed the significance of such studies.

Reflecting concern over the safety of GMOs, the UN's Food Safety Agency, representing 101 nations worldwide, in 1999 ruled unanimously to continue a 1993 European moratorium on Monsanto's genetically engineered hormonal milk (rBGH). This ban was not reported in the American media, further indicating the extent of Monsanto's influence in the media.

Award-winning journalists Steve Wilson and Jane Akre both were fired when they tried to expose the cover-up of such studies as well as the ban on growth hormones in Europe. According to the Goldman Environmental Prize website, "As investigative reporters for the Fox Television affiliate in Tampa, Florida, [Wilson and Akre] discovered that while the hormone had been banned in Canada, Europe and most other countries, millions of Americans were unknowingly drinking milk from rBGH-treated cows. The duo documented how the hormone, which can harm cows, was approved by the government as a veterinary drug without adequately testing its effects on children and adults who drink rBGH milk. They also uncovered studies linking its effects to cancer in humans. Just before broadcast, the station cancelled the widely promoted reports after Monsanto, the hormone manufacturer, threatened Fox News with 'dire consequences' if the stories aired. Under pressure from Fox lawyers, the husband-and-wife team rewrote the story more than eighty times. After threats of dismissal and offers of six-figure sums to drop their ethical objections and keep quiet, they were fired in December, 1997."

The addition of unsafe, even toxic, chemicals to food and water may be attributed to laxity and greed on the part of producers, but when coupled with the public statements of leading globalists concerning the desire to reduce the human population, which will be discussed later, it takes on a much darker aspect.

CODEX ALIMENTARIUS

ONE WOULD THINK THAT a good diet with plenty of vitamins might help prevent disease and malnutrition, but even here the New World Order may interfere.

The World Health Organization (WHO) was founded in 1948 with the goals of setting global standards of health and helping governments to strengthen national health programs. The WHO and the United Nations' Food and Agriculture Organization (FAO) work together in committees, conferences, and commissions. One of their most significant joint efforts is the Codex Alimentarius (Latin for "food code") Commission, which sets standards for food commodities, codes for hygiene and technology, pesticide evaluations, and limits on pesticide residues. It also evaluates food additives and veterinary drugs and sets guidelines for contaminants. Approximately 170 nations accept its standards and codes.

In recent years, controversy had grown over the application of food standards to traditional vitamins and mineral supplements. A major cause for concern by nutritionists is that the Codex Alimentarius list is recognized by the World Trade Organization (WTO). It is feared that the WTO will use Codex Alimentarius standards in disputes over the classification of vitamins as food.

Such fears are not irrational since in 1996 the German delegation to the Codex Alimentarius Commission advocated a ban on herbs, vitamins, and minerals sold for preventative or therapeutic reasons and advanced a position that supplements should be classified as drugs with attendant restrictions and physician prescriptions. Though the commission agreed, there was an aftermath of such public protest that passage of the new classifications was postponed. As protests waned in mid-2005, the commission quietly adopted guidelines for vitamin and mineral food supplements, allowing member countries to regulate dietary supplements as drugs or other categories. Although the new classifications do not yet ban supplements outright, they do subject them to labeling and packaging requirements, set criteria for the setting of maximum and minimum dosage levels, and require that safety and efficacy are considered when determining ingredient sources.

Should supplements become as inaccessible as prescription drugs, John Hammell, founder of International Advocates for Health Freedom (IAHF), believes that the average consumer will lose out on the benefits of simple remedies like herbs, vitamins, minerals, homeopathic remedies, and amino acids. "The name of the game for Codex Alimentarius is to shift all remedies into the prescription category so they can be controlled exclusively by the medical monopoly and its bosses, the major pharmaceutical firms," said Hammell.

Despite government denials that this could occur, the Codex Alimentarius proposals are today law in Norway and Germany, where the entire health-food industry has literally been taken over by the drug companies. Hammell explained that in these countries, vitamin C above 200 mg is illegal as is vitamin E above 45 IU, vitamin B_1 over 2.4 mg, and so on. "The same is true of ginkgo and many other herbs, and only one government-controlled pharmacy has the right to import supplements as medicines which they can sell to health food stores, convenience stores or pharmacies," he added.

Opponents paint the Codex Alimentarius Commission as a "shady, secretive organization [that is] the thinly-veiled propaganda arm of the international pharmaceutical industry that does everything it can to promote industry objectives whilst limiting individual options to maintain health (which would diminish members' profits)."

Behind the Codex Alimentarius Commission is the UN and the WHO. According to critics, both organizations are working for multinational pharmaceutical corporations and international banks whose owners support reducing the human population through such means as reducing the availability of necessary minerals in the human diet. This, in turn, could increase the occurrences of various debilitating diseases such as cancer and diabetes, the number three cause of death in adults in the United States.

Citing a study at the University of Vancouver Medical School, naturopathic physician and author Dr. Joel D. Wallach indicated that vanadium, a soft white metallic element found in certain minerals, could replace insulin in adult onset diabetics, a condition representing 85 percent of all diabetics.

In a 2005 speech, Wallach said, "I've seen it work on hundreds and hundreds of people. Now to me this is criminal. If you write to Hills Packing Company that makes Science Diet dog food . . . high tech foods for animals . . . and say, 'How many minerals, exactly, is in Science Diet dog food?' They'll write back there's 40 minerals. You write Checkerboard Square in St. Louis, Ralston Purina, and say 'Just how many minerals are in your rat pellets for laboratory rats?' They'll say there are 28 minerals. I'll give anybody . . . a crisp new $100 bill if you can find me a human infant formula in a grocery store that has more than 11 [minerals]. . . . So dogs get 40 minerals, rats get 28 minerals, and human infants get 11. Is that fair? No! Doesn't matter if you're talking about SMA, Similac, Isomilk, ProSoyB. In fact, that's why they call Similac, Similac, because it lacks everything."

While efforts in the United States to curtail vitamins and supplements have been stymied by public opposition, proponents found another ally in the Federal Trade Commission (FTC), which has now made Codex a trade issue. At the Uruguay Round of the General Agreement on Tariffs and Trade (GATT) (which created the World Trade Organization), the United States agreed to submit its laws to the international standards, which included the Codex Alimentarius Commission's standards for dietary supplements. What this means is that now Codex Alimentarius is enforced by the WTO, whose international standards could supersede domestic laws without the American people's consent or vote in the matter.

According to Hammell, if a country disagrees with or refuses to follow Codex standards, the WTO can apply pressure by withdrawing trade privileges and imposing crippling trade sanctions.

The WTO was established with the understanding it was to push the world toward greater economic integration. However, according to many, the WTO has ended up politicizing trade by putting the stamp of officialdom on some very bad policies and promotes further loss of American sovereignty to supranational organizations. According to Llewellyn H. Rockwell Jr., president and founder of the Ludwig Von Mises Institute, "The WTO has the power to order Congress to change any U.S. law the WTO deems a 'barrier to free trade.' If Congress does not obey the WTO, then American businesses and consumers will face trade sanctions. Congress has already changed America's tax laws in response to WTO commands. It is possible that the WTO will force America to adopt the restrictive regulations of foods and dietary supplements endorsed by the UN's CODEX commission."

Despite centuries of human experience with healing herbs and vitamins, today's corporate medicine industry, especially the pharmaceutical giants that can be traced back to the Nazi I. G. Farben complex, has attempted to limit any healing agent to pharmaceuticals. Agents for this suppression of natural healing are the Food and Drug Administration and the Federal Trade Commission.

Legitimate standardized codes for dietary supplements, such as Codex Alimentarius, require expensive clinical studies, research, tests, and analysis well beyond the financial reach of all but the largest corporations. In other words, a huge mound of personal narratives supporting natural remedies would be useless against a few reports from well-paid corporate scientists. "In working to protect the business interests of vaccine manufacturers [the pharmaceutical corporations], both the FDA and FTC have declared all-out war against any products that might offer consumers options other than vaccines," said Mike Adams, *NaturalNews* editor and self-styled "health ranger," whose articles and books have attracted a worldwide audience of nearly a million people.

"The FDA's official position is that there is no such thing as any herb, any plant, any nutrient or any dietary supplement that has any beneficial effect on the human body. Thus, no herb, plant, nutrient or supplement can EVER be approved by the FDA to protect against influenza. As you've figured out, the whole game is rigged from the start. Herbs that have antiviral properties will never be approved as anti-virals. And, frankly, for the people running natural product companies to try to play the 'FDA game' is useless. You can never appease tyranny. Trying to 'conform' to the requirements of the FDA and FTC is like Jewish prisoners trying to conform to the wishes of Hitler. You've been condemned from the start!" said Adams.

FLUORIDATED WATER

How safe is drinking water?

Controversy over the addition of the chemical sodium fluoride to municipal drinking water supplies has raged since the early 1950s. It was a time when Nazi scientists were being settled within the United States under the auspices of Project Paperclip.

The *Reader's Digest Oxford Complete Wordfinder* defines fluoride merely as "any binary compound of fluorine." But fluorine was defined as a "poisonous pale yellow gaseous element of the halogen group."

Charles Eliot Perkins, a prominent U.S. industrial chemist, was sent by the U.S. government to help reconstruct the I. G. Farben chemical plants in Germany at the end of the war. In 1954, he wrote a letter to the Lee Foundation for Nutritional Research, stating that he had learned that the Nazi regime had used sodium fluoride as a means of "mass control." "I want to make this very definite and very positive," Perkins wrote. "The real reason behind water fluoridation is not to benefit children's teeth. . . . The real purpose behind water fluoridation is to reduce the resistance of the masses to domination and control and loss of liberty. Repeated doses of infinitesimal amounts of fluorine will in time gradually reduce the individual's power to resist domination by slowly poisoning and narcotizing this area of brain tissue, and make him submissive to the will of those who wish to govern him. . . . I say this with all the earnestness and sincerity of a scientist who has spent nearly 20 years' research into the chemistry, biochemistry, physiology and pathology of 'fluorine.' . . . Any person who drinks artificially fluoridated water for a period of one year or more will never again be the same person, mentally or physically."

Most people do not realize that fluoride is a key ingredient in Prozac and many other psychotropic drugs. Prozac, whose scientific name is fluoxetine, is 94 percent fluoride.

Though fluoride purportedly prevents tooth decay, it only has been shown to affect decay in children under twelve. Today, two-thirds of all municipal water and most bottled water in the United States contain sodium fluoride. Fluoride is a poisonous waste product of aluminum manufacture that accumulates in the human body. The use of aluminum cookware has been strongly linked to Alzheimer's disease, a progressive brain disorder that gradually destroys a person's memory and ability to learn, reason, and make judgments. A *Christian Science Monitor* survey in 1954 showed that seventy-nine of the eighty-one Nobel Prize winners in chemistry, medicine, and physiology refused to endorse water fluoridation. Nevertheless, every U.S. Public Health Service surgeon general since the 1950s has supported putting this rat poison ingredient into America's water supply.

The experts cannot decide where the truth lies in the fluoride controversy. Virginia dental surgeon and nutritionist Dr. Ted Spencer wrote, "A few years ago, I was asked by the head of our local health department to conduct a review of existing journal research on the toxicity of fluoride with emphasis on its cancer causing potential. I went to the National Medical Library and produced for him some 40 articles on the toxicity of fluoride. When we reviewed them, there was some discrepancy in whether or not fluoride was mutagenic . . . half of the articles said that it was and half said that it was not. But it cannot be both ways. . . . We wondered what was wrong."

Spencer discovered that fluoride has been banned in European nations such as Sweden, Norway, Denmark, Germany, Italy, Belgium, Austria, France, and the Netherlands. It is especially interesting to note that West Germany banned the use of fluorides in 1971, a time when it was still heavily occupied by Allied soldiers. "Apparently they could no longer silence the German scientists who had proved that fluoridation is a deadly threat to the population," wrote Eustace Mullins, a former Library of Congress staffer and World War II veteran who wrote numerous books on conspiracy topics including medicine, finance, and politics.

Despite Europe's bans, America continues to pursue fluoridating all water supplies and ignoring studies like those of Dr. Dean Burk, the chief chemist emeritus of the U.S. National Cancer Institute. Burk stated, "In point of fact, fluoride causes more human cancer death, and causes it faster, than any other chemical." Dr. Perry Cohn of the New Jersey Department of Health discovered a correlation between osteosarcoma—a principal childhood cancer—and fluoridation. After creating a 2005 survey in seven New Jersey counties, Cohn found the incidence of osteosarcoma in boys under the age of ten was 4.6 times higher in fluoridated areas than in nonfluoridated areas. The incidence of cancer was 3.5 times higher in the ten to nineteen age group and over twice as high in the twenty to forty-nine age group.

Studies indicate that every major city using fluoridated water has experienced an increase in the rates of cancer. "Not a fair trade for good looking teeth," commented Dr. Spencer, adding, "All allopathically-trained dentists are very familiar with the ADA [American Dental Association] and other 'authoritative' positions on fluoride. They rarely mention its toxic potential or the few studies revealing increased tooth decay after fluoride use."

Spencer also referred to studies that suggest fluoride causes unscheduled DNA synthesis, sister chromatid exchanges, and mutagenic effects on cells. "These terms may not bother some people at all, but they mean that there will be an increase in cancer after the ingestion of fluoride," Spencer wrote. Although each person must decide for themselves the dangers of fluoride, Spencer did point to several studies with convoluted titles that conjure images of grotesque science experiments: "Sodium Fluoride-induced Chromosome Aberrations in Different Stages of the Cell Cycle," "Chronic Administration of Aluminum Fluoride or Sodium Fluoride to Rats in Drinking Water: Alterations in Neuronal and Cerebrovascular Integrity," and "Toxin-Induced Blood Vessel Inclusions Caused by the Chronic Administration of Aluminum and Sodium Fluoride and Their Implications in Dementia."

Given the massive amounts of money being paid by the pharmaceutical corporations to the corporate mass media, it is highly doubtful that many Americans will learn of the results of these studies any time soon. The entire history of fluoride in America is one of deceit and conspiracy. In 1946, a Wall Street attorney and former counsel to the Aluminum Company of America (now known by the acronym Alcoa) named Oscar Ewing was appointed by President Truman to head the Federal Security Agency. Ewing became in charge of not only the U.S. Public Health Service but also the Social Security Administration and the Office of Education.

Congressman A. L. Miller, a physician turned Republican politician, accused Ewing of being placed in a highly paid position by Alcoa, a Rockefeller syndicate, to promote fluoridation. Miller stated, "The chief supporter of the fluoridation of water is the U.S. Public Health Service. This is part of Mr. Ewing's Federal Security Agency. Mr. Ewing is one of the highly paid lawyers for the Aluminum Company of America."

Other opponents were less kind. Leaflets handed out in New York City boldly stated, "Rockefeller agents order fluoride-(rat-) poisoning of nation's water. Water fluoridation is the most important aspect of the cold war that is being waged on us—chemically—from within, by the Rockefeller-Soviet axis. It serves to blunt the intelligence of a people in a manner that no other dope can. Also, it is genocidal in two manners: it causes chemical castration and it causes cancer, thus killing off older folks. ... This committee [Ewing's study of fluoride] did no research or investigation on the poisonous effects of water fluoridation. They accepted the falsified data published by the U.S.P.H.S. [U.S. Public Health Service] on the order of boss Oscar Ewing, who had been 'rewarded' with $750,000 by fluoride waste producer, Aluminum Co." Suspiciously, it was also reported that Ewing told fellow senators not to drink fluoridated water.

HEALTH-CARE BLUES

IF THE NIGHTMARES OF natural-health advocates come to pass, a sick person soon will have no recourse but to seek professional medical assistance, which may not exist, according to recent reports.

This nation's health-care system is in a shambles. Health-care costs are moving beyond 16 percent of gross domestic product and the U.S. health-care system is sometimes 100 percent more expensive than anywhere else, yet Americans do not live as long as citizens in other nations. Every citizen in these countries is covered by a health-care plan, whereas in America, 15 percent of the population—about 47 million people—are uncovered at any given time. Fifty percent of bankruptcies in the United States are due to medical bills, and many workers avoid changing jobs for fear of losing medical coverage, especially when they have preexisting conditions.

Many factors contribute to the poor state of health care in America, including malpractice anxiety for physicians, which leads to defensive practice. Also at play is the lack of coverage for preventive and mental-health care, which could serve as a prophylactic for expensive emergency care later on. More troubling is the profiteering of insurance and drug companies—a system that rewards physicians for overprescribing drugs. In her book *Overtreated: Why Too Much Medicine Is Making Us Sicker and Poorer,* Shannon Brownlee explains that a serious part of the health-care issue is the lack of clinical research needed to guide physicians' decisions. According to Brownlee, up to 80 percent of health decisions involve ambiguity—the variability of diagnosis and available treatments—which leads to unnecessary treatments and costs.

But don't blame the doctors for the failures of the American health industry.

In 2009, the *Wall Street Journal* reported that an increasing number of doctors, including specialists, were either opting out of Medicare entirely or not accepting patients with Medicare coverage, blaming low reimbursement rates and complaining that the burden of bureaucratic paperwork was not worth the effort. Dr. Michael E. Truman, a Texas family physician in practice for nearly forty years, explained, "Over the past several years, I've noticed that reimbursements for services I provide are being cut or staying the same while the cost of business has escalated a great deal. Current reimbursements from Medicare are 35 percent below what most other insurance carriers pay. . . . I have no idea what they are going to do this year, but if rates are lowered 25 percent, most doctors will start limiting the number of Medicare patients they see because reimbursement is below their cost for doing business. I haven't seen anything in the new health proposals that will remedy this problem."

Truman said most large insurance companies refuse to increase reimbursements to match inflation. "We have very little to say about it except not to see their patients and that means closing our office," he said. With decreasing reimbursement, doctors will be forced to start seeing forty to fifty patients a day, which means the patients will pay the price. "They will get about five minutes of the doctor's time. With so little time with the patient, the doctors will be ordering more tests to cover their ass and turning care over to their nurse practitioners.

"When I went into practice in 1972, we didn't have any PPO's or HMO's. No one stood at our front door and collected part of our fee before the patient ever got in the office. We now have to subsidize big salaries for the insurance CEO's and who knows who else. . . . They are getting rich off every doctor in practice today and insurance premiums are going up every year to the point that many of my patients can't afford their insurance anymore and they are now paying cash. Most of the insurance companies today are nothing but parasites, offering no vitality to medical care, just sapping whatever life is left out of it."

With more and more doctors dropping out of insurance plans, soon "there is no guarantee that you will be able to see a physician no matter what coverage you have," said Marc Siegel, an internist and associate professor of medicine at the NYU Langone Medical Center. "Of course, we're promised by the Obama administration that universal health care insurance will avoid all these problems. But how is that possible when you consider that the medical turnstiles will be the same as they are now, only they will be clogged with more and more patients? The doctors . . . will be even more overwhelmed."

Deserting doctors may be the least of the health-care problems facing a zombie nation. Analysts estimate that the Obama administration's proposed universal health-care program may cost upward of $2 trillion over a ten-year period. There is difficulty in even funding existing programs. In a 2009 article for FrontPageMag.com, Mackinac Center for Public Policy associate Tait Trussell warned that "we are totally unprepared fiscally even for existing programs. Neither Social Security nor Medicare is ready for the onslaught of the 78 million Americans who will stop paying into retirement programs, and who instead will begin to draw on benefits government has promised them. The first line of baby boomers began signing up for early retirement under Social Security last year [2008]. Soon the 78-million-person tsunami of seniors will expect to be covered by Medicare."

But, just like the FDIC and Social Security, there is no stockpile of funds to fulfill government promises of health care. Payroll taxes supplying trust funds for these programs already are inadequate. According to the nonpartisan Congressional Budget Office, the Obama budget plan will increase federal spending 25 percent faster than revenues during the next ten years. "Incredibly, this is almost modest, dollar-wise, compared to the current unfunded liability for Social Security and Medicare" noted Trussell. "It totals $101.7 trillion in today's dollars. This is more than seven times the 2008 gross domestic product (GDP), our total economy, according to calculations by the National Center for Policy Analysis. These enormous figures to fund Social Security and Medicare seem too huge to even want to be acknowledged by some policy-makers."

In February 2009, John C. Goodman, president of the Dallas-based National Center for Policy Analysis, outlined the coming costs for government programs: "In 2012, Social Security and Medicare will need one out of every ten general income tax dollars to make up for their combined deficits. By 2020, the federal government will need one out of every four income tax dollars to pay for these programs. By 2030, the midpoint of the Baby Boomer retirement years, it will require one of every two income tax dollars. So it is clear that the federal government will be forced either to scale back everything else it's doing in a drastic way or raise taxes dramatically."

Goodman added, "If health-care consumers are allowed to save and spend their own money, and if doctors are allowed to act like entrepreneurs—if we allow the market to work—there is every reason to believe that health care costs can be prevented from rising faster than our incomes. Otherwise, prepare for the tax tsunami."

Is it possible that the globalists foresee this looming tax tsunami only too well and are siphoning every dime out of the U.S. economy before it hits? Such calamity could provide the very excuse they need to gain total control of not only the U.S. economy but also the economies of the nations who support the U.S. dollar.

Over and beyond the stretched-thin health-care industry and approaching financial chaos, even more medical horrors loom on the horizon.

THE MYCOPLASMA ATTACK

The victims of the neurodegenerative/systemic degenerative disease Myalgic Encephalomyelitis/Fibromyalgia are ill with a very real physical disease deriving from a sub-viral particle developed from the Brucellosis bacterial toxin.
—DONALD W. AND WILLIAM L. C. SCOTT,
authors of *The Brucellosis Triangle*

IN RECENT HORROR MOVIES, tiny microorganisms infect humans and turn them into flesh-eating zombies. Often, the virus has been accidentally loosed from a covert government laboratory. Although it doesn't seem like a pathogen exists for transforming a normal person into a cannibalistic zombie, there are a number of man-made germs and toxins that have been in development since before World War II that can devastate the human body.

NAZI AND JAPANESE BIOLOGICAL WARFARE

IN THE WAKE OF World War II, thousands of die-hard Nazis were arriving in the United States, thanks to a technology-for-immunity swap arranged between Hitler's right-hand man, Martin Bormann, and America's Wall Street elite, which included John J. McCloy and his protégé, Allen Dulles.

According to Dr. Len G. Horowitz's research, "The WHO [World Health Organization] was heavily funded and influenced by the Rockefeller family, along with the United Nations and the World Bank ... [and] the fact that John D. Rockefeller's business managers and lawyers, John Foster and Allen Dulles, had created the partnership between the world's largest oil conglomerate and I. G. Farben—Germany's leading industrial organization prior to World War II. . . ." Before the war, attorney McCloy had represented the I. G. Farben drug combine. In *The Rise of the Fourth Reich,* it was detailed how the Dulles brothers and their prewar work for Schroeder, Rockefeller & Company, City National Bank chairman John J. McCloy, and Union Banking Corporation director Prescott Bush acted as principal agents for Hitler's Germany. It might also be noted that the UN building in New York City sits on Rockefeller-donated land.

McCloy, who served as high commissioner in postwar Germany, also was chairman of the Ford Foundation, Chase Manhattan Bank, the Salk Institute, E. R. Squibb & Sons, and the powerful Council on Foreign Relations, described in the *New York Times* as a group that "fixes major goals and constitutes itself a ready pool of manpower for the more exacting labors of leadership." In his 1989 *Times* obituary, McCloy was termed "chairman of the Establishment."

Though U.S. laws were in place to forbid postwar Germans from conducting research on chemical warfare, these were largely ignored as John McCloy hired experts as "consultants" and helped fund German industries to produce chemical warfare materials for the American military. At the same time, Allen Dulles was named director of the CIA. Prior to the war he had served as legal representative of the Nazi Shroeder Bank and then during the war as an officer for the Office of Strategic Services (OSS), where he supervised army intelligence translator Henry Kissinger, who would go on to become secretary of state under President Richard Nixon. It was Dulles as head of the CIA who expunged many Paperclip scientists' Nazi backgrounds.

During this time, Wernher von Braun, long considered the father of our NASA space program, and other top rocket scientists entered the country, along with Walter Emil Schreiber, the chief of Nazi medical science who had supervised the sterilization of men using surgery, X-rays, and drugs and had overseen the exchange of humans and mice as recipients of a deadly typhus virus. Despite being described as "the prototype of an ardent and convinced Nazi," Schreiber worked for a decade in the chemical division of the U.S. European Command and for a time at the Air Force School of Aviation Medicine in Texas.

Another German immigrant, Kurt Blome, told U.S. military interrogators in 1945 that he had been ordered in 1943 to experiment with plague vaccines on concentration camp prisoners. Blome went on to work for the U.S. Army Chemical Corp. These Nazis were joined at Fort Detrick by Japanese general Ishii Shiro, the man in charge of the infamous Unit 731, the Japanese biological research and development unit responsible for the deaths of three thousand people, including American prisoners.

It was the work of such enemy researchers that was continued and expanded in the United States following World War II that may have resulted in many recent health disasters.

MYCOPLASMAS AND PRIONS

IN THE EARLY 1940s, Nazi medical scientists had managed to isolate the bacterial toxin from *Brucella* bacteria (usually known as Brucellosis or undulant fever and mostly found in mammals, especially cows) and form it into a crystalline form or agent.

Brucellosis is an ancient bacteria and was selected because it was insidious, very difficult to detect, and present in almost every organ or system of the human body. When activated by the crystalline agent, brucellosis stimulates various diseases that prompt a variety of symptoms, including debilitating fatigue, high fever, shivering, aching, drenching sweats, headache, backache, weakness, and depression. Damage to major organs is possible, leading to ailments such as multiple sclerosis, arthritis, and heart disease.

The Paperclip medical scientists coming to America brought with them this toxin, known as a mycoplasma—a distinct type of bacteria lacking a cell wall. A U.S. government report dated January 3, 1946, carried a section entitled "Production and Isolation, for the First Time, of a Crystalline Bacterial Toxin." The Nazi bug had been reduced to a crystalline form, creating an artificial virulent disease agent derived from the original bacteria.

This crystalline bacterial agent could be dispensed by aerial spraying or by infected insects. The agent also did not respond to most antibiotics, including penicillin. Acting as a parasite, it stimulated both bacterial and viral diseases and, because it attached to specific cells without killing them, was virtually undetectable by conventional medical diagnosis techniques. Such diseases are considered untreatable and usually fatal, because they mostly affect the brain or neural tissue.

These subviral bacterium particles have various names. They have been termed "prions" by Nobel Prize winner Dr. Stanley B. Prusiner; "stealth viruses" by Dr. John Martin of the Center for Complex Infectious Diseases; "amyloids" by the late Dr. Carleton Gajdusek, winner of the 1976 Nobel Prize in Medicine for his work on mysterious epidemics at the National Institutes of Health (NIH); and "Mycoplasma/Brucellosis" by Donald Scott and Garth Nicolson.

According to a paper by Stanley Prusiner, prions are unprecedented infectious pathogens that cause fatal neurodegenerative diseases by the entirely novel mechanism of altering proteins in the body. "Prion diseases may present as genetic, infectious, or sporadic disorders, all of which involve modification of the prion protein (PrP)," wrote Prusiner.

Paperclip scientists working on these infectious organisms were based primarily in laboratories at Fort Detrick, Maryland; Cold Spring Harbor, New York; and Edgewood Arsenal, Maryland. "It was here and in hundreds of other laboratories throughout America that immediately after World War II our former enemies' scientists were brought in under Operation Paperclip to continue their research and development of some of the most horrible weapons of mass destruction known to mankind," noted molecular researchers Garth and Nancy Nicolson in their 2005 book *Project Day Lily.*

The husband and wife molecular researchers noted there are two hundred species of *Mycoplasma.* Most are innocuous and do no harm. Only four or five are pathogenic. "*Mycoplasma fermentans* (incognitus strain) probably comes from the nucleus of the *Brucella* bacterium. This disease agent is not a bacterium and not a virus; it is a mutated form of the *Brucella* bacterium, combined with a visna virus, from which the mycoplasma is extracted," they said. "[T]he little mycoplasma also lost some of its genetic information, such as the genes that encode the thick cell wall and other genes that code for certain enzymes in metabolic pathways. Thus it is smaller than the most common bacteria, and without the distinctive cell walls found in most bacteria it can take on a variety of morphologies. It must hide inside animal or human cells to survive, and although originally thought to be fairly fragile, the little mycoplasma was hardier than anyone had ever imagined."

Although considered primitive by bacteriological standards, the mycoplasma actually evolved from bacteria that contained cell walls but lost its ability to make its own cell wall, probably because it no longer needed it when hiding inside hosts' cells and tissues. "But it made up for the loss of some of its genetic information by having evolved with other genetic sequences that allowed it to enter and colonize cells just like viruses. . . . [But] it was not a virus because it retained the genetic and biochemical remnants of bacteria. Like a virus, however, it damaged cells by interfering with some of the cells' biochemical cycles, and it encoded some nasty molecules that caused invaded cells to slowly self-destruct and die," said the Nicolsons, noting that important targets inside cells were the mitochondria, cellular "batteries" that produce energy and the DNA.

The Nicolsons explained that biological warfare research conducted between 1942 and now has created more deadly and infectious forms of mycoplasma. Continuing the work of Nazi scientists, researchers in the United States "weaponized" the mycoplasma by reducing the pathogen to a synthesized crystalline form. They later tested it on an unsuspecting public in North America.

According to the Nicolsons, the U.S. military's fascination with building this kind of biological weapon lies in the fact that the "creature will hide inside cells and cause unbelievable havoc. It will destroy the mitochondria, eventually sending cells into an unrelenting death program, and in the process gene expression will go crazy and surrounding cells will become damaged. This bug will then escape from its dying host cell and go to other places to eventually colonize every organ. And because pieces of the cellular membrane are dislodged when this little mycoplasma leaves its cellular hiding places, its victims should also be presented with an array of autoimmune symptoms similar to those found in various degenerative illnesses. It may even mimic some neurodegenerative diseases. It's beautiful, because it should cause diseases such as multiple sclerosis and rheumatoid arthritis, but no one will ever guess that they are caused by an infection. Most physicians . . . will never figure this out. . . . What a delightful weapon!"

Several researchers, including the Nicolsons, Dr. Leonard G. Horowitz, Dr. Joseph S. Puleo, and authors of *The Brucellosis Triangle,* Donald W. and William L. C. Scott, have linked this mycoplasma pathogen to a host of increasingly common neurosystemic diseases, such as Alzheimer's, bipolar disorder, Crohn's colitis, chronic fatigue syndrome, Creutzfeldt-Jakob, diabetes, dystonia, fibromyalgia, Huntington's, lupus, Lyme disease, multiple sclerosis, myalgic encephalomyelitis, Parkinson's disease, and even schizophrenia. Some strains of *Mycoplasma* are now being blamed for cancer and AIDS. According to the former chief virologist for the pharmaceutical company Merck Sharp & Dohme, the late Dr. Maurice Hilleman, this disease agent is now carried by everybody in North America and possibly most people throughout the world.

Mycoplasma researchers claim many people today suffering from various neurological diseases are actually ill with brucellosis. However, because the disease toxin pathogen has been isolated from the source bacterium in a crystalline form, there is no blood or tissue test that will confirm this fact.

Weaponized mycoplasmas generate ammonias that are deposited into the infected cell nuclei. "These nasty 'beasts' intertwine with the genetic machinery and are intra-cellular rather than inter-cellular. Other infectious agents are involved in the afflicted individual. These agents are usually mosaics of naturally occurring bacteria and viruses, and the effect upon the afflicted individual depends upon the individual's genetic predisposition and immunological make-up," stated Garth Nicolson. "Each person is affected differently by the infection, but all afflicted individuals share a constellation of symptoms.

"We have a survey that describes 120 signs and symptoms," added Nancy Nicolson. "In the case of the pathogenic mycoplasmas that we investigated, we found the HIV-1 envelope gene associated with the mycoplasma. This gene renders the mycoplasma more deadly. I have always wondered how many people that have been diagnosed as HIV positive actually have the chimeric—a mosaic of the mycoplasma bacterian and HIV?" Reportedly there are ten strains of HIV. HIV-1 promotes AIDS by compromising the immunization system, whereas HIV-2 does not promote AIDS. The other eight HIV strains are included in the biowarfare arsenal. The pathogenic mycoplasma can promote a non-HIV AIDS that mimics the symptoms of AIDS. "No one will talk about this!" said Nancy Nicolson. "The mycoplasmas have been genetically engineered with pieces of genetic material from other pathogens such as brucella. The mycoplasmas are often co-factors with the Lyme disease microorganism. All these emerging diseases correlate to bio-warfare experiments conducted during the Cold War that went seriously awry. Remember the US did approximately 208 open air tests on the US population without their knowledge or consent over a 30 years period."

It is possible that the crystalline disease toxin from the pathogens is one of the *Mycoplasma* species—a technological feat accomplished by U.S. military biochemical researchers working with Nazi Paperclip scientists. In 1946, the director of the War Research Service, George W. Merck, reported the possibility of using crystalline toxins to Secretary of War Robert P. Patterson. It should be noted that the War Research Service initiated America's biological weapons program, and Merck went on to become president of the Merck & Company pharmaceutical firm. Although Merck died in 1957, his early knowledge of the disease toxin means it could have been passed along to his colleagues at Merck Pharmaceutical. That Merck was involved in such research can be seen in a *New England Journal of Medicine* article that noted that a study of the hepatitis B vaccine, used extensively in gay and drug-addict communities, was supported "by a grant from the Department of Virus and Cell Biology of Merck, Sharp and Dohme Research Laboratories, West Point, VA."

After extensive study, researchers Donald W. and William L. C. Scott concluded that those suffering from chronic fatigue syndrome and fibromyalgia are actually victims of "man-altered versions of brucellosis emanating from the 'triangle'—that is, the areas around Fort Detrick, Washington, D.C., New York City's East Side and Long Island's federal Animal Disease Center, and Cold Spring Harbor Laboratory." These locations are often mentioned in biological warfare literature. Fort Detrick and Cold Spring Harbor, especially, were centers of Nazi Paperclip research activity.

According to the Scotts' report, this pathogen was tested during the summer of 1984 at Tahoe Truckee High School in California via the air duct system. Individual rooms were fitted with an independent recycling air supply system and the teachers' lounge was designated as the infection target. Within months, seven of eight teachers assigned to this room became very ill.

Tahoe Truckee High School was only one of several locations where the specially designed pathogens were tested. Some pathogens were distributed by aerosol sprays and others were spread through contaminated mosquitoes. The Scotts reported that, during the 1980s, one hundred million mosquitoes a month were bred at the Dominion Parasite Laboratory in Belleville, Ontario. From there, the mosquitoes were tested by both Canadian and U.S. military authorities after being infected with brucellosis. Some observers believe the 1999 outbreak of human encephalitis in New York City, due to what was designated West Nile virus, may have been the result of these infected mosquitoes.

Additionally, the Scotts also claim that unsuspecting victims were tested by both the military and CIA and monitored by the National Institutes of Health and the Centers for Disease Control. Encouraged by what they thought was a successful test, military leaders reportedly passed the brucellosis bioagent to Saddam Hussein, who in the mid-1980s was fighting a protracted war against Iran with the aid of the CIA. With the approval of Vice President George H. W. Bush in 1985, Saddam received "a startling array of biological pathogens . . . the essential raw material for a disabling weapon." This included shipments of both *Brucella abortus,* biotypes 3 and 9, and *Brucella melitensis,* biotypes 1 and 3. These toxins continued to be sold to Saddam through May 2, 1986, as "shipments number 21 and 22 from [the American Type Culture Collection] ATCC in Rockville, Maryland."

In a 2005 article entitled "Molecular Terrorism," Gary Tunsky credited both the Scotts and the Nicolsons with creating a growing public awareness of the mysterious and debilitating effects of mycoplasma infection.

"Chances are if you feel sick and tired and your doctor is unable to make a definite diagnosis because lab tests, blood chemistry profiles and tissue cultures fail to reveal any disease pathogen, you might very well be infected with Mycoplasma," suggested Tunsky.

"Since Mycoplasma cannot be successfully treated with the usual short course duration of antibiotics due to their intracellular location, slow proliferation rate and inherent resistance to most antibiotics, the few Mycoplasma experts that specialize in this field are recommending six-months to one year of non-stop treatments using strong antibiotics such as Cipro and Doxycycline," he added. "However, if a patient does not want to destroy their body and immune system with Cipro and Doxycycline, a total overhaul of every cell from head to toe using a multi-faceted, non-toxic, holistic treatment approach is absolutely necessary to overcome Mycoplasma infections naturally. This is why vitamins and nutritional supplementation are so important in the therapy."

Tunsky said the reason so many Americans are caught up in a medical merry-go-round of being bounced from one doctor to the next without ever receiving a proper diagnosis is that mainstream medical doctors are not trained to find hard-to-detect pathogens. "Since mycoplasma hides intra-cellularly and invades multiple organs and systems, it manifests a vast array of symptoms throughout the whole body, making a correct diagnosis virtually impossible for a mainstream doctor's linear, magic bullet mentality," he explained. Such inability to make a quick and simple diagnosis lies behind the mysterious malady that struck members of the U.S. military in the Persian Gulf War of 1990–91.

GULF WAR SYNDROME

AFTER SADDAM OBTAINED A stockpile of the brucellosis, it was discovered that this contagious designer bacteria had mutated and become airborne. And it was too late. According to the Scotts, Saddam used his toxins on American troops during the Persian Gulf War. This attack by mycoplasma, exacerbated by the impaired immunization systems caused by untested vaccines, the depleted uranium used in antitank shells, and oil well fires, combined in a toxic mixture resulting in the illness known as Gulf War syndrome. "Researchers could only look dumbly on when 100,000 veterans returned from the Gulf War presenting all of the brucellosis symptoms. . . . And the Pentagon could only take up the tried and tested myth that the veterans were not really sick at all. They only imagined they were," the Scotts explained.

Troops initially were told that no such infection existed and that the problem was mostly in their minds. But over the years, authorities were forced to admit that something had triggered severe illness in many Gulf War veterans. Curiously, French troops who served in the Gulf War did not receive the same mix of vaccines as the British and Americans and did not suffer from Gulf War syndrome. Apparently their undamaged immunization systems were able to withstand the mycoplasma attack.

A 1993 staff report to Senator Donald W. Riegle Jr., entitled "Gulf War Syndrome: The Case for Multiple Origin Mixed Chemical/Biotoxin Warfare Related Disorders," contrasts the relationship between the high rate of Gulf War illnesses among troops exposed to direct agent attacks and the much lower rates among those exposed only to the indirect fallout from coalition bombings of Iraqi chemical, biological, and nuclear targets. Because the U.S. military was not likely to reveal one of its most secret biochemical weapons or face liability by admitting that it had been sold to Saddam Hussein, the report concluded that vaccines "were to blame for the troops' illnesses." However, the report also hinted at the possibility of other causes, stating, "While other possible causes of the Gulf War Syndrome, such as petrochemical poisoning, depleted uranium exposure, and regionally prevalent diseases, have been discussed, no other explanation proves as compelling."

Although Riegle's report was completed in September 1993, it was not made available until April 1997, when the American Gulf War Veterans Association was finally able to obtain a copy. Not only were service members being forced to take untested vaccines, many veterans were not receiving adequate medical care due to missing medical records. The Senate Veterans Affairs Committee Report 103-97 issued on December 8, 1994, showed that the military medical records of 51 percent of 150 Gulf War veterans surveyed were either missing or inaccurate. Clearly, something other than mere negligence must have been at play if so many medical records were missing or inaccurate.

In 2009, Gulf War infection due to man-made mycoplasma seemed to be repeating itself. In mid-August, three Canadian soldiers were quarantined at a hospital in Quebec City, Canada, after returning from Kandahar, Afghanistan. The soldiers were infected with a drug-resistant "superbug" formally titled *Acinetobacter baumannii,* but dubbed by the American troops "Iraqibacter." Fearing they too may have contracted this bug, two civilian patients who were in contact with the soldiers were also isolated. "This isn't the first case we've had. We've received military patients returning from Afghanistan with this bacterium since 2007," said a hospital spokesperson. In a 2007 report, Wound Care Canada wrote that incidences of this strain have increased in U.S. military hospitals. America's CDC has issued a report stating that an increase of *Acinetobacter baumannii* in military hospitals treating U.S. troops serving in Iraq, Kuwait, and Afghanistan was noticed as far back as 2002.

Following the Gulf War and the misrepresentations of the government, the mycoplasma spread to the civilian population whereupon many people began suffering from debilitation and tiredness. Once it was known that the contagion was spreading into the general population, top officials with the National Institutes of Health and Centers for Disease Control as well as the Defense Department and the Department of Health and Human Resources claimed the disease was connected to the Epstein-Barr virus. They labeled it "chronic mononucleosis," and it has now become known as chronic fatigue syndrome (CFS). Like the veterans before them, victims of this ailment initially were told it was merely a psychological condition.

Yet by 2010, the CDC had acknowledged CFS as a long-term debilitating and complex disorder characterized by profound fatigue that is not improved by bed rest and that may be worsened by physical or mental activity. The CDC estimated more than one million people in the United States are affected by the syndrome and that there are "tens of millions of people with similar fatiguing illnesses who do not fully meet the strict research definition of CFS."

One victim, Dr. Martin Lerner of William Beaumont Hospital, told his peers in the American Society of Microbiology that the mysterious disease left his heart damaged, and that he suspected that CFS was caused by viral infection. Lerner, who founded the Treatment Center for Chronic Fatigue Syndrome in Beverly Hills, Michigan, created the Energy Index Point Score in hopes it would become a standard measurement tool to evaluate the degree of disability for CFS patients. Lerner has connected the Epstein-Barr virus, human herpes virus-6, and cytomegalovirus and similar infections to CFS. These are the very debilitating diseases studied by Donald and William Scott, who concluded that the victims of such neurodegenerative and systemic diseases "are ill with a very real physical disease deriving from a sub-viral particle developed from the brucellosis bacterial toxin."

The idea that a man-made biological weapon may be responsible for the ill health of millions of Americans is horrifying enough. Is it possible that such a catastrophic circumstance is the result of a conscious plan by the globalists?

DEPOPULATION EFFORTS

RESEARCHERS NOW BELIEVE THAT virtually everyone in North America—and perhaps the world—carries the crystalline pathogen, although no symptoms will become apparent until the latter stages of some serious disease. Many conspiracy theorists believed in early 2009 that something within the swine flu vaccinations would trigger the pathogen.

Swine flu, officially a new strain of the H1N1 influenza virus, was first identified in the spring of 2009 following an outbreak in Mexico. Oddly, although the strain contains a combination of genes from swine, avian (bird), and human influenza viruses, it cannot be spread by eating pork or pork products, leading many suspicious persons to suspect that swine flu is of human manufacture.

Some theorists also believed that the spread of the health-destroying mycoplasma toxin fits well with the agenda of the wealthy elite who have long supported eugenics and have been looking for ways to cull the human herd of "useless eaters." Many cite a classified study made by the U.S. National Security Council under Henry Kissinger in 1974, entitled "National Security Study Memorandum (NSSM) 200: Implications of Worldwide Population Growth for U.S. Security and Overseas Interests." This study, also known as the Kissinger Report, stated that population growth in the so-called lesser-developed countries (LDCs) represented a serious threat to U.S. national security. The study was adopted as official policy in November 1975 by unelected president Gerald R. Ford.

In a 1981 interview concerning overpopulation, former ambassador to South Vietnam and Chairman of the Joint Chiefs of Staff Maxwell Taylor, after advocating population reduction through limited wars, disease, and starvation, blithely concluded, "I have already written off more than a billion people. These people are in places in Africa, Asia, and Latin America. We can't save them. The population crisis and the food-supply question dictate that we should not even try. It's a waste of time."

As if he were reading from Taylor's script, England's Prince Philip was quoted in *People* magazine as saying, "Human population growth is probably the single most serious long-term threat to survival. We're in for a major disaster if it isn't curbed—not just for the natural world, but for the human world. The more people there are, the more resources they'll consume, the more pollution they'll create, the more fighting they will do. We have no option. If it isn't controlled voluntarily, it will be controlled involuntarily by an increase in disease, starvation and war." Years later,

Philip mused, "In the event that I am reincarnated, I would like to return as a deadly virus, in order to contribute something to solve overpopulation."

In the early 1970s, Associate Supreme Court Justice Ruth Bader Ginsburg may have echoed the views of Ivy League intellectuals when she said she believed the *Roe v. Wade* abortion decision was predicated on the Supreme Court majority's desire to diminish "populations that we don't want to have too many of." She added that it was her expectation that the right to abortion created in *Roe* "was going to be then set up for Medicaid funding for abortion."

Where did Ginsburg get the idea that American policy-making elites were interested in decreasing undesirable populations? Some researchers suggested that Ginsburg, at some point, became acquainted with the writings of John Holdren or of like-minded people in the most militant branch of the population control movement. In 1977, Holdren was a young academic who helped antinatalist guru Paul Ehrlich and his wife, Anne, write *Ecoscience: Population, Resources, Environment.*

Holdren's work states, "If some individuals contribute to general social deterioration by overproducing children, and if the need is compelling, they can [could] be required by law to exercise reproductive responsibility—just as they can be required to exercise responsibility in their resource-consumption patterns. . . ." Expressing the desire for "a Planetary regime" by controlling all human economic activity and interactions with the environment, the authors suggested the "power to enforce the agreed limits" on population growth by whatever means necessary. This includes involuntary sterilization, abortion, or even mass involuntary sterilization through the infiltration of sterilizing agents into public water supplies.

Internet blogger and radio host William Norman Grigg pointed out that amid the Obama administration's efforts to impose centralized "universal" health care, John Holdren sits as Barack Obama's "science czar," in which he counsels the president on the role of science in public policy. "This relationship has a certain Strangelovian undercurrent, given Holdren's enthusiasm for eugenicist and totalitarian methods of population 'management,'" he noted.

Prolific author G. Edward Griffin, best known for his book on the Federal Reserve, *The Creature from Jekyll Island,* also voiced concern over Holdren's thoughts on martial law and depopulation. Noting Holdren's advocacy of forced abortions and putting sterilization chemicals in the water supply, Griffin stated that Holdren discussed the possibility of reducing the population by insidious means. "He was not concerned with the ethical or freedom issues involved with these measures, only their practicality. Now we find this same man, an academic expert on population reduction, at the right hand of the President of The United States, advocating mass vaccination against the Swine Flu using vaccines that half of the medical profession believes are unsafe. . . . Remember, all of those who hold power in the governments of the world today are collectivists [self-styled globalists], and the guiding rule of collectivism is that individuals and minorities must be sacrificed, if necessary, for the greater good of the state or of society. Of course, those who rule will decide what the greater good is and who is to be sacrificed," Griffin said.

This, of course, is the basic problem with population control. The idea of limiting the burgeoning Earth population is probably desirable, as the increasing number of humans as well as their waste is placing a strain on the planet. The rub comes with the question of who will decide which segments of the population must forgo childbearing for the good of the majority. So far, it is the wealthy elite—the globalists—who have taken the lead in creating ways of holding down population growth through eugenics, drugs, and birth-control measures.

Former assistant secretary of housing Catherine Austin Fitts agreed with Griffin that one of the globalists' goals is depopulation. "Perhaps it is the goal of a swine flu epidemic as well, whether bio-warfare or hype around a flu season," she warned. "I keep remembering my sense of urgency leaving the Bush administration in 1991. We had to do something to turn around the economy and gather real assets behind retirement plans and the social safety net. If not, Americans could find themselves deeply out on a limb. I felt my family and friends were in danger. They did not share my concern. They had a deep faith in the system. As my efforts to find ways of reengineering government investment in communities failed to win political support, Washington and Wall Street moved forward with a debt bubble and globalization that was horrifying in its implications for humanity.

"Overwhelmed by what was happening, I estimated the end result. My simple calculations guessed that we were going to achieve economic sustainability on Earth by depopulating down to a population of approximately 500 million people from our then current global population of 6 billion [by 2009, 7.7 billion]. I was . . . used to looking at numbers from a very high level. To me, we had to have radical change in how we governed resources or depopulate. It was a mathematical result."

Fitts noted than some government budget analysts have concluded that the nation can no longer afford previously assumed social safety nets like Social Security and Medicare. "That is, unless you change the actuarial assumptions in the budget—like life expectancy," she said. "Lowering immune systems and increasing toxicity levels combined with poor food, water and terrorizing stress will help do the trick. A plague can so frighten and help control people that they will accept the end of their current benefits (and the resulting implications to life expectancy) without objection. And a plague with proper planning can be highly profitable. Whatever the truth of what swine flu is or vaccines rushed into production without proper testing and peer review, it is a way to keep control in a situation that is quickly shifting out of control."

MANUFACTURED AIDS

ADDING TO FEARS OVER conscious efforts to involuntarily reduce the human population are growing concerns that some killer plagues are man-made. To this day, many citizens still believe that acquired immune deficiency syndrome (AIDS) was created by bioengineers working in the United States after the immigration of Nazi eugenicists. It was in 1983 that AIDS was publicly recognized as a deadly and rapidly spreading disease. When the CDC called AIDS "a peculiar biological curiosity among

New York City homosexuals," suspicion grew around the world that AIDS was the product of germ warfare experiments designed to destroy undesirables. One theory was that it was developed between 1969 and 1972 in U.S. laboratories, then released in Africa by unsuspecting WHO workers in 1975 in doses of the smallpox vaccine. It was believed by some that AIDS came to the United States in 1978 in hepatitis B vaccine laced with human immunodeficiency virus (HIV).

"Don't Discount Conspiracy Theories on Origins of AIDS," stated a headline in Kenya's newspaper, the *Daily Nation*. In a December 2009 article, it was noted that thirty-three million persons worldwide and thirteen million sub-Saharan Africans have died of AIDS. Writer Angeyo Kalambuka noted, "Soon after the US State Department published the Global 2000 Report for the President in 1980 advising that the world population must be reduced by 2 billion people by the year 2000, Thomas Ferguson of the Office of Population Affairs elaborated in the *Executive International Review* that 'the quickest way to reduce population is through famine, like Africa, or through disease, like in the Black Death . . . population reduction is now our primary policy objective'."

The belief that AIDS was manufactured by the United States is supported by a record of hearings before a subcommittee of the House Committee on Appropriations in 1969.

In his testimony to the subcommittee in 1969, Dr. D. M. MacArthur, deputy director for research and technology at the Department of Defense, said, "Within the next five to ten years, it would probably be possible to make a new infective microorganism which could differ in certain important aspects from any known disease-causing organisms. Most important of these is that it might be refractory to the immunological and therapeutic processes upon which we depend to maintain our relative freedom from infectious disease." In other words, he meant a type of germ that would neutralize the normal human immunization system. MacArthur told the congressmen that tentative plans for the development of this organism had already been drawn up between the Pentagon and the National Academy of Sciences' National Research Council and that the project would cost $10 million. Interestingly, MacArthur admitted that such a program was "highly controversial" and that there were many "who believe such research should not be undertaken lest it lead to yet another method of massive killing of large populations."

"Mycoplasmas will forever be at the heart of the U.S. biological warfare program," stated attorney Boyd Graves, a graduate of the U.S. Naval Academy and director of AIDS Concerns for the Common Cause Medical Research Foundation in Ontario, Canada. Graves had produced a timeline flowchart that correlated more than twenty thousand scientific papers and fifteen years of progress reports concerning a secret federal virus development program, which he claimed proves the man-made origins of AIDS. "The 1971 flowchart makes it perfectly clear . . . [and] provides absolute evidence of the United States' intent to kill its own citizens and others," declared Graves.

Graves told one interviewer, "No one in this U.S. government has downloaded the 1971 flowchart. . . . There is a substantial basis in U.S. law and fact for the allegation and the conclusion that the United States intentionally made HIV/AIDS with the purpose for use as a population control weapon; a quiet and silent holocaust of people of color, toward the development of a New World Order."

The depopulation views of Maxwell Taylor, Henry Kissinger, and others are echoes of the words from a 1996 full-page ad by Negative Population Growth, Inc. (NPG), that was published in *Foreign Affairs,* the official publication of the Council on Foreign Relations:

"We need a smaller population in order to halt the destruction of our environment, and to create an economy that will be *sustainable* [original emphasis] over the very long term. We are trying to address our steadily worsening environmental problems without coming to grips with their root cause—overpopulation. All efforts to save our environment will ultimately be futile unless we not only halt U.S. population growth, but reverse it, so that our population can eventually be stabilized at a *sustainable* [original emphasis] level, far lower than it is today."

According to the ad by NPG, the population level being sought by the globalists was described as "a U.S. population in the range of 125 to 150 million, or about its size in the 1940s." According to U.S. Census statistics in mid-2009, the population stood at 307,229,513. Thus, more than half the current population needs to disappear to reach the level envisioned by the globalists.

Such globalist thinking continues today. On May 5, 2009, some of America's leading billionaires met in a private Manhattan home just a week before the annual meeting of the secretive Bilderbergers in Greece. Calling themselves the "Good Club," attendees included Bill Gates, David Rockefeller Jr., Warren Buffett, George Soros, New York mayor Michael Bloomberg, Ted Turner, and Oprah Winfrey. According to John Harlow of the *Sunday Times,* the group agreed with Gates that human overpopulation was a priority concern. "Another guest said there was 'nothing as crude as a vote' but a consensus emerged that they would back a strategy in which population growth would be tackled as a potentially disastrous environmental, social and industrial threat," wrote Harlow.

So to achieve the globalists' dream of a U.S. population of no more than 150 million, the current population would have to be halved. What's to happen to more than 150 million Americans?

Apparently, those with great wealth and power have decided to take overpopulation into their own hands. And these individuals were connected to the same families and corporations that funded communism in Russia and then National Socialism in prewar Germany.

"Today with AIDS, mad cow disease, chronic fatigue, and the rest, history is apparently repeating," noted Dr. Len Horowitz. "In fact, even the message is the same. The millions of Holocaust victims were told they were going into 'showers' for 'public health' and 'disinfection.' That's why we are being told to get vaccinated. Virtually nothing has changed, not even the message."

As an indication that nothing has substantially changed in the ruling hierarchy of the globalists, on February 8, 2009, President Barack Obama's national security adviser, General James L. Jones, opened a speech to the Forty-fifth Munich Conference on Security Policy in Germany by admitting that he takes his "daily orders from Dr. Kissinger, filtered down through General Brent Scowcroft and Sandy Berger."

Donald and William Scott believed a high-level agenda to reduce the population went even above Henry Kissinger: "The Washington corner of the brucellosis triangle with its military, NIH, Treasury and Justice [Department] components have had their ties to and have largely taken their directions from the New York corner dominated by the Rockefeller interests. And the Rockefeller interests through the agency of the CFR [Council on Foreign Relations], the Rockefeller Institute/University, the Cold Spring Harbor Laboratory, the Rockefeller Foundation and the Chase Manhattan Bank have constituted a vast machine of power and baleful influence whose parts have meshed together in an effort to maintain that power."

DEAD MICROBIOLOGISTS

ONE REASON WHY MORE doctors don't want to look more closely at the mycoplasma pandemic may be that work in the field of microbiology appears to be hazardous to one's healthy. By mid-2009, nearly one hundred scientists around the world had died, many under suspicious circumstances. Most of them were microbiologists.

Researcher Mark J. Harper compiled a list of scientists in some way connected to the study of viruses or vaccines. "While some of these deaths may be purely coincidental and seem to pose no connection, many of these deaths are highly suspicious and appear not to be random acts of violence. Many are just plain murders," commented Harper.

While not everyone on this list died an unnatural death, the sheer number and scope is breathtaking. With this many dead, couldn't this mean that someone, somewhere, wants to get rid of those who see through the conspiracy of fraudulent pandemics and might produce effective antidotes?

A full list of these names and dates of death can be found on several Internet websites, including Mark Halper's site http://www.puppstheories.com/forum/index.php?showtopic=6521.

DR. RIFE'S DISCOVERY

INSIDIOUSNESS OF CONTROL HAS become so pervasive that a remarkable scientist was professionally discredited and ruined for claiming to discover a cure for dangerous diseases. Today Dr. Royal Raymond Rife's suppressed technology is making a worldwide comeback despite the opposition of the medical establishment.

In the 1930s, Rife demonstrated the ability of specific radio-wave frequencies to disrupt viral and bacteria cells. Every biochemical compound, including single-cell organisms, oscillate with a unique frequency vibration. Because germs are carbon-based life forms, they are susceptible to disruption by radio frequencies. When the amplitude, or resonance, of the frequency is intensified, the cell can be shattered and destroyed. By increasing the intensity of a frequency, Rife increased the natural oscillations of one-celled bacteria and viruses until they distorted and disintegrated from structural stresses. A crude analogy to this effect is a glass shattering when a singer sounds a high note.

Rife's work with pathogens began as a result of his invention of the "Universal Prismatic Microscope," which was more effective in studying organisms than electron microscopes because those devices killed specimens by bombarding them with radiant energy. Using specially ground quartz prisms in an elongated microscope tube, Rife not only was able to view live specimens but also his view was amplified up to sixty thousand times. He became the first human to see and photograph live viruses and to note that they evolved and changed form just as other organisms.

A 1944 report from the Smithsonian Institution entitled "The New Microscope," by Dr. R. E. Seidel (report #3781), stated, "Under the Universal Microscope disease organisms such as those of cancer . . . and other disease may be observed to succumb when exposed to certain lethal frequencies. . . ." This was strong support for claims that Rife's frequency therapy actually worked to destroy diseases.

Following decades of research, Rife isolated the frequencies of numerous disease cells, including cancer, and by broadcasting them back to the cells in an intensified form was able to shatter the original disease cell. This technology does not harm normal healthy cells or tissue. There has not been one documented case of a person harmed by a Rife-type device.

In fact, there are narratives of many successes. A Special Research Committee of the University of Southern California confirmed that Rife frequencies were reversing many ailments, including cancer. By 1934, Rife had isolated a virus that incited cancer cells and stopped it by bombarding it with radio frequencies. He was successful in killing both carcinoma and sarcoma cancers in more than four hundred tests on animals and in using his frequencies to cure sixteen cancer patients diagnosed as terminal by conventional medicine.

Soon enough, the established medical community realized that this device not only would wreck the pharmaceutical industry, but damage medical practices in general. Cures meant fewer visits to the doctor. Opposition immediately came from Dr. Thomas Rivers of the Rockefeller Institute, who had not even seen Rife's equipment in operation. Rivers claimed evolved forms of viruses did not exist. Conflict broke out between those persons who had seen viruses changing into different forms beneath Rife's microscopes and those who had not.

"Because his microscope did not reveal them, Rivers argued that there was 'no logical basis for belief in this theory [evolving forms of viruses],'" explained national radio commentator Jeff Rense. "The same argument is used today in evaluating many other 'alternative' medical treatments; if there is no precedent, then it must not be valid. Nothing can convince a closed mind. Most had never actually looked though the San Diego microscopes [of Rife] . . . air travel in the 1930's was uncomfortable, primitive, and rather risky. So, the debate about the life cycle of viruses was resolved in favor of those who never saw it. Even modern electron microscopes show frozen images, not the life cycle of viruses in process,"

Overworked and underfunded, Rife and his associates were easy targets for attack. The health authorities made false claims against him, altered his test procedures so that his demonstrations would fail, and made impossible demands on him.

In 1934, Rife declined an offer to partner with Morris Fishbein, then head of the American Medical Association. "We may never know the exact terms of this offer. But we do know the terms of the offer Fishbein made to Harry Hoxsey for control of his herbal cancer remedy," reported Rense. "Fishbein's associates would receive all profits for nine years and Hoxsey would receive nothing. Then, if they were satisfied that it worked, Hoxsey would begin to receive 10% of the profits. Hoxsey decided that he would rather continue to make all the profits himself. When Hoxsey turned Fishbein down, Fishbein used his immensely powerful political connections to have Hoxsey arrested 125 times in a period of 16 months. The charges (based on practicing medicine without a license) were always thrown out of court, but the harassment drove Hoxsey insane."

Rife's troubles turned more serious. His lab was ransacked on several occasions, but no suspect was ever caught. He was also harassed by health officials. Baseless and costly lawsuits were brought against him resulting in his bankruptcy. The suits, some filed by persons with connections to pharmaceutical corporations, ultimately failed. The USC's Special Research Committee was disbanded, Rife was marginalized, and his device today is available only as a costly research instrument employed by a few doctors and private citizens. Rife died a broken man in 1971.

Although Rife's work has been confirmed by scientists and researchers outside the United States, the conventional medical community still ignores the benefits of this technology and continues to prosecute those who do. Those Americans who have confirmed or endorsed various areas of Rife's work include Dr. Edward C. Rosenow Sr., former chief of bacteriology at the Mayo Clinic; Dr. Arthur I. Kendall of Northwestern Medical School; Dr. George Dock of the Los Angeles County Medical Association library; Dr. Alvin Foord, professor of pathology at the University of Southern California; Rufus Klein-Schmidt, president of USC; Dr. Milbank Johnson, director of the Southern California AMA; Whalen Morrison, chief surgeon for the Santa Fe Railway; Dr. George Fischer of Children's Hospital, New York; Karl Meyer, with the Hooper Foundation; and many others.

Barry Lynes, a California investigative reporter, learned of the Rife story through John Crane, who had worked at Rife's side from 1950 until

Rife's death in 1971. Initially skeptical of the claims of the healing benefits of Rife technology, after studying the documentation held by Crane, Lynes became outraged by the injustices that had wrecked Rife's life's work. Lynes's 1987 book on Rife and his work, entitled *The Cancer Cure That Worked! Fifty Years of Suppression,* became an underground favorite and sparked renewed interest in Rife's work. Starting in 1995, San Diego manufacturer James Folsom marketed and distributed Rife-type devices when he took over the Royal Rife Research Society. He claimed to have hundreds of testimonials that his devices improved physical symptoms, and in many cases led to remission in cancer. According to Folsom, he had no dissatisfied customers.

Folsom was raided by the FDA in 2003 during Operation Cure All, a campaign that targeted various companies in the alternative-health market. Although Folsom's equipment was confiscated, Folsom heard no more about it for years. But then in October 2007, just days before his vulnerability would have ended under a statute of limitations, Folsom was arrested and charged with several felonies, including selling a class III medical instrument without a license. Folsom argued he did not need a license because his equipment was a class I biofeedback device. These devices were exempt and had been used for more than seventy years with no known harm or side effects. Regardless, the FDA claimed these biofeedback devices were under its jurisdiction over medical devices under a 1976 law. That law allowed for the prosecution of selling high-voltage medical devices. However, it should be noted that Folsom's machine at that time could be powered by a nine-volt battery.

Despite being offered a plea bargain that would allow him to plead guilty to a misdemeanor and make him pay a $250 fine and suffer one year of unsupervised probation, Folsom decided to go to trial.

According to U.S. Attorney Karen Hewitt, Folsom's business generated more than $8 million in revenue over its years in operation. Assistant U.S. Attorney Melanie Pierson said the case was the largest involving illegal medical devices that she had seen in twenty years working as a prosecutor in San Diego County.

The trial was held in U.S. District Court, where no discussion of the effectiveness of the Rife's technology was allowed. Originally, Folsom wanted to produce stacks of testimonials from satisfied customers but Melanie Pierson objected. Folsom then tried to assert that for more than seventy years, no harmful effects had been documented from the devices. This too was not allowed. Folsom then tried to argue that none of his customers had been dissatisfied, but to no effect. Aside from prosecutors and government officials, the only witnesses at the trial were twenty-four of Folsom's friends and fellow device distributors. They unanimously testified to Folsom's good character and clean business practices. Incredibly, the prosecution used this testimony against Folsom, claiming that, in fact, Folsom was such a brilliant fraud that even his peers and customers weren't aware they'd been defrauded. Prosecutors claimed Folsom used the false name "Jim Anderson" to avoid being caught by the FDA, and that he gave buyers the false impression that the FDA had approved the devices for "investigational purposes." Folsom admitted he had used the name as a salesman at a different company but had used his real name on all official and government correspondence.

In February 2009, a U.S. federal jury in San Diego convicted Folsom of twenty-six felony counts for selling Rife devices under the name of "Nature-tronics," "AstroPulse," "Biosolutions," "Energy Wellness," and "Global Wellness." Folsom, sixty-eight, faced more than 140 years in prison, literally a life sentence at his age, and $500,000 in fines. He is being held in the federal government's Western Region Detention Facility in San Diego, now managed by the private firm GEO Group, Inc.

A few weeks after Folsom's trial and conviction, the FDA issued a news release announcing that manufacturers of twenty-five types of medical devices marketed prior to 1976 must submit safety and effectiveness information to the agency so that it may evaluate the risk level for each device type. Supporters of Folsom said the FDA's decision to scrutinize such preexisting technology was most likely the result of his trial.

One Folsom supporter stated, "Jim stood on his principles for his innocence and to clear the Rife name. It was an impossible task. Jim was up against an endless supply of money through the FDA and an unjust system. Research has found since Jim's conviction that our judicial system is more of a money machine than Big Pharma & the Medical Industrial Complex."

Observers saw Folsom's conviction as a blow against those supporting Rife technology. They also predicted that those interested in the technology would have to go to foreign websites such as http://www.rife.de/, a site in Germany where the sales and use of Rife-type devices are legal.

With the FDA seeking to require prescriptions for everyday vitamins and suppressing potentially useful medical technologies like Rife's, not to mention the new government-controlled national health-care plan, it would appear as though there is a conscious effort to prevent the public from acquiring healthful alternatives to chemical drugs.

But why would the government harm us with untested vaccines and the suppression of potentially healthful therapies? Wouldn't such actions also adversely affect the health of the global elite? Some researchers believe the answer may be that the inner-core globalists already utilize such technology or something even more advanced. Is it possible they can cure themselves of the same illnesses they allow to be inflicted on others? The globalist elites may not be worried that their eugenics plans will touch their families. They may believe they can protect their own DNA with race-specific pathogens. If they should contact some dire affliction, might they easily eliminate it with just a few short hours of frequency technology or advanced antidotes for immunization—therapies cloaked from the general population.

The possibility of holding such publicly denied therapies that might end disease and halt or regress the aging process would prove a most effective enticement in the recruitment of minions to aid in the advancement of their globalist agendas.

DRUGGING THE POPULATION

There will be, in the next generation or so, a pharmacological method of making people love their servitude, and producing dictatorship without tears, so to speak, producing a kind of painless concentration camp for entire societies, so that people will in fact have their liberties taken away from them, but will rather enjoy it, because they will be distracted from any desire to rebel by propaganda or brainwashing, or brainwashing enhanced by pharmacological methods. And this seems to be the final revolution.

—ALDOUS HUXLEY, 1961

BIG PHARM

DRUGS ARE BIG BUSINESS. Only five biopharmaceutical companies— Novartis, GlaxoSmithKline, MedImmune, the Australian firm CSL, and Sanofi-Pasteur—have been awarded massive contracts by the U.S. Department of Health and Human Services (HHS) to develop and produce more than 195 million doses of swine flu vaccine. This is in addition to the seasonal flu vaccine.

According to Dr. Joseph Mercola, an osteopathic physician and author of sixteen books on health and alternative medicine, including two *New York Times* bestsellers, "CSL has contracts to supply $180 million worth of bulk antigen to the U.S. MedImmune will supply 40 million doses of its live attenuated nasal spray swine flu vaccine for more than $450 million. And Sanofi-Pasteur is providing more than 100 million doses of monovalent swine flu vaccine, a $690 million order."

About half of the world's largest pharmaceutical corporations are not American, but rather European. Among the top ten pharmaceutical companies are the American companies Pfizer, Merck, Johnson & Johnson, Bristol-Myers Squibb, and Wyeth (formerly American Home Products). The rest of the top pharmaceutical companies are the British companies GlaxoSmithKline and AstraZeneca; the Swiss companies Novartis and Roche; and the French company Aventis (which in 2004 merged with another French company, Sanafi Synthelabo, putting it in third place). These corporations essentially function alike, but their drug prices in America are much higher than in other nations' markets. For example, a bottle of one thousand aspirin costs less in Mexico than a bottle of five hundred across the border in the United States and, obviously, no company will sell a product without making a profit.

To give some indication of the money involved in the modern drug business, the legal pharmaceutical market totaled $712 billion globally in 2007, of which about $80 billion was for psychiatric drugs. According to several authorities, including Harvard psychiatrist Dr. Peter R. Breggin; Bruce Wiseman, national president of the Citizens Commission on Human Rights; geneticist Dr. Thomas Roeder; Dr. Hyla Cass, a former assistant clinical professor of psychiatry at UCLA School of Medicine; and David Healey and David B. Menkes, both of the North Wales Department of Psychological Medicine, psychiatric drugs may be the culprit behind many homicides, suicides, and school shootings.

Even worse, the $80 billion doesn't even include the illegal drug market.

A former editor in chief of the *New England Journal of Medicine*, Dr. Marcia Angell, wrote in the *New York Review of Books,* "The combined profits for the ten drug companies in the Fortune 500 ($35.9 billion) were more than the profits for all the other 490 businesses put together ($33.7 billion). Over the past two decades the pharmaceutical industry has moved very far from its original high purpose of discovering and producing useful new drugs. Now primarily a marketing machine to sell drugs of dubious benefit, this industry uses its wealth and power to co-opt every institution that might stand in its way, including the US Congress, the FDA, academic medical centers, and the medical profession itself."

In her 2004 book *The Truth About the Drug Companies: How They Deceive Us and What to Do About It,* Dr. Angell argues that the current power of the pharmaceutical industry can be directly traced to the industry's phenomenal growth during the Reagan years, with George H. W. Bush and his globalist supporters in command following Reagan's wounding during an assassination attempt in March 1981.

"The watershed year was 1980," she noted. "Before then, it was a good business, but afterward, it was a stupendous one. From 1960 to 1980, prescription drug sales were fairly static as a percent of US gross domestic product, but from 1980 to 2000, they tripled. They now stand at more than $200 billion a year. Of the many events that contributed to the industry's great and good fortune, none had to do with the quality of the drugs the companies were selling."

The success of Big Pharm has more to do with marketing than with the effectiveness of its drugs. Dr. Michael Wilkes, professor of medicine and vice dean for medical education at the University of California, Davis, joined other critics in describing a recent phenomenon called "disease-mongering," an activity in which large drug corporations attempt to convince healthy people they are sick and need drugs in order to boost sales.

"Most pharmaceutical companies devote huge amounts of money to prevent, control and cure diseases," he added. "When their profits don't match corporate expectations, they 'invent' new diseases to be cured by existing drugs."

"Countless examples of disease-mongering are driven by the pharmaceutical industry's drive to sell drugs," wrote Dr. Wilkes. "Conditions such as female sexual dysfunction syndrome, premenstrual dysphoric disorder, toenail fungus, baldness and social anxiety disorder (a.k.a. shyness) are a few places where the medical community has stepped in, thereby turning normal or mild conditions into diseases for which medication is the treatment."

Ironically, though Big Pharm invents new diseases, they rarely invent a new drug. Surprisingly, most new and important drugs brought to market in recent years were based on taxpayer-funded research at universities, small biotechnology companies, or the National Institutes of Health (NIH). In fact, most supposedly "new" drugs are merely a variation of older drugs.

"If I'm a manufacturer and I can change one molecule and get another twenty years of patent rights, and convince physicians to prescribe and consumers to demand the next form of Prilosec, or weekly Prozac instead of daily Prozac, just as my patent expires, then why would I be spending money on a lot less certain endeavor, which is looking for brand-new drugs?" asked Dr. Sharon Levine, associate executive director of the Kaiser Permanente Medical Group.

"What's true of the eight-hundred-pound gorilla is true of the colossus that is the pharmaceutical industry. It is used to doing pretty much what it wants to do," wrote Dr. Marcia Angell. "The most important of these laws [that relax restrictions on pharmaceutical corporations] is known as the Bayh-Dole Act, after its chief sponsors, Senator Birch Bayh (D-Ind.) and Senator Robert Dole (R-Kans.). Bayh-Dole enabled universities and small businesses to patent discoveries emanating from research sponsored by the National Institutes of Health, the major distributor of tax dollars for medical research, and then to grant exclusive licenses to drug companies. Until then, taxpayer-financed discoveries were in the public domain, available to any company that wanted to use them. But now universities, where most NIH-sponsored work is carried out, can patent and license their discoveries, and charge royalties. Similar legislation permitted the NIH itself to enter into deals with drug companies that would directly transfer NIH discoveries to industry. . . . Thus, when a patent held by a university or a small biotech company is eventually licensed to a big drug company, all parties cash in on the public investment in research."

Under this system, research paid for by public money became a commodity to be sold for profit by privately owned companies. Dr. Angell provides examples of the large consulting fees paid by pharmaceutical corporations to individual faculty members and to NIH scientists and directors. These fees allow for globalist pharmaceutical corporations to further intrude into the nation's medical education.

The lucrative connection between Big Pharm and medical schools and hospitals has brought about a definite corporate-friendly atmosphere. "One of the results has been a growing pro-industry bias in medical research [in both schools and hospitals]—exactly where such bias doesn't belong," stated Dr. Angell.

She noted that the huge amounts of money flowing from Big Pharm began to change the ethos of medical schools and teaching hospitals. Such nonprofit institutions began to view themselves as partners of industry. Faculty researchers were encouraged to obtain patents on their work, which were then assigned to their universities. The schools then sold the right to Big Pharm and shared in royalties. Many medical schools and teaching hospitals even created technology transfer offices to capitalize on faculty discoveries.

Dr. Angell also noted the excessive salaries for pharmaceutical executives. Take, for instance, the whopping $74,890,918 salary paid to Charles Heimbold Jr. in 2001, the former chairman and CEO of Bristol-Myers Squibb. This does not count his $76,095,611 worth of unexercised stock options. At the same time, the chairman of Wyeth made $40,521,011 in 2001, not counting his $40,629,459 in stock options.

DTC ADS

SELLING IS THE NAME of the game. Drug advertising is now ubiquitous in all major media outlets. Despite spending 7.1 percent less on direct-to-consumer (DTC) drug advertising in the third quarter of fiscal 2008, a Nielsen Media Research report showed that pharmaceutical firms still spent about $4.8 billion on DTC advertising for television, radio, and print ads in magazines and newspapers.

Here's how the top few drugs worked out in sales per advertising dollar spent:

The cholesterol drug Lipitor earned $34.09 for each ad dollar spent.
The asthma drug Advair Diskus earned $27.98 per ad dollar.
The heartburn remedy Nexium earned $44.92 per ad dollar.
The allergy drug Singulair earned $45.24 per ad dollar.
The allergy medication Zyrtec (now available without prescription) earned $33.86 per ad dollar.

DTC advertising more than tripled between 1997 and 2005, growing from $1.3 billion to $4.2 billion since the U.S. Food and Drug Administration eased restrictions governing these types of drug ads.

It has been estimated that $8 billion of the $235 billion spent by consumers on prescription drugs in 2008 came from DTC advertising. And the 2008 decline in DTC advertising—a first in recent U.S. history—was offset by launch campaigns on drugs such as Cialis, Abilify, Nasonex, and Plavix.

While TV ads show visuals of happy people, idyllic countrysides, laughing children, and playful pets, a droning audio voice rapidly skips through possible side effects. The pain medication Vioxx was heavily advertised by its maker but later recalled when it was shown the drug increased the risk of heart attack in some people. "The fact that it was so heavily marketed magnified its ultimate damage," said Michael Russo, a health-care proponent for the public advocacy group California Public Interest Research Group (CalPIRG).

Perhaps Big Pharm cares more for promoting their drugs than developing something better and safer. Published estimates predict that whereas the drug industry spent about $57.5 billion on U.S. marketing in 2004, it spent only $31.5 billion on research and development. Percentage-wise, of the $235.4 billion in U.S. sales in 2004, promotion consumed 24.4 percent of sales dollars while R&D only took 13.4 percent.

"Although some academic studies suggest that DTC advertising can help people who need to start taking drugs and others to remain compliant with existing treatment regimens, the lack of fair balance in many DTC ads that promote drug benefits and downplay risks is what is driving legislation to curb its use," stated a comment posted on BioJobBlog.com, a website dedicated to bioindustry employment. " . . . Interestingly, about ten years ago, a friend who works for a major pharmaceutical company told me that she always waits five years before using a newly approved drug. At the time, I thought it was an odd thing for her to say since she had been in the business for over 15 years. However, over the past five years or so, several high-profile drugs that were heavily promoted by DTC advertising had to be withdrawn from the market. To that end, while DTC advertising may be 'great for business,' it may not always be in the best interest of American consumers who use prescriptions."

The site also noted that DTC advertising is allowed in only two countries—New Zealand and the United States.

The ever-increasing predominance of DTC drug advertising has prompted several members of Congress to introduce legislation to curtail the ads. Legislators were disgusted with tax deductions for drug marketers using DTC advertising and commercials offering products that gave four-hour erections during prime-time television hours.

Not only did drug advertising trouble the public, but so did the disproportion of actual drug costs to retail sale price. In 2003, the website ThePeoplesVoice.org posted this chart of the actual price of active ingredients used in some of the most popular drugs sold in America.

Brand Name	Consumer Price per 100	Cost of General Active Ingredients per 100 tab/cap	Percent Markup
Celebrex 100 mg	$130.27	$0.60	21,712%
Claritin 10 mg	$215.17	$0.71	30,306%
Keflex 250 mg	$157.39	$1.88	8,372%
Lipitor 20 mg	$272.37	$5.80	4,696%
Norvasc 10 mg	$188.29	$0.14	134,493%
Paxil 20 mg	$220.27	$7.60	2,898%
Prevacid 30 mg	$44.77	$1.01	34,136%
Prilosec 20 mg	$360.97	$0.52	69,417%
Prozac 20 mg	$247.47	$0.11	224,973%
Tenormin 50 mg	$104.47	$0.13	80,362%
Vasotec 10 mg	$102.37	$0.20	51,185%
Xanax 1 mg	$136.79	$0.024	569,958%
Zestril 20 mg	$89.89	$3.20	2,809%
Zithromax 600 mg	$1,482.19	$18.78	7,892%
Zocor 40 mg	$350.27	$8.63	4,059%
Zoloft 50 mg	$206.87	$1.75	11,821%

Fortunately, the government has acted in response to the growing public awareness of Big Pharm malfeasance. In September 2009, Pfizer Inc., the world's largest drug manufacturer, was ordered to pay a record $2.3 billion civil and criminal penalty after the government found the firm guilty of unlawful prescription drug promotions. Prosecutors charged the company with promoting four prescription drugs, including the pain-killer Bextra (taken off the market in 2005), after studies indicated that the drugs increased the chances of heart attack and had been used as a treatment for medical conditions different from those for which federal regulators had approved.

A spokesman for the Justice Department said the fine, which included both criminal and civil penalties, was the largest criminal fine in U.S. history.

Authorities noted that this was the fourth settlement involving false and misleading advertising claims in the past ten years. They called Pfizer a "repeat offender" and said the company's conduct would be monitored for the next five years.

Previously, Pfizer was accused of inviting doctors to all-expense-paid meetings at resorts as consultants. U.S. attorney for Massachusetts Mike Loucks said, "They were entertained with golf, massages, and other luxuries." He added that Pfizer continued to violate the same laws with other drugs even while negotiating the Bextra settlement with Justice Department attorneys.

New York attorney general Andrew Cuomo told the media, "Pfizer ripped off New Yorkers and taxpayers across the country to pad its bottom line. Pfizer's corrupt practices went so far as sending physicians on exotic junkets as well as wining and dining health care professionals to persuade them to prescribe the company's drugs for patients in taxpayer-funded programs."

Another Big Pharm giant's consolidation efforts point to high-level connections with both the globalists and the Nazis. GlaxoSmithKline (GSK), the second-largest pharmaceutical company in the world after Pfizer, was founded in London in 1880 by two American pharmacists—Henry Wellcome and Silas Burroughs—as Burroughs Wellcome & Company. Glaxo Laboratories, originally a baby food manufacturer, went multinational in 1935. After the postwar acquisition of other companies, including Meyer Laboratories, Glaxo merged with Burroughs Wellcome in 1995. The new name of the company was GlaxoWellcome. In 2000, after merging with

SmithKlineBeckman, the firm became GlaxoSmithKline.

The original Burroughs Wellcome drug firm was wholly owned by Wellcome Trust, whose director was the British lord Oliver Franks, a man described as "one of the founders of the post-war world." Franks was ambassador to the United States from 1948 to 1952 and was also a director of the Rockefeller Foundation and its principal representative in England. He was a director of the Kurt von Schroeder Nazi Bank, which at one time handled Hitler's personal bank account. Franks also was a director of the Rhodes Trust, which was used in the late 1800s by the African diamond magnate Cecil Rhodes to create his Round Table Groups, a forerunner of the Council on Foreign Relations. As a Rhodes director, Franks was in charge of approving Rhodes scholarships such as the one awarded Bill Clinton in 1968.

According to former intelligence officer Dr. John Coleman, members of Rhodes's Round Tables, armed with immense wealth gained from control of gold, diamonds, and drugs, fanned out over the world to take control of fiscal and monetary policies and political leadership in all countries where they operated. This conspiratorial network was confirmed by President Clinton's academic mentor, the Georgetown University historian Carroll Quigley, who wrote, "There does exist, and has existed for a generation, an international Anglophile network which operates, to some extent, in the way the radical Right believes the Communists act. I know of the operations of this network because I have studied it for 20 years and was permitted for two years, in the early 1960s, to examine its papers and secret records. I have no aversion to it or to most of its aims and have, for much of my life, been close to it and to many of its instruments. . . . [I]n general my chief difference of opinion is that it wishes to remain unknown, and I believe its role in history is significant enough to be known."

While Franks is known as director of the trust that owned a large drug company, most people do not know the extent of the Rockefellers' influence over modern medicine and drugs.

According to Eustace Mullins, the drug industry is controlled by a Rockefeller "Medical Monopoly," largely through directors on pharmaceutical boards representing Rockefeller entities. "The American College of Surgeons maintained a monopolistic control of hospitals through the powerful Hospital Survey Committee, with members Winthrop Aldrich and David McAlpine Pyle representing the Rockefeller control."

Winthrop Aldrich, whose sister was married to John D. Rockefeller Jr., served as president and board chairman of Chase National Bank from 1930 to 1953. He also served on the Committee on the Cost of Medical Care (CCMC), which was started by Dr. Alexander Lambert, the personal physician to Teddy Roosevelt and a president of the AMA beginning in 1910. According to Dr. Charles C. Smith, a physician who researched the activities of the committee and published a report in 1984: "He [Dr. Lambert] obviously was to be the needed 'figurehead.' . . . The full time staff was headed by Harry H. Moore of Washington, who in 1927 published 'American Medicine and the People's Health' while a member of Public Health Service. His main tenets were the need for a *system* [original emphasis] to distribute medical care and an insurance plan to pay for it."

So early in the twentieth century, administrators and economists were deciding the future of America's health care. Moore was aided by C. Rufus Rorem, who received a PhD in economics at the University of Chicago, which was founded and funded by Rockefeller. Rorem, according to Smith, was more concerned about hospital prepayment than in health care. Following his work for the CCMC, Rorem went on to become executive director of the Blue Cross Plan Commission between 1936 and 1946.

"I think the most important principle spawned by this Committee was not at all what was planned," wrote Dr. Charles C. Smith Jr., who authored a medical history study paper on the committee. A minority on the committee fruitlessly recommended that government competition in the medical practice be discontinued. They also argued in opposition to corporate medicine being financed through intermediary agencies, such as health maintenance organizations (HMOs). Allegedly, these types of organizations exploit the medical professions and fail to provide high-quality health care.

"The tenor of the [CCMC] report was such that one can read into it the seeds of everything that led to the health care system we have today. . . . So at last we find ourselves, as always, in a health care crisis,"

Dr. Smith wrote in 1984. This health-care crisis continues today.

"Rockefeller's General Education Board has spent more than $100 million to gain control of the nation's medical schools and turn our physicians to physicians of the allopathic school, dedicated to surgery and the heavy use of drugs," wrote author Mullins, who spent more than thirty years researching the Rockefeller medical monopoly.

Recalling how John D. Rockefeller Sr.'s father, William "Big Bill" Rockefeller, once tried to sell unrefined petroleum as a cancer cure, Mullins wrote, "This carnival medicine show barker would hardly have envisioned that his descendents would control the greatest and most profitable Medical Monopoly in recorded history."

Mullins reported that the German chemical company I. G. Farben and its subsidiaries in the United States through the Rockefeller interests (such as the cartel between Rockefeller's U.S.-based Standard Oil Co. and I. G. Farben as revealed in a 1941 investigation by the government) were responsible for trying to build a monopoly by suppressing discoveries of its own drugs. From 1908 to 1936, I. G. Farben withheld its discovery of sulfanilamide, an early sulfa drug, until the firm had signed working agreements with the important drug firms of Switzerland, Sandoz and Ciba-Geigy. After years of a working relationship, these two firms were finally joined in 1996 to form one of the largest corporate mergers in history—Novartis.

During the first half of the twentieth century, the Nazi drug cartel, I. G. Farben, along with the drug companies controlled through Rockefeller interests, dominated the development, production, and distribution of numerous drugs, including substances that are downright dangerous.

ASPARTAME

AMERICANS AREN'T JUST BEING affected by chemicals in pharmaceutically produced drugs. One of the many controversial chemicals now being used by millions of Americans is aspartame, an additive sugar substitute found in most diet soft drinks and more than five thousand foods, drugs, medicines, and most sugar substitutes such as NutraSweet, Equal, Metamucil, and Canderel.

When heated to more than 86 degrees Fahrenheit, aspartame releases free methanol, which breaks down into formic acid and formaldehyde in the body. Keep in mind the human body temperature is 98.6 degrees and that formaldehyde is a deadly neurotoxin. The remaining formaldehyde from free methanol then breaks down into formic acid—the venom of ant stings.

In 1987, Dr. Louis J. Elsas, a professor of pediatrics and director of the Division of Medical Genetics at Emory University, testified before the U.S. Senate Committee on Labor and Human Resources about phenylalanine, one of the two amino acids in aspartame. He said, "In the developing fetus such a rise in maternal blood phenylalanine could be magnified four to six fold by the concentrative efforts of the placental and fetal blood brain barrier and this concentration kills such cells in tissue culture. The effect of such an increased fetal brain concentrations in vivo would probably be much more subtle and expressed as mental retardation, microcephaly, or potential certain birth defects." When Dr. Elsas told the senators about phenylalanine in 1987, infant autism rates were 1 in 1,500. Today they are 1 in 150 and rising. It would appear that certain drugs are wrecking our newborn children.

Dr. Madelon Price, a professor of neurobiology at Washington University, said, "Aspartic acid (aspartate) has been known to be a neurotoxin for 30 years [now 40 years]. Rodents that have ingested too much aspartame as infants are stunted as adults, obese and have sexual and reproductive dysfunctions."

Until the Reagan administration, the Food and Drug Administration had refused to approve the use of aspartame. The FDA's own toxicologist, Dr. Adrian Gross, told Congress that aspartame can contribute to or even cause seizures, brain tumors, and brain cancer, and violated the Delaney clause, which forbids putting anything in food that is known to cause cancer. "And if the FDA violates its own laws, who is left to protect the public?" he asked.

Dr. H. J. Roberts with the Palm Beach Institute for Medical Research devoted an entire chapter of his book *Aspartame Disease: An Ignored Epidemic* to aspartame interaction with drugs such as Coumadin, Dilantin, antidepressants, and other psychotropic agents as well as Inderal, Aldomet, hormones, and insulin. Roberts said aspartame interacts with all cardiac medication, and even noted drug reactions after a person stopped using aspartame products. "The issue of sudden death related to aspartame and its breakdown products has been raised a number of times, particularly among previously well individuals using such products . . . including pilots and drivers, and athletes. . . ." He added, "The need for clinicians and corporate-neutral investigators to evaluate the contributory role of aspartame in cardiopulmonary disorders and sudden death, and drug interactions with aspartame, is underscored by the frequency of persons dying unexpectedly being categorized as 'death due to causes yet to be determined.' "

Dr. Betty Martini, a twenty-two-year veteran in the medical field and founder of Mission Possible International, has worked with doctors around the world to remove aspartame from food, drinks, and medicine. She recounted how pharmaceutical interests subordinated public welfare:

> *Donald Rumsfeld was CEO of Searle, that conglomerate that manufactured aspartame. For 16 years the FDA refused to approve it, not only because it's not safe but because they wanted the company indicted for fraud. Both U.S. prosecutors [for the FDA] hired on with the defense team and the statute of limitations expired. They were Sam Skinner and William Conlon. Skinner went on to become Secretary of Transportation squelching the cries of the pilots who were now having seizures on this seizure-triggering drug, aspartame, and then Chief of Staff under President Bush's father. Some of these people reach high places. Even Supreme Justice Clarence Thomas is a former Monsanto attorney. (Monsanto bought Searle in 1985, and sold it a few years ago.)*

Yet even with friends in high places, the FDA still refused to allow NutraSweet on the market. Termed a deadly neurotoxic drug masquerading as an additive by opponents, aspartame interacts with antidepressants and also interacts with vaccines and other toxins and unsafe sweeteners like Splenda. "Both being excitotoxins, the aspartic acid in aspartame and MSG, the glutamate people were found using aspartame as the placebo for MSG studies, even before it was approved. The FDA has known this for a quarter of a century and done nothing even though it's against the law. Searle went on to build a NutraSweet factory and had $9 million worth of inventory," said Martini. Donald Rumsfeld was on President Reagan's transition team and the day after Reagan took office, Dr. Arthur Hull Hayes, the man who would approve aspartame, was appointed as FDA commissioner.

Former Searle salesperson Patty Wood Allott supported the idea that Rumsfeld was behind the approval of aspartame by stating that in 1981 Rumsfeld told company employees "he would call in all his markers and that no matter what, he would see to it that aspartame be approved [that] year." FDA commissioner Hayes had previously served in the U.S. Army Chemical Weapons Division and initially had approved aspartame only as a powdered additive. But in 1983, just before he left his position for a public relations job with Burson-Marsteller, the chief public relations firm for both Monsanto and Searle, Hayes approved aspartame for all carbonated beverages. Since that time he never spoke publicly about aspartame (Hayes died in February 2010).

Rumsfeld is merely one example of the cozy relationship between government and Big Pharm. A former CEO of Searle and a member of the Trilateral Commission—a globalist group designed to foster economic cooperation between the United States, Japan, and Europe—Rumsfeld is also a major stockholder in Gilead Sciences, a California biotech firm that owns the rights to Tamiflu. When the population was being threatened with the bird flu in 2005, CNN reported Rumsfeld's Gilead holdings at somewhere between $5 million to $25 million.

The incestuous relationship between Big Pharm corporate business and the government makes a mockery of American free enterprise. While it is free to decide which drugs to promote and distribute, it is also free to price them as high as the traffic will bear. Yet Big Pharm is dependent on government in the form of patent protection and FDA approval to protect its drug monopoly.

In a move bewildering to those who are not aware of the globalist agenda and its control, Congress expressly prohibited Medicare from negotiating lower drug prices through its bulk purchasing power. The excesses of the globalists' pharmaceutical corporations have prompted many Americans to seek price relief by traveling to Canada or Mexico to purchase drugs.

Dr. Angell said the pharmaceutical industry has moved very far from its original high purpose of discovering and producing useful new drugs and it now primarily a marketing machine to sell drugs of dubious benefit. Big Pharm "uses its wealth and power to co-opt every institution that might stand in its way, including the US Congress, the FDA, academic medical centers, and the medical profession itself [as] most of its marketing efforts are focused on influencing doctors, since they must write the prescriptions," she said.

Don't look for any real relief from either Democrats or Republicans. While campaigning in 2008 both then senators Barack Obama and Hillary Clinton pledged to fight the huge pharmaceutical and insurance industries. These promises echoed similar promises made by Mrs. Clinton during her husband's administration. Yet campaign contributions data showed that both Obama and Clinton were the largest recipients of Big Pharm donations in 2008 campaign funding. According to the Center for Responsive Politics, Obama received $1,425,501 from the health service sector and health maintenance organizations (HMOs), while Clinton came in second with $575,746 in contributions. Trailing both Obama and Clinton were $427,228 for John McCain and $186,700 for Mitt Romney.

Only an awakened American public can rein in the power of the globalist pharmaceutical monopoly, concluded Angell. "Drug companies have the largest lobby in Washington, and they give copiously to political campaigns. Legislators are now so beholden to the pharmaceutical industry that it will be exceedingly difficult to break its lock on them. . . . But the one thing legislators need more than campaign contributions is votes. That is why citizens should know what is really going on. Contrary to the industry's public relations, they don't get what they pay for. The fact is that this industry is taking us for a ride, and there will be no real reform without an aroused and determined public to make it happen."

DRUGGING THE KIDS

IN YEARS PAST, IF a child was acting up or caught staring out the window, he or she received a rap on the knuckles with a ruler and was told to stay with the rest of the class. Today, the child is sent to the school nurse, who often tells the parents the student has been diagnosed with attention-deficit/hyperactivity disorder (ADHD) and advises them to see a psychiatrist, who usually recommends the administration of Prozac (94 percent sodium fluoride), Ritalin, or Zoloft—psychotropic drugs that have been shown to produce psychosis in lab rats.

At least one state has put a stop to this practice. In 2001, the Connecticut House of Representatives voted 141–0 on a law prohibiting school personnel from recommending to parents that their children take Ritalin or other mood-altering drugs. One of the bill's primary sponsors, Republican state representative Lenny Winkler, quoted studies showing the number of children taking Ritalin nationally jumped from 500,000 in 1987 to more than 6 million by 2001. The bill also prohibited the state Department of Children and Families from taking children away from parents who declined to put their children on mood-altering drugs.

If the fact that unaware parents are being urged to drug their children is bad enough, consider that the effectiveness of the medication they're being asked to use has come under scrutiny. A 1999 study at the Human Development Center at the University of Wisconsin in Eau Claire found that thirteen "ADHD" children on medication performed progressively worse over four years on standardized tests than a group of thirteen normal children with similar IQs and other characteristics. Another study by Dr. Gretchen LeFever, an assistant professor of pediatrics and psychiatry at Eastern Virginia Medical School, revealed that while children in her community used the drug Ritalin two to three times more than the national rate, their academic performance in relation to their peers showed no improvement. Her persistence in questioning the rising incidence of drug use in schoolchildren was muted in 2005 when she was fired.

Alan Larson, a former secretary of the Oregon Federation of Independent Schools, criticized the expanding diagnosis of attention-deficit disorder (ADD), stating, "[T]he labeling of children with ADD is not because of a problem the kids have; it is because of a problem teachers who cannot tolerate active children have." Other questionable diagnoses include syndromes concerning children who are victims of obesity, junk food, lack of exercise, and inattentive parents. Clearly, some children have serious mental disorders, but these are relatively few compared with the number of currently medicated children.

There is also the possibility that some of the diagnoses that doctors give to children are for nonexistent diseases. In his 1991 book *Toxic Psychiatry*, psychiatrist Peter Breggin wrote: "Hyperactivity is the most frequent justification for drugging children. The difficult-to-control male child is certainly not a new phenomenon, but attempts to give him a medical diagnosis are the product of modern psychology and psychiatry. At first psychiatrists called hyperactivity a brain disease. When no brain disease could be found, they changed it to 'minimal brain disease' (MBD). When no minimal brain disease could be found the profession transformed the concept into 'minimal brain dysfunction.' When no minimal brain dysfunction could be demonstrated, the label became attention deficit disorder. Now it's just assumed to be a real disease, regardless of the failure to prove it so. Biochemical imbalance is the code word, but there's no more evidence for that than there is for actual brain disease."

Textbooks of psychological disorders blossomed in size after programs such as Project Paperclip brought German psychiatrists into the military and intelligence fields after World War II. In its 1952 *Diagnostic and Statistical Manual for Mental Disorders* (DSM), the American Psychiatric Association defined only 106 mental disorders. By the publication of *DSM-IV* in 1994, the number had grown to 374. Meanwhile, the number of child psychologists in U.S. schools grew from a mere 500 in 1940 to more than 22,000 by 1990. In 2006, the number of school psychologists, including clinical and consultation, had grown to 152,000, with an anticipated 176,000 by 2016.

The unscientific and political nature of psychiatry was noted in a resignation letter to the APA from Dr. Loren R. Mosher, former chief of the Center for Studies of Schizophrenia at the National Institute of Mental Health: " . . . why must the APA pretend to know more than it does? DSM IV is the fabrication upon which psychiatry seeks acceptance by medicine in general. Insiders know it is more a political than scientific document. . . . It is the way to get paid."

This growth of an immense and well-funded field of psychiatry is worrisome to those who recall that in both Nazi Germany and the Soviet Union, the incarcerations and, ultimately, the genocides practiced there all began innocuously as mental health programs. Persons who were considered defective, either physically or mentally, were the first victims of the Nazis, long before they turned to the Jews.

"Today, though psychiatry may still be suspect among the public, it has won over both government and the media. The profession and its treatments inundate talk shows, magazines and the front pages of our news papers," wrote Bruce Wiseman, the U.S. national president of the Citizens Commission on Human Rights and former chairman of the history department at John F. Kennedy University.

Lysergic acid diethylamide (LSD) was initially studied as an antipsychotic and antidepressant as well as a truth drug by the military. When LSD was outlawed in 1968 for its dangerous side effects, drug companies sought substitutes. They developed the antidepressant Prozac (fluoxetine), then Zoloft (sertraline), Effexor (venlafaxine), and Paxil (paroxetine).

These companies also developed the drug Ritalin. Long after the war, Dr. Helmut Remschmidt proposed "a genetic answer" to hyperactivity and was a leading proponent of the use of drugs such as Ritalin. Remschmidt was the director of the Society for Child and Adolescent Psychiatry and studied under Dr. Hermann Stutte, a man associated with Nazi psychiatrists involved in the German euthanasia program. Remschmidt received his doctorate from Robert Sommer, director of the Deutscher

Verband fur Psychische Hygiene, or the German Association for Mental Hygiene, the institution that in the late 1920s laid the psychiatric groundwork for the idea of "mental hygiene." The end result of this attempt at eugenics was the hands-on Nazi sterilization and euthanasia programs that led to the Holocaust. Long after the war, Remschmidt still proposed "a genetic answer" to hyperactivity and was a leading proponent of the use of drugs such as Ritalin.

Could it be that Ritalin is doing more harm than good? A 1986 edition of the *International Journal of the Addictions* listed 105 adverse reactions to Ritalin, including serious ones such as dangerously high blood pressure, aggressiveness, restlessness, hallucinations, unusual behavior, and suicidal tendencies. Investigative reporter Kelly Patricia O'Meara spent sixteen years working as a congressional staffer before writing investigative articles for *Insight Magazine.* Her reports on child vaccines and mood-altering drugs prompted congressional hearings. She wrote: "Thirty years ago the World Health Organization (WHO) concluded that Ritalin was pharmacologically similar to cocaine in the pattern of abuse it fostered and cited it as a Schedule II drug—the most addictive in medical use. The Department of Justice also cited Ritalin as a Schedule II drug under the Controlled Substances Act, and the Drug Enforcement Agency (DEA) warned that 'Ritalin substitutes for cocaine and d-amphetamine in a number of behavioral paradigms.'" O'Meara referenced a 2001 study at the Brookhaven National Laboratory that confirmed the similarities between cocaine and Ritalin, but found that Ritalin is more potent than cocaine in its effect on the dopamine system, an area of the brain many doctors believe is most affected by such narcotics.

Although Americans wonder why there has been a rash of school shootings and teen suicides in recent years, few take into account that virtually all of these killings have involved a student who was on—or was just coming off—mood-altering drugs. In five cases of school shootings between March 1998 and May 1999—including the tragedy at Columbine High School—the students involved with the shootings were medicated. Though it was downplayed by the media, Seung-Hui Cho, the gunman in the Virginia Tech shootings in April 2007, had been undergoing psychological counseling and possessed prescription psychoactive drugs.

In his book *Reclaiming Our Children,* psychiatrist and drug critic Dr. Peter Breggin argued that Eric Harris's violence at Columbine was caused by the prescription drug Luvox. "I also warned that stopping antidepressants can be as dangerous as starting them, since they can cause very disturbing and painful withdrawal reactions," said Dr. Breggin.

The claim that drugs are behind school shootings was echoed as far back as a 1999 article in *Health and Healing,* written by Dr. Julian Whitaker: "[V]irtually all of the gun-related massacres that have made headlines over the past decade have had one thing in common: they were perpetrated by people taking Prozac, Zoloft, Luvox, Paxil or a related anti-depressant drug."

A website called TeenScreenTruth.com is dedicated to gathering information off the Internet to help teens "connect the dots to see the revealing connections" between mood-altering drugs and teen violence. The site compiled a list of violent episodes dating as far back as 1985 when Steven W. Brownlee, an Atlanta postal worker on psychotropic drugs, killed two coworkers. Despite sealed medical records, the sheer totality of evidence pointing to psychiatric drugs as the culprit behind most school shootings, teen suicides, and other violent behavior is most compelling, if not overwhelming.

It would appear from the evidence that German drug science and German psychiatry have provided the foundation for today's schools where children increasingly are being steered to drugs for any complaint—from true antisocial behavior to merely daydreaming.

The effort by Big Pharm to mold education, physicians, politicians, and even health care in general to its will requires massive amounts of money. Such great sums are only available to the globalists with Nazi roots and well beyond the reach of even well-off Americans, thanks to a crumbling economy and never-before-seen debt.

PSYCHIATRY AND EUGENICS

BY APPLYING PSYCHOLOGICAL TECHNIQUES developed by the Germans, Big Pharm, the corporate mass media, and even education have been turned into tools for mind control. But before examining how this has occurred, one must first understand the history of psychology and psychiatry.

HISTORY OF PSYCHIATRY

PRIOR TO THE LATE 1800s, the mentally ill were treated little better than torture victims—chained to the walls of basements, cages, or dungeons; beaten; and subjected to "therapies" such as bloodletting, partial drowning, and primitive shock treatments. A change to these treatment methods came when, in the 1860s, German medical doctor Wilhelm Maximilian Wundt proposed the idea that man is simply a higher-order animal and that feeling and emotions may be studied and altered scientifically rather than through physical punishment. Wundt's work emphasized the physiological relationship of the brain and the mind. He explored the nature of religious beliefs, denied the human soul, and began to identify mental disorders and abnormal behavior, which led to the creation of the field of psychology. His *Lectures on the Mind of Humans and Animals* was published in 1863, and a year later, he was promoted to assistant professor of physiology at Heidelberg University. There his work continued and, with the support of German militarists and aristocrats, he became known as the "father of experimental psychology." Wundt, who found studies of the human soul incompatible with scientific empirical investigation, set out to explain what had previously been metaphysical matters in terms of mere animalistic and body chemical reactions.

Many believe that Wundt's studies, as well as other European studies of the human mind, were major influences on the Nazi eugenics programs, which ultimately led to some of the greatest horrors of the twentieth century. Thus, some of Germany's most learned men provided justification for Nazi euthanasia and extermination programs. "Hitler's philosophy and his concept of man in general was shaped to a decisive degree by psychiatry . . . an influential cluster of psychiatrists and their frightening theories and methods collectively form the missing piece of the puzzle of Hitler, the Third Reich, the atrocities and their dreadful legacy. It is the overlooked yet utterly central piece of the puzzle," wrote Dr. Thomas Röder and his coauthors, Volker Kubillus and Anthony Burwell, in their 1995 book *Psychiatrists—The Men Behind Hitler.*

Psychology, the scientific study of the human mind, and psychiatry, the study and treatment of mental disorders, go hand in hand and led to a viewpoint that certain people, endowed with better education, and presumably understanding, were more competent to judge the behavior of others.

As the field of psychiatry grew, so did its definitions, often to the point of absurdity. In 1871, a paper was published entitled "Psychical Degeneration of the French People," which purported that simply being French constituted a mental illness. "One of psychiatry's leading figures, Richard von Krafft-Ebing, added to his list of varieties of mental disorders 'political and reformatory insanity'—meaning any inclination to form a different opinion from that of the masses," stated Röder, Kubillus, and Burwell.

At the time of World War I, the attempt to bring respectability to the emerging psychiatric profession resulted in a certain bond that had been created between psychiatry and the aristocratic German government. The German military was particularly impressed with the "treatment" of Fritz Kaufmann's electroshock "therapy" because it helped minimize war neurosis or shell shock and quickly returned disturbed soldiers to the front. It was more of a disciplinary measure than true medical therapy. After being electrically shocked, most soldiers quickly agreed to return to service.

Psychiatry continued to grow in power even as its agenda continued to widen. Psychiatrist P. J. Möbius, who had lectured on the "psychological feeble-mindedness of the woman," pronounced, "The psychiatrist should be the judge about mental health, because only he knows what ill means."

The rush to isolate and "cure" mental defectives in Nazi Germany quickly was interpreted to include malcontents and dissidents opposed to Hitler's regime. This concept resulted in the Nazi Sterilization Act, which went into effect in July 1933, just six months after Hitler's ascension to power. This law provided for the compulsory sterilization of anyone deemed defective, deficient, or undesirable by the State. One of the leading and articulate authorities behind this act was Dr. Ernst Rüdin, a psychiatrist who in 1930 had traveled to Washington, D.C., to present a paper on "The Importance of Eugenics and Genetics in Mental Hygiene." It was well received by many Americans, especially among the globalists, who had come to embrace the racist and elitist views of the German philosophers, such as Georg Hegel, Friedrich Nietzsche, Martin Heidegger, and Rudolf Steiner. Just after Hitler took office in 1933, Rüdin, by then director of the Kaiser Wilhelm Institute, supported the Law for the Prevention of Genetically Diseased Children, the initial step toward the sterilization of those deemed "unworthy of life." Rüdin continued to be acknowledged as a leader in psychiatry. In 1992, the prestigious Max Planck Institute praised Rüdin for "following his own convictions in 'racial hygiene' measures, cooperating with the Nazis as a psychiatrist and helping them legitimize their aims through pertinent legislation."

Prescott Bush, the father and grandfather of two U.S. presidents, along with being a member of the secretive Skull and Bones fraternity, was among those Yale activists promoting the Mental Hygiene Society. This organization evolved into the World Federation of Mental Health, which included the prominent Montagu Norman, a former partner of Brown Brothers, governor of the Bank of England (1920–1944), and godfather to Nazi banker Hjalmar Schacht's grandson. Norman, himself a mental patient, appointed Brigadier General John Rawlings Rees, the former chief psychiatrist and psychological warfare expert for British Intelligence, as the director of the Tavistock Psychiatric Clinic.

Dr. John Rawlings Rees, as a cofounder of the World Federation for Mental Health, spelled out the federation's agenda before the annual general meeting of the National Council for Mental Hygiene on June 18, 1940: "We can therefore justifiably stress our particular point of view with regard to the proper development of the human psyche, even though our knowledge be incomplete. We must aim to make it permeate every educational activity in our national life. . . . We have made a useful attack upon a number of professions. The two easiest of them naturally are the teaching profession and the Church: the two most difficult are law and medicine.

"Public life, politics and industry should all of them be within our sphere of influence. . . . If we are to infiltrate the professional and social activities of other people I think we must imitate the Totalitarians and organize some kind of fifth column activity! If better ideas on mental health are to progress and spread we, as the salesmen, must lose our identity. . . . Let us all, therefore, very secretly be 'fifth columnists.' "

Beverly K. Eakman, an author and a commissioner for the Citizens Commission on Rights, wrote, "Colleagues [of Rees] such as Canadian Drs. Brock Chisholm and Ewen Cameron, 'progressive' U.S. educators like Edward Thorndike, James Earl Russell, John Dewey and Benjamin Bloom, and [a] bevy of foundations, associations and tax-supported 'research centers' became Rees' enablers. This cadre of like-minded and self-styled 'experts' first seized upon Russian Ivan Pavlov's 'classic conditioning'; followed that up with German psychologist Kurt Lewin's 'group dynamics,' Russian neuropsychologist Alexander Luria's 'disorganization of behavior,' and the U.S. psychologist B. F. Skinner's deprivation-based 'operant conditioning' coupled with U.S. social psychologist Elliot Aronson's 'cognitive dissonance.' Together, they created Rees' dream: 'a controlled psychological environment.' Today, the Department of Defense (DOD) has a new name for it: 'perception management' (PM), and the psychopharmaceutical industry has hit the jackpot."

Perception management to the Department of Defense simply means getting the public to respond as DOD officials wish without their realizing it, knee-jerk reactions, leaving reason behind much like subliminal advertising. An early yet clear use of this technique was the name change in 1947 from the "War Department" to the "Defense Department." "In so doing, the subject [of perception management] is thrust unawares into a twisted view of reality. In today's politically correct environment, this unorthodox technique is sold as intellectual and academic freedom," explained Beverly Eakman. "Similarly, encounter sessions (or 'therapy groups') are predicated on fostering emotional toughness. Facilitators lead participants to accept ideas and deportment they normally would not tolerate. What they actually get is 're-education,' Soviet-style. Schools of behavioral science, such as Esalen Institute and the Western Training Laboratory for Group Development, allude to consensus—group thinking—as being the objective. Encounter groups deliberately heighten peer pressure—isolating holdouts of a viewpoint and intimidating weaker individuals by ridiculing them, cursing at them, yelling at them, and ostracizing them until they 'cave.' Some even commit suicide.

"That's why NTL [the National Training Laboratories Institute], for example, carries a disclaimer which the applicant must sign prior to admission [stating] 'No person concerned about entering a stress situation should participate in NTL programs. . . . A small percentage of participants have experienced stress reactions in varying degrees. There is no means of predicting such reactions or screening out or otherwise identifying those predisposed to such reactions.'

"Now any thoughtful person, upon reading this, would realize that the very concept of psychological screening must be a sham. If psychologists are unable to predict or screen out individuals predisposed to become upset by NTL's daunting program, then how do they expect to 'screen' the entire population for mental illness? Yet just such an initiative was funded by Congress in 2002, with copycat bills set for launch in several states. Could our nation's leaders be looking to avert political dissent under the pretext of preventing emotional 'diseases'? Wouldn't be the first time. . . ."

EUGENICS

THE PERCEPTION MANAGEMENT ACTIVITIES of twisting semantics and promoting groupthink were psychological methods employed by the German Nazis that resulted in the deaths of millions of innocents as a matter of State policy, a holocaust in anyone's book. Although most

Americans are aware of the horrors inflicted by Hitler's Nazis on Europe's citizens in pursuit of their creating a "master race," and many see eugenics as a racist pseudoscience seeking to eliminate anyone whom a self-proclaimed elite views as undesirable, few realize that the theological and scientific basis for the Nazis' beliefs originated in the United States, particularly in California, long before the Nazis came to power in Germany.

In the late nineteenth century, the United States had joined fourteen other nations in passing various types of eugenics legislation. Thirty states had laws providing for the sterilization of mental patients and imbeciles. At least sixty thousand such "defectives" were legally sterilized.

In 1925, Justice Oliver Wendell Holmes, writing for the majority in a Supreme Court case, stated, "It is better for all the world, if instead of waiting to execute degenerate offspring for crime or to let them starve for their imbecility, society can prevent those who are manifestly unfit from continuing their kind."

Sir Francis Galton, English psychologist and father of the eugenics movement, defined eugenics as "the science of improving the stock [to] give more suitable races or strains of blood a better chance of prevailing speedily over the less suitable. . . ." In order to determine who was dirtying the gene pool requires extensive comprehensive statistics on the population. So in 1910, the Eugenics Records Office was established as a branch of the Galton National Laboratory in London, endowed by Mrs. E. H. Harriman, wife of U.S. railroad magnate Edward Harriman and mother of diplomat and early-day globalist Averell Harriman.

After 1900, the Harrimans, the family that gave Prescott Bush's family their start, along with the Rockefellers provided more than $11 million to create the privately owned Eugenics Records Office of Charles B. Davenport at Cold Springs Harbor, New York, as well as eugenics studies at Harvard, Columbia, and Cornell. The first International Congress of Eugenics was convened in London in 1912, with Winston Churchill as a director. Clearly, the concept of "bloodlines" was as significant to the British and American elite as it was to Hitler and the Nazis.

In 1932, when the Congress met in New York, it was the Hamburg

America Shipping Line, controlled by Harriman associates George Walker and Prescott Bush, that brought prominent Germans to the meeting. In attendance was Dr. Ernst Rüdin, aforementioned authority behind the Nazi Sterilization Act and member of the Kaiser Wilhelm Institute for Genealogy and Demography in Berlin. Rüdin was unanimously elected president of the International Federation of Eugenics Societies for his work in founding the Deutschen Gesellschaft fur Rassenhygiene, or the German Society for Racial Hygiene, a forerunner of Hitler's racial institutes. But, as stated previously, the groundwork for eugenics was laid in the United States.

California was considered the epicenter of the American eugenics movement, according to Edwin Black, author of *War Against the Weak: Eugenics and America's Campaign to Create a Master Race.* "During the Twentieth Century's first decades, California's eugenicists included potent but little known race scientists, such as Army venereal disease specialist Dr. Paul Popenoe, citrus magnate and Polytechnic benefactor Paul Gosney, Sacramento banker Charles M. Goethe, as well as members of the California State Board of Charities and Corrections and the University of California Board of Regents," wrote Black.

Black said that within the first twenty-five years of eugenics legislation, California sterilized 9,782 individuals, mostly women, many of whom were classified as "bad girls," or diagnosed as "passionate," "oversexed," or "sexually wayward." Some women were sterilized because of what was deemed an abnormally large clitoris or labia. In 1933 alone, Black found at least 1,278 compulsory sterilizations were performed, 700 of which were on women. He said California's two leading "sterilization mills" in 1933 were Sonoma State Home with 388 operations and Patton State Hospital with 363 operations. Other sterilizations were also performed in centers at Agnews, Mendocino, Napa, Norwalk, Stockton, and Pacific Colony.

Black noted, "Eugenics would have been so much bizarre parlor talk had it not been for extensive financing by corporate philanthropies, specifically the Carnegie Institution, the Rockefeller Foundation and the Harriman railroad fortune. They were all in league with some of America's most respected scientists hailing from such prestigious universities as Stanford, Yale, Harvard, and Princeton. These academicians espoused race theory and race science, and then faked and twisted data to serve eugenics' racist aims."

He described how the Rockefeller Foundation helped create the German eugenics movement and even funded the program that the infamous Nazi Dr. Josef Mengele worked in before he became the "Angel of Death" at Auschwitz.

"The grand plan was to literally wipe away the reproductive capability of those deemed weak and inferior—the so-called 'unfit,'" said Black. "The eugenicists hoped to neutralize the viability of 10 percent of the population at a sweep, until none were left except themselves. One solution offered was simply execution or euthanasia, as listed in a 1911 study funded by the Carnegies entitled 'Preliminary Report of the Committee of the Eugenic Section of the American Breeder's Association to Study and to Report on the Best Practical Means for Cutting Off the Defective Germ-Plasm in the Human Population.'" Interestingly enough, the most popular idea for euthanasia in the United States at that time was the employment of gas chambers.

Black concluded, "Hitler studied American eugenics laws. He tried to legitimize his anti-Semitism by medicalizing it, and wrapping it in the more palatable pseudoscientific facade of eugenics. Hitler was able to recruit more followers among reasonable Germans by claiming that science was on his side. While Hitler's race hatred sprung from his own mind, the intellectual outlines of the eugenics Hitler adopted in 1924 were made in America."

Despite much public renunciation of eugenics following the revelations of the Nazi racial extermination programs at the Nuremburg trials, work on population control continues right up to today under more politically correct names. Some conspiracy-oriented researchers see the fingerprints of eugenics theology in today's efforts to reduce the human population as previously discussed. Many of the same families and foundations that support birth-control organizations today were connected to the eugenics movement of the past. According to its Summary of Financial Activities ending in June 2008, Planned Parenthood ended the year with $966.7

million in revenues. Of this amount, $349.6 million came from unspecified government grants and contracts compared to $374.7 million from its health-care centers and $186 million in private contributions and bequests.

THE PSYCHOLOGY OF CONSERVATISM

In August 2003, the National Institute of Mental Health (NIMH) and the National Science Foundation (NSF) announced the results of a $1.2 million taxpayer-funded study. The conclusion was that people who believe in traditional values—such as monogamous marriage, balanced budgets, strict interpretation of the Constitution—are mentally disturbed. In studying what they called "the psychology of conservatism," the researchers wrote that the core of political conservatism is a resistance to change and a tolerance for inequality that promote fear and uncertainty. This results in psychological factors commonly linked to conservatism, such as "fear and aggression, dogmatism and intolerance of ambiguity, uncertainty avoidance, a need for cognitive closure and terror management." In their paper entitled "Political Conservatism as Motivated Social Cognition," the authors concluded that political conservatism stems from the need to satisfy various psychological needs, but admitted that it is unlikely that conservative ideology can be ascribed to a "single motivational syndrome."

The researchers also admit in the paper that their term "motivated social cognition" refers to "a number of assumptions about the relationship between people's belief and their motivational underpinnings." They compared Hitler, Mussolini, and President Ronald W. Reagan as "right-wing conservatives," saying they all shared a resistance to change and the acceptance of inequality.

So, in addition to being identified in FEMA materials as potential terrorists, thanks to psychobabble funded by NIMH and the NSF, constitutionalists are now in danger of being diagnosed with a mental disorder.

And what should be done with political conservatives suffering from "motivated social cognition"? Dr. José M. R. Delgado, a former professor of neuropsychiatry at Yale University Medical School and a man who has been connected with the CIA's MK-Ultra mind-control experiments, has recommended, "We need a program of psychosurgery for political control of our society. The purpose is physical control of the mind. Everyone who deviates from the given norm can be surgically mutilated. The individual may think that the most important reality is his own existence, but this is only his personal point of view. . . . Man does not have the right to develop his own mind. . . . We must electronically control the brain. Someday armies and generals will be controlled by electronic stimulation of the brain." Dr. Delgado will enjoy living in the Orwellian digitally controlled New World Order.

DRUG THE WOMEN AND CHILDREN FIRST

IT IS VERY POSSIBLE that the globalists are now trying to control the minds of American citizens by funding research facilities and by supporting specific legislation. The U.S. Preventative Services Task Force has urged routine screening of all American teenagers for depression, and politicians were ready to step up to the plate. Just three months into 2009, Congress was introduced to eight bills on widespread mental health screening.

In 2007, legislation entitled the Postpartum Mood Disorders Prevention Act was introduced; it called for the mental screening of mothers for signs of depression. Such screening for depression may soon become state law in Illinois. Similar legislation has already been adopted or at least introduced in several other states. In 2009, this mass screening scheme was brought up again as the Melanie Blocker Stokes Mom's Opportunity to Access Health, Education, Research, and Support for Postpartum Depression Act of 2009, otherwise known simply as the Mother's Act. This law was reintroduced into both bodies of the new

Congress in January 2009, after the 2007 bill died in the Senate in 2008.

Critics see the Mother's Act as an insidious plan that essentially would allow for infants, pregnant women, and nursing mothers to be drugged even more than usual. The legislation would also allow Child Protective Services to take children from parents with fewer restrictions. This legislation is similar to bills Congress has declined to pass for eight years.

"The true goal of the promoters of this act is to transform women of child bearing age into life-long consumers of psychiatric treatment by screening women for a whole list of 'mood' and 'anxiety' disorders and not simply postpartum depression," stated investigative journalist Evelyn Pringle writing for the political newsletter *Counterpunch.* "Enough cannot be said about the ability of anyone with a white coat and a medical title to convince vulnerable pregnant women and new mothers that the thoughts and feelings they experience on any given day might be abnormal."

Any woman who has gone through pregnancy knows that there are accompanying periods of ups and downs. Moods swings while carrying a child have been part of the experience since the beginning of time. Except for the few exceptional cases of true clinical depression, to many it appeared unnecessary to subject normal healthy women to a regimen of psychiatric drugs at the first sign of a bad day.

Another concern for parents is autism. A recent U.S. government study stemming from the 2007 National Survey of Children's Health reported that autism rates climbed 200 percent between 2001 and 2009. This new estimate indicated about 673,000 American children have autism.

"This is an alarming increase in a disease that many ill-informed doctors and scientists still brush off as being genetic in origin," said Mike Adams, editor of *NaturalNews,* a widely read natural-health source. "But genes can't explain such a rapid increase in the number of children being diagnosed with the disease. Clearly, some other factor is at work, and many parents suspect vaccines are one of the primary contributing factors."

In 1998, the Dawbarns Law Firm of Norfolk, England, along with Freeth Cartwright of Nottingham, filed lawsuits against three manufacturers of measles, mumps, and rubella (MMR) vaccine after parents reported more than fifteen hundred instances of perceived side effects following the administration of MMR vaccines, introduced there in 1988. The firms succeeded in obtaining legal aid for the children, and management of the cases was transferred ultimately to Alexander Harris solicitors of London in 1999. Despite assurances from British health officials denying any connection between side effects and the vaccines, cases were set for trial in the High Court to decide the preliminary issue of whether the vaccine caused symptoms of autism and bowel problems among the claimants. The cases were funded under the English legal aid system and supported by twenty-seven experts who prepared reports supporting the children's case. The parents believed their children were normal before being vaccinated, and saw nothing but the vaccinations to account for the changes in their children. The cases stalled and have not proceeded after legal aid was withdrawn in August 2003.

The June 2009 issue of *Toxicological & Environmental Chemistry* included a paper that concluded the routine administering of childhood vaccines containing a mercury substance called thimerosal could cause "significant cellular toxicity in human neuronal and fetal cells."

"This latest study confirms that damage [from the mercury-based preservative thimerosal] does occur in human neuronal and fetal cells, even at low concentrations," wrote Dr. Joseph Mercola, in the comments sections of his natural health newsletter. Mercola, owner of the Illinois Natural Health Center, said, "[R]ates of autism in the U.S. have increased nearly 60-fold since the late 1970s, rising right along with the increasing number of vaccinations added to the childhood vaccination schedule. Although autism may be apparent soon after birth, most autistic children experience at least several months, or even a year or more of normal development—followed by regression, defined as loss of function or failure to progress. Typically, by the age of three, at which time the child has received at least 24 of their scheduled vaccinations, symptoms of autism are fully apparent, affecting their communication and social skills, and impairing the child's ability to play, speak and relate to the world."

Many people feel the drugs and vaccines being administered to children are not fully tested or guaranteed safe. They feel children are being used as guinea pigs for Big Pharm.

Mike Adams also spoke out against the exploitation of young children for drug testing and claimed that it amounts to nothing less than chemical child abuse. "So-called 'bi-polar disorder' was wholly invented by psychiatrists with strong financial ties to drug companies," Adams wrote on his website. "The purpose of this disease is not to help children, but to sell drugs to anyone and everyone, including toddlers."

He added, "I often wonder when the rest of the country will wake up and notice that the mass-drugging of our nation's children has gone too far. Why isn't the mainstream media giving this front-page coverage? Why aren't lawmakers demanding an end to the chemical abuse of our children? Why isn't the FDA halting these trials on toddlers out of plain decency? You already know the answer: Because they're all making money from this chemical assault on our nation's children. The doctors, hospitals, drug companies, psychiatrists and mainstream media all profit handsomely from the sales of mind-altering drugs to children. Ethics will never get in the way of old-fashioned greed."

Adams said children should be given "some sunshine, play time and some time with nature" instead of drugs. "[You then] get balanced, healthy children. It's no secret, it's just common sense.

"But psychiatry has no common sense," argued Adams, "and no one in the industry dares mention that most so-called mental disorders are really just caused by nutritional imbalances. Because to admit to the truth about the mental health of children would be to render their careers irrelevant. And no psychiatrist is going to commit career suicide by admitting that bipolar disorder was just made up, or that toddlers need good food, not expensive drugs. Just like conventional doctors, psychiatrists have to protect their egos and revenue streams, and that means convincing parents that little Johnny has a brain chemistry imbalance and he'll have to take psychotropic drugs for life. The parents, as gullible as ever, naively go along with the scam, usually after being frightened into compliance by a psychiatrist who warns them what might happen to little Johnny if they *don't* [original emphasis] drug him. 'He might commit suicide,' they're sternly warned."

Bipolar disorder is a psychiatric diagnosis describing persons, usually children, who display a wide range of emotions, who experience exuberant highs and depressing lows. Other do not see such behavior as a disorder, but rather the normal ups and downs of the growth process. There is no scientific means to confirm a diagnosis of bipolar disorder.

David Healy, a former secretary of the British Association for Psychopharmacology and author of *Mania,* a book on bipolar disorder, said this disorder is somewhat of a mythical entity. "The problems that currently are grouped under the heading 'bipolar disorder' are akin to problems that, in the 1960s and 1970s, would have been called 'anxiety' and treated with tranquilizers or, during the 1990s, would have been labeled 'depression' and treated with antidepressants," said Healy in a 2009 interview in *Psychology Today.* Referring to what he described as "biobabble," Healy said this refers to "things like the supposed lowering of serotonin levels and the chemical imbalance that are said to lie at the heart of mood disorders. . . . This is as mythical as the supposed alterations of libido that Freudian theory says are at the heart of psychodynamic disorders. While libido and serotonin are real things, the way these terms were once used by psychoanalysts and by psychopharmacologists now—especially in the way they have seeped into popular culture—bears no relationship to any underlying serotonin level or measurable chemical imbalance or disorder of libido. What's astonishing is how quickly these terms were taken up by popular culture, and how widely, with so many people now routinely referring to their serotonin levels being out of whack when they are feeling wrong or unwell.

"In the case of bipolar disorder the biomyths center on ideas of mood stabilization. But there is no evidence that the drugs stabilize moods. In fact, it is not even clear that it makes sense to talk about a mood center in the brain. A further piece of mythology aimed at keeping people on the drugs is that these are neuroprotective—but there's no evidence that this is the case and in fact these drugs can lead to brain damage."

Some historians believe Vincent Van Gogh suffered from bipolar disorder. Fortunately, there were no synthetic psychiatric drugs available in his day to dull him down and prevent him from completing his works of art, today considered masterpieces.

FLU AND OTHER SWINISH IDEAS

I'm going to get [the swine flu vaccination] if that helps at all. But I'll tell you, my
wife is not going to immunize our kids.

—DR. MEHMET OZ, vice chair and professor of surgery
at Columbia University and host of *The Doctors,* when asked on CNN
if his family would be inoculated with the swine flu vaccine

THE UNITED STATES WAS once an industrial fountainhead, spewing
forth streams of consumer goods such as automobiles, televisions, and
refrigerators in international trade. Now, America is merely a nation of
zombies working in the service industry.

Today, America's largest consumer goods industries are health care and
legal drugs.

Not feeling well?

Just take a pill, even if you think you don't need one.

BIG PHARM PAYS OFF

IN *SURVIVING AMERICA'S DEPRESSION EPIDEMIC,* Dr. Bruce
Levine explained how the pharmaceutical industry's psychological drug
cartel works: "Mental health treatment in the United States is now a
multibillion-dollar industry and all the rules of industrial complexes
apply. Not only does Big Pharma have influential psychiatrists ... in their
pocket, virtually every mental health institution from which doctors, the
press, and the general public receive their mental health information is
financially interconnected with Big Pharma. The American Psychiatric
Association, psychiatry's professional organization, is hugely dependent
on drug company grants, and this is also true for the National Alliance for
the Mentally Ill and other so-called consumer organizations. Harvard and
other prestigious university psychiatry departments take millions of dol-

lars from drug companies, and the National Institute of Mental Health funds researchers who are financially connected with drug companies."

Sometimes the money goes to the people right at the top of these organizations. Dr. Charles Nemeroff, chairman of Emory University's psychiatry department, was one of several academics who came under investigation by the Senate Finance Committee for failing to disclose millions of dollars in income from pharmaceutical corporations.

According to Senate Finance Committee reports, Nemeroff was paid more than $960,000 by Paxil maker GlaxoSmithKline, from 2000 through 2006. Yet Nemeroff listed less than $35,000 on his Emory disclosure forms. Apparently, Nemeroff had earnings that totaled $2.8 million from speaking and consulting arrangements with drug companies between 2000 and 2007, but only disclosed a fraction of that amount. Compare that amount to the fact that Emory University's entire department of psychiatry received only $25,000 in 2008 from drug manufacturer Eli Lilly, according to the first quarter report of that firm. After the controversy, Nemeroff stepped down as department chairman. In late 2009, Nemeroff was named chairman of the psychiatry department at the University of Miami.

Nemeroff joined other prominent psychiatrists who recently have been exposed for extensive conflicts of interest due to millions in undisclosed funding from pharmaceutical corporations. There was also concern over Big Pharm funding selected advocacy groups.

"The majority of the public may or may not be familiar with these so-called mental health advocacy organizations, such as the National Alliance on Mental Illness (NAMI), Children and Adults with Attention Deficit Hyperactivity Disorder (CHADD) or the myriad of bipolar, depression or ADHD 'support groups' that are inundating the Internet. But they need to be," advised the Citizens Commission on Human Rights (CCHR). "These are groups operating under the guise of advocates for the 'mentally ill,' which in reality are heavily funded pharmaceutical front groups—lobbying and working on state and federal laws which affect the entire nation—from our elderly in nursing homes to our military, pregnant women and nursing mothers and schoolchildren."

Another issue that is as troublesome as that of Big Pharm lining the pockets of academics is the revolving door between government drug "experts" and Big Pharm executives. In 2009, former CDC chief Dr. Julie Gerberding became president of Merck's vaccine division. Thus, the former chief of the top public disease agency now oversees the $5 billion Merck division that markets the vaccines for cervical cancer, chickenpox, and, of course, H1N1 swine flu. *NaturalNews* editor Mike Adams noted, "The CDC . . . has been running defense for Merck for many years, downplaying vaccine side effects and insisting that Merck's vaccines are safe. Now that the president of Merck's vaccine division and the former chief of the CDC are one and the same, it brings up obvious questions of whether there was some level of ongoing collusion between the CDC and Merck and how deeply Dr. Gerberding might have been involved."

Adams, who advocated a law prohibiting government public health officials from ever working for pharmaceutical corporations, added, "There's just too much risk of cross-contamination of influence, which is why we have the corruption and collusion problems we're seeing today with the FDA, FTC, and CDC, all of which seem to be operating as marketing extensions of the pharmaceutical industry."

ADJUVANTS AND SQUALENE

IN MID-2009, THE WORLD Health Organization (WHO) and the CDC predicted a death-dealing onslaught of swine flu, a curious mixture of older human influenza viruses mixed with strains of avian (bird) flu and swine flu designated H1N1, a subtype of the influenza A virus. Both the WHO and the CDC talked seriously about instituting mandatory inoculations during the swine flu scare of 2009.

Even in the midst of the flu scare, critics were accusing pharmaceutical corporations of manipulating the WHO in an effort to sell swine flu vaccine so as to recoup the millions of dollars they had invested in researching and developing pandemic vaccines following the bird flu scares of

2006 and 2007. In early 2010, Dr. Wolfgang Wodarg, the president of the Health Committee of the Council of Europe, accused the pharmaceutical lobbies and the governments involved of a "great campaign of panic" based on the swine flu. A German epidemiologist by profession, Wodarg won unanimous approval from the Health Committee of the Council of Europe for a commission of inquiry into what he described as a "massive operation of disinformation."

Others had even deeper suspicions. Could it be that the swine flu epidemic was manufactured? There is strong evidence that this man-made disease comes from post–World War II–era biological warfare experimentation, as discussed in the section on mycoplasmas.

"It is obvious that the vaccine manufacturers stand to make billions of dollars in profits from this WHO/government-promoted pandemic," said Dr. Russell Blaylock, a board-certified neurosurgeon, author, and lecturer. "Novartis, the maker of the new pandemic vaccine, recently announced that they would not give free vaccines to impoverished nations—everybody pays. One must keep in mind that once the vaccine is injected, there is little you can do to protect yourself—at least by conventional medicine. It will mean a lifetime of crippling illness and early death. There are much safer ways to protect oneself from this flu virus, such as higher doses of vitamin D_3, selective immune enhancement using supplements, and a good diet."

In an article titled "The Vaccine May Be More Dangerous Than Swine Flu," Blaylock examined the swine flu pandemic from both 1976 and 2009 and pointed out that Novartis made an agreement with WHO for a pandemic vaccine. "What is terrifying is that these pandemic vaccines contain ingredients, called immune adjuvants, that a number of studies have shown cause devastating autoimmune disorders, including rheumatoid arthritis, multiple sclerosis and lupus. Animal studies using this adjuvant have found them to be deadly. A study using 14 guinea pigs found that when they were injected with the special adjuvant, only one animal survived. A repeat of the study found the same deadly outcome," reported Blaylock.

The adjuvant Blaylock mentioned in his article is called squalene, a chemical that may have something to do with Gulf War syndrome, the mysterious illness that afflicted many Gulf War veterans. Squalene is an unsaturated organic compound that acts as an intermediary in the production of cholesterol. Squalene occurs normally in the human body but at low levels in blood plasma and at elevated levels in viral influenza.

Squalene was initially derived for commercial use from shark liver oil. Today a synthetic form of squalene is used in a number of pharmaceuticals. For years, the Department of Defense denied the presence of squalene in the anthrax vaccine. However, the FDA tested several samples of the vaccine and found the compound throughout in varying levels.

Citing the Military Vaccine Resource Directory website, Dr. Anders Bruun Laursen, who has written extensively on vaccines in general and squalene in particular, noted, "The average quantity of squalene injected into the US soldiers abroad and at home in the anthrax vaccine during and after the Gulf War was 34.2 micrograms per billion micrograms of water. According to one study, this was the cause of the Gulf War syndrome in 25% of 697,000 US personnel at home and abroad." Laursen said these values were confirmed by Professor Robert F. Garry in testimony before the House Subcommittee on National Security, Veterans Affairs, and International Relations in 2002. Garry was the man to first discover the connection between the Gulf War syndrome and squalene. Squalene was subsequently banned from use by the Pentagon and by a federal court judge in 2004.

The Constitution of the United States makes it clear that the sanctity of the individual person is inviolate except under a court order following due process. Article Four of the Bill of Rights states, "The right of the people to be secure in their persons, house, papers, and effects, against unreasonable searches and seizures, shall not be violated. . . ." Articles Nine and Ten clearly restrain any act of the federal government against the states or the people.

Forcing people to submit to an involuntary injection is an egregious form of restricting freedom. How can we call someone free if that person cannot determine for themselves what may be injected into their own body? There are many people who believe that they should be able to object to mandatory inoculations. These people should consider the

State Emergency Medical Powers Act and PATRIOT Acts I, II, and III, BARDA (Biomedical Advanced Research and Development Authority). These acts make it legal for mandatory vaccinations or druggings to take place without exemptions.

Regarding the dangers of swine flu inoculations, Dr. Laursen said that many people's fears are with the adjuvants in the vaccines—in particular squalene—which "in all probability was responsible for the Gulf War syndrome." There is also a great deal of fear over the virus antigen's condition (dead, attenuated, live) and "a deeply rooted mistrust in our politicians and the vaccine producers' motives and morals. . . ."

Laursen said one vaccine allotment contained 10.68 mg of squalene per 0.5 ml. "This corresponds to 2.136.0000 microgrammes pr. billion microgrammes of water, i.e. one million times more squalene per dose than [noted in the Military Vaccine Resource Directory]. There is [every] reason to believe that this will make people sick to a much higher extent than in 1990/91. This appears murderous to me."

Laursen said he contacted the medical authorities in Denmark, where the government has ordered mass vaccinations, only to discover they knew nothing about the composition of one vaccine called Pandremix, manufactured by GlaxoSmithKline (GSK). "Then I addressed the Danish Medicinal Agency. They admitted that the Pandremix vaccine from GlaxoSmithKline does contain squalene and thimerosal," noted Laursen. "They have not rejected my remark that the squalene concentration is dangerous. In contrast, the AstraZeneca MedImmune nasal vaccination avoids squalene side effects."

Although in the past the FDA has banned squalene, this ban may have been ignored during the rush to develop a swine flu vaccine after President Obama declared swine flu a "national emergency" in late October 2009. "Clearly, bypassing the FDA requirements for safety testing of these new adjuvants and the vaccines which contain them puts the entire population at risk for serious, possibly life-threatening side effects, particularly any of the 12,000 paid trial participants (6,000 children) who are unfortunate enough to be randomized into the adjuvant containing groups," warned Laursen. "My advice: If you are forced to be vaccinated against the harmless swine flu (H1N1), demand a vaccination with the AstraZeneca nasal vaccine MedImmune, thereby avoiding squalene side effects."

Actually getting the swine flu vaccination can be a painful ordeal. According to information supplied by the vaccine manufacturer Novartis, reactions to the vaccine's injection site may include pain that may limit limb movement, redness, swelling, warmth, ecchymosis (bleeding), and induration (loss of feeling).

Possible side effects to the swine flu vaccine may include hot flashes/flushes, chills, fever, malaise, shivering, fatigue, asthenia (loss of strength), facial edema (excess moisture), immune system disorders, hypersensitivity reactions (including throat and/or mouth edema), cardiovascular disorders, vasculitis (blood vessel inflammation), syncope shortly after vaccination (temporary loss of consciousness), digestive disorders, diarrhea, nausea, vomiting, abdominal pain, blood and lymphatic disorders, metabolic and nutritional disorders, loss of appetite, arthralgia (joint pain), myalgia (muscle pain), myasthenia (muscle weakness), nervous system disorders, headache, dizziness, neuralgia, paraesthesia (tickling or numbness), febrile convulsions, Guillain-Barré Syndrome, myelitis (including encephalomyelitis and transverse myelitis [inflammation of the spinal cord or bone marrow]), neuropathy (abnormalities in the nervous system, including neuritis), paralysis (including Bell's Palsy [facial paralysis]), respiratory disorders, dyspnea (shortness of breath), chest pain, cough, pharyngitis (throat inflammation), rhinitis (nose inflammation), Stevens-Johnson syndrome (a life-threatening skin condition), pruritus (skin itching), urticaria (skin eruptions), and rashes. "In rare cases, hypersensitivity reactions have lead to anaphylactic shock and death," stated Novartis literature on the vaccine.

The extremeness of these side effects may explain why several physicians publicly issued warnings against the swine flu vaccine. Some pointed to some unsavory history regarding vaccines. These warnings were not lost on thoughtful Americans, who also questioned the effectiveness of the vaccine. A September 2009 poll by the University of Michigan's C. S. Mott Children's Hospital indicated that out of 1,678 parents, 60 percent decided against vaccinating their children. About half of the parents who objected to the H1N1 flu shot expressed concern about possible side effects of the vaccine. Only 40 percent said they would agree to an inoculation against the swine flu.

Almost half of those polled indicated they did not expect their kids to become infected or did not believe in the seriousness of the flu pandemic. Dr. Matthew Davis, University of Michigan professor of pediatrics and internal medicine and the poll's director, noted differences along racial and ethnic lines in parents' responses. More than half of Latino parents said they would bring their kids to get vaccinated, whereas only 38 percent of white parents and 30 percent of African American parents said they would do so.

A September 2009 Canadian study from researchers at the British Columbia Centre for Disease Control and Laval University also called into question the effectiveness of the swine flu vaccine. The study indicated that people vaccinated against seasonal flu are twice as likely to catch the swine flu. The lead researchers from the study were prevented from speaking in public until their study is reviewed and published.

Despite skepticism over the study's results (which contradict previous governmental assurances that swine flu inoculations are safe), several provincial Canadian health agencies announced that they were suspending seasonal flu vaccinations.

"It has confused things very badly," said Dr. Ethan Rubinstein, head of adult infectious diseases at the University of Manitoba. "And it has certainly cost us credibility from the public because of conflicting recommendations. Until last week, there had always been much encouragement to get the seasonal flu vaccine." He said the study methodology appeared sound. "There are a large number of authors, all of them excellent and credible researchers [of this study]. And the sample size is very large—12 or 13 million people taken from the central reporting systems in three provinces. The research is solid."

Many people were objecting to the hype over the pandemic. Though it was not reported much at all, the results of a mid-2009 survey of Hong Kong health-care workers indicated that more than half of the doctors and nurses questioned would decline the swine flu vaccine if they were offered inoculation. In fact, an initial study of 2,225 health-care specialists in the Hong Kong public hospital system showed that only 28.4 percent indicated an "overall willingness to accept pre-pandemic H5N1 vaccine." The most prevalent reasons that the health-care workers declined shots were a fear of side effects and doubts about the vaccine's efficacy. Only after the media started spreading fear about the flu and after the WHO raised the pandemic alert level to Phase 5 did a second survey show the above percentage rise to 47.9. The most common reasons that respondents gave for why they would accept the vaccine were "wish to be protected" and "following health authority's advice."

Apparently, some American workers won't even get a choice when it comes to vaccinations. Albany Medical Center spokesman Gregory Mc-Garry confirmed that "corrective action" might be taken against workers who did not follow orders to get a flu shot by October 16, 2009. Under emergency regulations adopted by the State Hospital Review and Planning Council in August 2009, officials for Capital Region hospitals in New York State threatened disciplinary action and even termination if all workers (including janitors, food service workers, doctors, and nurses) refused to take the vaccination shots. Elmer Streeter, a spokesman for St. Peter's Hospital in Albany, told newsmen in August, "There are very few exceptions. We will be requiring [flu shots] of all our employees as a condition of employment." Local news reports stated that workers first would be suspended for five days if they refuse the shot. After another five days, they would face possible termination.

Despite the fact that President Barack Obama declared swine flu a national emergency and despite the WHO's classification of the disease as a worldwide pandemic, serious researchers and some mainstream news outlets, such as CBS, reported that the counting of swine flu victims was widely overestimated and that many people diagnosed with the flu did not have it at all.

According to CBS reporter Sharyl Attkisson, "In late July [2009], the CDC [Centers for Disease Control] abruptly advised states to stop testing for H1N1 flu, and stopped counting individual cases. The rationale given for the CDC guidance to forego testing and tracking individual cases was: why waste resources testing for H1N1 flu when the government has already confirmed there's an epidemic? . . . CBS News learned that the decision to stop counting H1N1 flu cases was made so hastily that states weren't given the opportunity to provide input."

CBS requested state-by-state information on swine flu victims, but the news organization was stalled for some time by the CDC. Everyone was shocked when figures for state flu cases were finally released. "The vast majority of cases were negative for H1N1 as well as seasonal flu, despite the fact that many states were specifically testing patients deemed to be most likely to have H1N1 flu, based on symptoms and risk factors, such as travel to Mexico," CBS reported.

Even real cases were hyped by a compliant corporate media. One headline in September 2009 stated: "H1N1 Flu Infects Over 250 Georgetown Students." Yet a closer investigation at Georgetown University showed the number of sick students came only from "estimates" made by counting students who went to the Student Health Center with flu symptoms, students at the emergency room, and even those who called the H1N1 hotline or the Health Center's doctor on call, not from laboratory tests.

In early February 2010, the whole stressful pandemic of swine flu was unraveling, with vaccines being returned to manufacturers unsold. Sanofi Pasteur in Swiftwater, Pennsylvania, issued a nationwide recall of its H1N1 vaccine after it was discovered to have a lack of potency. The discovery was made after allotments had been shipped to all fifty states.

THE KANSAS CITY PANDEMIC OF 1921

THE SWINE FLU PANDEMIC may be just the latest occurrence in a history of instances where powerful organizations exaggerated the dangers of a disease in order to profit from scaring the population. In the early 1920s in Kansas City, Missouri, a citizen's watchdog group called "The Advertiser's Protective Bureau" successfully prosecuted the Missouri state chapter of the AMA, the Jackson Medical Society, for unduly spreading fear about a smallpox pandemic when none existed. The bureau reported:

"In the fall of 1921, the health of the city was unusually good, but slow for the doctors. So the Jackson Medical Society met and resolved to make an epidemic in the city. According to the minutes of this meeting, 'a motion was made and seconded, that a recommendation be made by the committee, to the board of health, that an epidemic of smallpox be declared in the city . . . it was moved and seconded that a day be set aside, termed Vaccination Day, on which physicians would be stationed at all schools, clinics, public buildings and hospitals to vaccinate "free of charge" . . . it is further recommended that wide publicity be given, stating that vaccination is a preventive of smallpox, and urging the absolute necessity of vaccination for every man, woman, and child in the city.' "

Dr. A. True Ott, a naturopathic medical doctor and talk-show host who specializes in health and medicine issues, researched this case and noted the Jackson Medical Society's propaganda blitz was highly successful. "Over a million previously healthy and happy American citizens were hypnotized and terrorized into placing the vaccine toxins into their bloodstreams. All public school children in the region were vaccinated while at school! Parents who dared question the vaccination of their children were ostracized and publicly vilified. The court record on this case is very clear. In the weeks and months following the 'mass vaccinations' the area's hospital beds were filled to over-flowing with vaccine-induced smallpox cases. Tens of thousands of people became ill, and many hundreds of innocents died, and many more were permanently crippled. Of course, the newspapers then trumpeted how wise the medical establishment was to promote the vaccines, stating how much worse the death toll would have been without the vaccination campaign."

Evidence presented in court showed there was no epidemic at any time, either in Kansas City or the state. However, the Jackson Medical Society produced large quantities of posters, flyers, newspaper stories, and ads featuring lurid pictures of children covered with massive smallpox sores and open wounds. The Advertiser's Protective Bureau later proved that these photographs came from British newspapers.

According to Dr. Ott, "While the Protective Bureau won the criminal court case the American People lost. The case should have made front-page headlines around the nation, showing the Modus Operandi of certain corrupt 'medical practitioners'—how, by means of fraud, treachery, and trickery, [the Jackson Medical Society] made millions of dollars in windfall profits while thousands of innocent, trusting, and naive Americans suffered and died. The entire sordid affair, with all its damning details, was kept out of the American Press. John D. Rockefeller's AMA [American Medical Association], with its millions of dollars of influence made sure of that!"

The polio vaccine of the 1950s is yet another instance in which a vaccine has hurt Americans more than it has helped. Prior to the polio vaccine, parents were deathly afraid their children would contract polio, an infectious viral disease often resulting in paralysis or permanent disability, such as suffered by President Franklin D. Roosevelt. The population was greatly relieved with the discovery and distribution of the Jonas Salk polio vaccine beginning in the mid-1950s. But after millions of Americans and others around the world were given the new vaccine, scientists discovered the vaccine contained a cancer-causing monkey virus called Simian vacuolating virus 40 (SV-40), a virus closely related to human immunodeficiency virus (HIV), and one that was born through the manufacture of the polio vaccine from infected monkey glands. SV-40 has been connected to brain tumors, bone cancers, lung cancers, and leukemia. It can be transmitted from mother to child in the womb as well as through sexual intercourse.

There has been a good deal of documentation over how pervasive this disease has become in the American population. Yet very little of this story has been brought to the attention of the public by the corporate mass media. Conspiracy-minded researchers are suspicious that no samples of pre-1962 polio vaccine can be found. Although more than ten million people were inoculated with potentially contaminated batches of vaccine, there is now no way to determine if they were exposed to the SV-40 virus, which can lie dormant in the human body for years before causing tumors and cancer.

And one should not forget the swine flu scare of 1976, when President Gerald R. Ford and some forty million Americans dutifully took swine flu shots.

And what was the death toll from that flu? Exactly one—the poor soldier who started the scare in the first place. The soldier died after his body reacted to an experimental vaccine while he was completing a "forced march" during training at Fort Dix, New Jersey. Others in the country had received the same experimental vaccine, and several deaths were reported. Just as disturbing, hundreds of others who were vaccinated suffered from Guillain-Barré syndrome (GBS), a debilitating response to the immune system that causes lupus or paralysis in the extremities and the facial muscles. Guillain-Barré is one of the world's leading causes of non-trauma-induced paralysis.

Court cases against the government and the vaccine manufacturers stacked up in the years following the 1976 scare. In July 2009, the media reported that Health and Human Services Secretary Kathleen Sebelius had taken steps to prevent a recurrence of lawsuits similar to those from 1976 by signing an order granting legal immunity to vaccine makers. This order was issued under provisions written into a 2006 law for public health emergencies.

Paul Pennock, a New York plaintiff's attorney on medical liability cases, was critical of the grant of immunity. He stated, "If you're going to ask people to do this for the common good, then let's make sure for the common good that these people will be taken care of if something goes wrong."

Though some may argue that liability is not an issue to consider in vaccination cases, the case of Lance Corporal Josef Lopez of Missouri is an appropriate rebuttal. After being deployed to Iraq for just nine days, Lopez ended up paralyzed in a coma and unable to breathe on his own. Had he been shot? Had his truck come too close to a roadside bomb? No. Lopez suffered a violent reaction to a smallpox vaccine administered by the military. Three years later, Lopez still had to wear a urine bag, walked with a limp, suffered short-term memory loss, and was taking fifteen pills a day to control leg spasms.

Yet when Lopez applied for GI benefits, the Veterans Administration rejected him, claiming that benefits are "for traumatic injury, not disease, not illness, not preventative medicine." Stephen Wurtz, the VA's deputy assistant director for insurance, said administrators were simply trying to follow the intent of Congress. "It has nothing to do with not believing these people deserve some compensation for their losses." VA officials were unable to say how many claims have been rejected because of vaccine-related injuries. The Military Vaccine Agency, which is in charge of troop vaccinations, did not respond to repeated requests for comment from a reporter.

Despite Lance Corporal Josef Lopez's debilitating reaction to the flu vaccine, the Defense Department announced on September 1, 2009, that swine flu vaccinations were mandatory for all military personnel, including health-care workers, deploying troops, those serving on ships and submarines, and new enlistees at the top of the list.

"Any place where we take a lot of people, squash them all together . . . and put them under stressful conditions will get the vaccine first," stated army lieutenant colonel Wayne Hachey, director of preventive medicine for Department of Defense health affairs. The vaccination program was to begin in early October 2009, and millions of doses had been readied.

Despite the fact that only twenty swine flu deaths were reported in Mexico by September 1, 2009, the U.S. corporate mass media continued a blitz of coverage on what was described as a pending pandemic. "That's not an epidemic. This has all the markings of a propaganda campaign benefiting the huge pharmaceutical firms producing vaccines. It's more than monetary motives that are driving this push. There seems to be a long-term agenda of making people totally dependent upon government money and actions to manage health," wrote Joel Skousen of *World Affairs Brief,* a long-running Internet news roundup service.

During the height of the swine flu scare, the Centers for Disease Control earmarked $16 million for an "outreach" program in major metropolitan areas that was aimed at garnering support for the swine flu inoculations. At the same time, major TV networks such as ABC and NBC were refusing to air ads that warned of the dangers of the vaccines or that criticized President Obama's health-care plan.

"Your biggest threat is first, schools (if you have children) and second, the workplace if they task employers to demand compliance of their employees. I oppose these measures as a matter of personal liberty and also due to the long history of vaccine contamination with immune damaging adjuvants like squalene and mercury," warned Skousen. "Of course, public schools, incubator of all things contagious, are back in session in September. Newscasters fret that 'the swine flu vaccine won't be ready for schoolchildren until mid-October,' clearly implying that an 'all schoolchildren vaccination campaign' is coming. All of this hype is aimed at priming everyone with sufficient fear so they will clamor for the vaccine—which could be very dangerous to your health. If history is any indicator, you won't see a dramatic rise in swine flu cases until the vaccine is administered—vaccines often carry some live virus 'by mistake.' "

As the school year began across the nation, schools prepared for what the Associated Press described as "the most widespread school vaccinations since the days of polio." The National School Boards Association told the AP that three-quarters of the districts in a recent survey agreed to allow vaccinations in school buildings, and according to an AP poll, almost two-thirds of the parents queried said they would give permission to have their child vaccinated if the vaccines were offered for free through the school.

South Carolina school superintendent Jim Rex said his state planned at least one vaccination clinic in each of the state's eighty-five school districts. South Dakota planned to offer both regular and swine flu vaccinations in many schools, said South Dakota state health secretary Doneen Hollingsworth. In mid-September 2009, more than seven hundred health and school officials participated in the National Association of County & City Health Officials' online seminar about how to run school flu vaccinations.

Despite all the media hype, official hand wringing, and experts predicting a deadly repeat of the 1918 killer pandemic, as of this writing, the swine flu appeared to be just another scam to increase profits for the pharmaceutical corporations and a failed attempt to see how much public control could be garnered by the globalist fascists.

FLU FEARS

WITH THE HISTORY OF the false smallpox epidemic in Kansas City as an indication of corporate malfeasance, one should look at who profits from pandemics. With swine flu, there should be public scrutiny of Baxter International, the giant worldwide pharmaceutical conglomerate that was given millions to develop a swine flu vaccine. In 2008, 44 percent of its total profits ($5.3 billion) came from pharmaceuticals and vaccines. In 2010, several websites were claiming that President Obama, as a senator in 2005, bought $50,000 worth of stock shares in two companies, one being Baxter. Apparently, in March 2005, Senator Obama attached an amendment to the Foreign Relations Committee Authorization Act (S. 600) authorizing $25 million for international efforts to combat the avian influenza. On April 28, 2005, Obama introduced the AVIAN Act (S. 969), a comprehensive bill addressing the threat of an avian flu pandemic. Interestingly enough, major outbreaks of the avian flu took place in 2006 and 2007, which prompted some to wonder how Obama could have known about the problem in 2005.

Baxter has been the center of several controversies, one of which was the adulteration of an avian flu vaccine with a pathogen. In late February 2009, a batch of the usual seasonal flu vaccines from a Baxter lab in Austria was contaminated with live H5N1 avian flu viruses (which has a 60 percent kill rate) and shipped to subcontractors in several countries. Luckily, some cautious researchers in the Czech Republic decided to inject the vaccine into laboratory ferrets to observe any side effects. The ferrets all died. Baxter officials quickly said the offending vaccines were destroyed and that "preventive and corrective" measures had been instituted.

Baxter's distribution of the adulterated flu vaccine caused concern. To many it illustrated how sloppy corporate handling of deadly viruses could break out into a full-blown public disaster, and others even saw this as an attempt to spread a pandemic for which the company could provide an antidote . . . for a price, of course.

Christopher Bona, Baxter's director of global bioscience communications, confirmed that the "experimental virus material" contained live avian flu virus but explained this was the result of "just the process itself, [and] technical and human error in this procedure."

One great fear of combining seasonal flu virus with a virulent avian flu virus—a process called reassortment—is that such mixing could produce new hybrid bird-human viruses with dire consequences for the human population.

The Czech media publicly questioned if Baxter's distribution of the deadly virus might have been a conspiracy to initiate a multination pandemic—a charge that may not be that absurd. According to routine laboratory protocols for vaccine makers, it is virtually impossible to accidentally mix a deadly live virus with a vaccine.

Mike Adams, editor of *NaturalNews* and a former trial tester for pharmaceutical companies, wrote, "Baxter is acting a whole lot like a biological terrorism organization these days, sending deadly viral samples around the world. If you mail an envelope full of anthrax to your senator, you get arrested as a terrorist. So why is Baxter—which mailed samples of a far more deadly viral strain to labs around the world—getting away with saying, essentially, 'Oops'?"

It seems Baxter has a long history of problems and controversies with its operations as well as its products. Beginning in the mid-1990s, more than a half-dozen persons in the United Kingdom tried to sue Baxter, Bayer, and four other pharmaceutical firms in the United States, claiming all had shipped blood contaminated with the HIV virus to Britain. The suits were continuing in 2007 after an American judge ordered the case moved to the United Kingdom.

In 2008, Baxter was charged with the distribution of contaminated doses of the Chinese-produced drug heparin, a blood thinner that is used in kidney dialysis. The heparin was provided to Baxter by Scientific Protein Laboratories of Waunakee, Wisconsin, and emanated from its plant in Changzhou City, China. The company is Baxter's main supplier of the active pharmaceutical ingredient in heparin.

In 2009, Baxter's subsidiary, Baxter Healthcare Corporation, settled a suit over excessive Medicaid billing in Kentucky for $2 million. Following an investigation, that state's attorney general, Jack Conway, had charged Baxter with charging the Kentucky Medicaid program inflated average wholesale prices for its intravenous solutions bearing no relationship to prices the firm charged its customers. This created an artificial large gap between Baxter's published prices and the real prices. At times this difference exceeded 1,300 percent, causing the Kentucky Medicaid program to pay substantially more for Baxter's drugs than their actual cost.

On August 15, 2001, two elderly patients in Spain died within hours of receiving dialysis from Baxter products. Eventually fifty-one more patients would die; though the cause was unclear, the company issued a worldwide recall of Baxter's two lines of filters, the sole common link between all the equipment used by the patients. Harry Kraemer, the company president at the time, apologized for the errors, shut down the factory producing filters, alerted competitors of the issue, and took a 40 percent pay cut along with a 20 percent cut for other executives. The company's earnings dropped by $189 million as a result of the issues. The company took quick action to reduce the impact of the event and prevent future recurrence and as a result suffered minimal damage to its reputation.

Despite Baxter's troubled past, at the end of 2009 the company remained one of the top contenders for making the swine flu vaccine. This is perhaps due to the fact that in 2008, Baxter was the first pharmaceutical company to announce the development of a swine flu vaccine. What is suspicious about the timing of Baxter's 2008 announcement is that the company applied for a patent on several viruses, including swine flu, on August 28, 2007, nearly two years before the disease was said to have suddenly appeared in Mexico. The fortuitous timing of Baxter's patent claim provides much grist for the mills of conspiracy theorists.

One of Baxter's competitors was Novartis Pharmaceuticals. In 2006, Novartis acquired the Chiron vaccine company, which at the time was embroiled in controversy after Britain's Medicines and Healthcare Products Regulatory Agency suspended the company's license to make the influenza vaccine Fluvirin in 2004. Both firms had agreements with the World Health Organization to produce a pandemic vaccine.

Author, lecturer, and neurosurgeon Dr. Russell Blaylock warned: "The

Baxter [swine flu] vaccine, called Celvapan, has had fast track approval. It uses a new vero cell technology, which utilizes cultured cells from the African green monkey. This same animal tissue transmits a number of vaccine-contaminating viruses, including the HIV virus."

Adjuvants are oil-based additives placed in vaccines that prompt the body to create antibodies against the targeted virus. Adjuvants can cause extreme inflammation. Animals injected with such adjuvants develop painful, incurable autoimmune diseases like multiple sclerosis, rheumatoid arthritis, or systemic lupus. Blaylock said that, after reviewing a number of studies on the adjuvant MF-59, which contains squalene, he noticed something interesting: "Several studies done on human test subjects found MF-59 to be a very safe immune adjuvant. But when I checked to see who did these studies, I found—to no surprise—that they were done by the Novartis Pharmaceutical Company and Chiron Pharmaceutical Company, which have merged. They were all published in 'prestigious' medical journals. Also, to no surprise, a great number of studies done by independent laboratories and research institutions all found a strong link between MF-59 and autoimmune diseases."

It is necessary to note that Daniel Vasella, chairman and CEO of Novartis, has regularly attended the secretive Bilderberg meetings since 1998. One would be foolhardy to believe that sheer coincidence could explain that, just two months after the 2009 Bilderberg meeting in Athens, the U.S. government gave Novartis $690 million to manufacture swine flu vaccines. It should be no secret how such deals are accomplished, considering the globalists in government service who attend Bilderberg meetings, such as Barack Obama, Hillary Clinton, and others.

It might also be noted that Novartis came from the 1996 merger of Ciba-Geigy and Sandoz Laboratories, both originally German entities and part of the massive I. G. Farben chemical cartel. The full name is Interessengemeinschaft Farbenindustrie Aktiengesellschaft, or the Syndicate of Dyestuff-Industry Corporations. It was the world's greatest chemical/drug combine from its inception in 1925 and a major supporter of the Nazi regime until broken up by the Allies at the end of World War II. This, once again, establishes a clear link between the old German Nazis, who desired to clean up the human gene pool by killing off undesirables, and the giant pharmaceutical houses of today, run by globalists who also desire to trim the human herd.

POT BUSTS ARE HIGH

IN EARLY 2009, WITH the economy stumbling and corruption constantly being revealed in high places, our nation's lawmen were on the job. The National Organization to Reform Marijuana Laws (NORML) reported that FBI statistics showed marijuana arrests were at an all-time high (no pun intended). Since 1965 marijuana arrests climbed from a mere 2 per hour to 100 per hour in 2008. In 2008, the FBI's *Uniform Crime Report* stated that police had arrested a record 872,721 persons for marijuana violations in 2007, the largest total number of annual arrests for cannabis ever recorded by the FBI. Marijuana arrests composed nearly 47.5 percent of all drug arrests in the United States, with almost three in four of those arrested under age thirty.

According to NORML executive director Allen St. Pierre, of those arrested for marijuana violations, approximately 89 percent (775,138 Americans) were only charged with possession. "These numbers belie the myth that police do not target and arrest minor cannabis offenders," said St. Pierre. "This effort is a tremendous waste of criminal justice resources that diverts law enforcement personnel away from focusing on serious and violent crime, including the war on terrorism. . . . The remaining 97,583 individuals [11 percent] were charged with 'sale/manufacture,' a category that includes all cultivation offenses, even those where the marijuana was being grown for personal or medical use."

Often those arrested are guilty of victimless crimes. One twenty-year-old Texas man used to make money selling small amounts of pot. He had a deal with local deputies—if he was caught, he had to share some of his stash before the cops left him alone. But the twenty-year-old made a mistake when he was stopped in an adjoining Texas county. After being arrested for possession of a small quantity of weed, the man was offered probation, but could not afford the probation fees and was forced to plead guilty. He now carries a felony record for the rest of his life with all the restrictions that implies.

Allen St. Pierre noted that annual marijuana arrests have nearly tripled since the early 1990s, while arrests for cocaine and heroin declined in the same period. St. Pierre concluded, "Arresting hundreds of thousands of Americans who smoke marijuana responsibly needlessly destroys the lives of otherwise law abiding citizens [and] increased enforcement of marijuana laws is being achieved at the expense of enforcing laws against the possession and trafficking of more dangerous drugs. . . . Enforcing marijuana prohibition costs taxpayers between $10 billion and $12 billion annually and has led to the arrest of nearly 20 million Americans. Nevertheless, nearly 100 million Americans acknowledge having used marijuana during their lives. It makes no sense to continue to treat nearly half of all Americans as criminals for their use of a substance that poses far fewer health risks than alcohol or tobacco. A better and more sensible solution would be to tax and regulate cannabis in a manner similar to alcohol and tobacco."

Observers of the modern drug scene are amazed at how it only took fourteen years for our great-grandparents to realize that alcohol prohibition not only was not working but, in fact, was creating a worse problem than the alcohol (police and judicial corruption, the rise of bootleggers and crime syndicates, and the large number of otherwise innocents caught up in the criminal justice system). In 1933, both the Prohibition laws and the constitutional amendment were amended and alcohol became controlled and taxed, but legal. It is interesting to note that laws against marijuana came into being just as Prohibition was on the way out. Perhaps this changeover is explained by the fact that Harry Anslinger, assistant Prohibition commissioner in the Bureau of Prohibition, was about to lose his job. Fortunately, he successfully lobbied Congress to pass antimarijuana laws just before being appointed as the first commissioner of the Treasury Department's newly formed Federal Bureau of Narcotics.

Conspiracy researchers have long suspected that the disparity between pot arrests and those for cocaine and heroin may be explained by allegations that the harder drugs, such as cocaine and heroin, cannot be easily produced at home and are imported by people working within the U.S. government to fund off-the-books operations. Such allegations stretch all the way back past the Vietnam War, but this is a story for another day.

NaturalNews editor Mike Adams has pointed out that all commercial hemp used by Americans for textiles, nutritional supplements, soaps, and ropes is imported from China, Canada, India, Chile, and many other countries. "Meanwhile, Americans farmers suffer under increasing debt and decreasing revenues from stalled crop prices. . . . The DEA makes no differentiation between industrial hemp and marijuana. To the DEA, it's all the same crop (never mind that smoking industrial hemp will only make you vomit, not high) and anyone caught planting hemp will be arrested and prosecuted using the same laws that were really only intended to halt hard-core street drug pushers. As anyone who isn't smoking crack has already figured out (and even a few who are), America's drug policy is a scandalous failure. Not only has the so-called 'War on Drugs' utterly failed to stop the flow of recreational drugs in America, it has criminalized struggling farmers who seek to grow industrial hemp as a profitable, renewable crop that's in high demand across multiple industries. The War on Drugs has accomplished one thing, though: It has filled the nation's prisons with small-time 'offenders' who got caught with an ounce or two of weed in their pockets. America's drug policy, it seems, is a boon for the prison industry, but a curse upon our nation's farmers," wrote Adams. The arrests for marijuana and the fines that come with the arrests are a boon for the global socialist fascists intent on tagging every citizen with a computer number.

Jeffrey A. Tucker, editor of the website for the Ludwig von Mises Institute (a research and educational center of classical liberalism and libertarian political theory), was taken aback in late 2009 when he read in his local Alabama newspaper about the arrest of twenty-five persons on methamphetamine-related charges. He thought it was amazing that all these people had meth labs in his hometown. But what caught his attention more were the published photos of the accused—old people, young people, long-haired and short-haired people, and every other type. "A cross section of rural America," thought Tucker. He noticed that "The arrests stem[med] from a three-month-long drug investigation that targeted individuals who were purchasing over the legal amount of pseudoephedrine [Sudafed], according to a release from the Lee County Sheriff's Office."

Only one of those arrested was charged with the unlawful manufacturing of a controlled substance. "This, we might presume, is the man with the meth lab; though we don't know for sure," said Tucker, noting that prior to 2005, "one could buy as many Sudafed packages as you did Big Mac sandwiches, and the police didn't care. Now, your 30-day allotment is nine grams. So this seems like it would be enough, but what if you are buying for two people or an entire family, or lose some, or give them away to a friend, or they fall to the back of the cabinet, or you're out of town?

"To me, this illustrates how regulations and rationing have a way of changing the subject from principles to practicalities," mused Tucker. "What if there were a rule that said that you can only purchase 30 Triple Whoppers from Burger King per person per month? Would we say, 'Oh, no one needs more than that?' Perhaps we would, but that is not the point. The point is that this is a violation of rights. Rationing of all types represents an egregious imposition on our right to choose. It weighs down daily life with arbitrary threats and increases the role of coercion in society— and this is true whether or not we actually bump up against the limits."

Tucker pointed out that even despite the legislated banning of certain substances, the black market always found ways of proliferating and selling drugs. "Whereas hundreds or thousands of pills used to be required to make meth, Bush's tough drug laws have led to new innovations: like the shake-and-bake method, which uses a legal number of pills and allows the user to make the stuff while driving. Yikes. That seems much more dangerous than texting while driving," he observed. "Keep in mind that all this insanity is a result of the laws themselves. People are still using the drug, but they are now risking their lives to do so. In other words, the laws are not working, except to make meth production and use even more dangerous."

The real horror, to Tucker, is the prohibition, "which has brought about a dark despotism that everyone pretends not to notice." He says, "To put it simply, this is an outrage, and it is even more disgusting that the local press is glad to play along with it. Here we have a nice illustration of how the police are used in an age of arbitrary law and despotic consumption controls. You become a criminal merely for buying today what was legal yesterday. And then society avoids you. You might be a druggie, and the suspicion alone is enough justification for you to be robbed of all rights and utterly smashed as a human being."

Tucker was merely voicing what many Americans feel but are afraid to speak out about in the growing police state that is the modern United States. But slowly and with great resistance at the federal level (the level controlled by globalist fascists who profit from the flowing of drug money through their banking system), the people are reaching for a new vision of their country. By 2010, fourteen states had legalized the use of marijuana for medical use and more were moving in that direction. Of course, this will eventually beg the question, "Hey, if my neighbor can smoke pot legally because he has glaucoma, how come I get busted if I smoke some?" The times they are a-changing.

But elsewhere, the screws are tightening. Suppression of citizens' rights through mandatory vaccinations of uncertain safety, unjust laws, unhealthy food and water, militarized law enforcement, unprovoked foreign wars, and crippling debt, both private and public, are only a part of the globalist fascists' long-term agenda to turn once-free Americans into subservient and controlled zombies.

DUMBED-DOWN EDUCATION

The aim of totalitarian education has never been to instill convictions but to destroy the capacity to form any.

— HANNAH ARENDT, social philosopher

EDUCATION CAN SHAPE THE brain of any person, even a zombie. Despite public claims to the contrary, education in America is, by almost every criterion, turning younger generations into dumbed-down and ignorant zombies.

OKLAHOMA SCHOOL STUDY

IN 2009, THE OKLAHOMA Council of Public Affairs (OCPA) enlisted a national research firm, Strategic Vision, to study student knowledge of civics. The test was taken from ten questions used by the U.S. Citizenship and Immigration Services. Candidates for U.S. citizenship must answer six questions correctly in order to become citizens. According to immigration service data, approximately 92 percent of those who take the citizenship test pass on their first try. The Oklahoma students did not do as well. The results indicated that only one in four Oklahoma public high school students could name the first president of the United States. Only about 3 percent of the thousand students surveyed would have passed the citizenship test. OCPA spokesman Brandon Dutcher said this is not just a problem in Oklahoma. According to Dutcher, Arizona students exhibited similar results.

Matthew Ladner, vice president of research at the Goldwater Institute, commented that "The results of this survey are deeply troubling. Despite billions of taxpayer dollars and a set of academic standards that cover all of the material, Oklahoma high school students display an overwhelming ignorance of the institutions that undergird political freedom."

Those surveyed were high school students who already had completed multiple classes in social studies and history. Theoretically, if they had failed those classes, they would not have been moved into high school. But the current educational philosophy dictates that self-esteem is more necessary than knowledge, and therefore these students were simply passed on to the next grade regardless of their aptitude for the material.

Here are the ten questions used in the Oklahoma study:

What is the supreme law of the land?
What do we call the first ten amendments to the Constitution?
What are the two parts of the U.S. Congress?
How many justices are there on the Supreme Court?
Who wrote the Declaration of Independence?
What ocean is on the east coast of the United States?
What are the two major political parties in the United States?
We elect a U.S. senator for how many years?
Who was the first President of the United States?
Who is in charge of the executive branch?

Recently, at least twenty-six states adopted stringent high school exit exams in an effort to promote increased learning. However, according to the *New York Times,* "As deadlines approached for schools to start making passage of the exams a requirement for graduation, and practice tests indicated that large numbers of students would fail, many states softened standards, delayed the requirement or added alternative paths to a diploma."

The dumbed-down condition of the schools is puzzling to many people since never before in history has a student population had access to such a wide variety and depth of educational resources. Yet, at the same time, never in the history of the world have students as a whole been less informed about the world largely due to a fixation on technology and self-interest. Those citizens who grew up between the 1950s and 1970s may recall that resources for knowledge were radios and TV sets, the daily newspaper, some magazines, the library, and an occasional visit to a museum.

Students now have the Internet, which places at their disposal the contents of nearly ten thousand American public libraries; TV screens everywhere (in airports, restaurants, clubs, and waiting rooms); and bookstores, both chains and individually owned. Yet despite this glut of resources, when the National Association of Scholars compared a test of current college seniors to a 1955 Gallup survey of high school students, the researchers who conducted the survey found no improvement in knowledge.

"Why is it that the older American students get, the worse they perform?" asked Mark Bauerlein, a professor of English at Emory University and a director of research and analysis for the National Endowment for the Arts. Bauerlein is the author of the national bestseller *The Dumbest Generation: How the Digital Age Stupefies Young Americans and Jeopardizes Our Future.* "[T]hat a nation as prosperous and powerful as the United States allows young citizens to understand so little about its past and present conditions, to regard its operative laws and values so carelessly, and to patronize the best of its culture so rarely is a sad and ominous condition."

Because students are actually studying only a small percentage of their total weekly time, Bauerlein posited that the debasement of education cannot be blamed on schooling alone, but instead on the surrounding culture—socializing, games, even spending habits. Bauerlein argued that the education of the young has been subverted by a culture of conformity, peer pressure, and popular culture enhanced by burgeoning technology. "Once youths enter the digital realm, the race for [their] attention begins, and it doesn't like to stop for a half-hour with a novel or a trip to the museum," Bauerlein wrote. "Digital offerings don't like to share, and tales of Founding Fathers and ancient battles and Gothic churches can't compete with a message from a boyfriend, photos from the party, and a new device in the Apple Store window."

THE VIDEO GENERATION

IN 2005, THE KAISER Family Foundation sponsored a report entitled *Generation M: Media in the Lives of 8–18 Year-Olds,* which theorizes that various media distract kids from serious study. Using a national representative sample of more than two thousand third through twelfth graders who completed detailed questionnaires, the report found students spending more time with "new media" such as computers, the Internet, and video games, without cutting time on "old media," like TV, print, and music. Often, students "multitasked" by using more than one medium at a time: for example, working on the computer while watching TV and texting via cell phones.

Some observers believed that multitasking actually made participants' minds sharper with increased mental activity. However, Mark Bauerlein noted that buried in the Kaiser report was a disturbing statistic: "While eight to eighteen-year-olds with high and low grades differed by only one minute in TV time (186 to 187 minutes), they differed in reading time by 17 minutes, 46-to-29—a huge discrepancy in relative terms . . . that suggests TV doesn't have nearly the intellectual consequences that reading does."

Bauerlein noted that years of TV and computer screen watching prime younger Americans for multitasking and interactivity at a deep cognitive level. "Perhaps we should call this a certain kind of intelligence, a novel screen literacy," stated Bauerline. "It improves their visual acuity, their mental readiness for rushing images and updated information. At the same time, however, screen intelligence doesn't transfer well to non-screen experiences, especially the kinds that build knowledge and verbal skills. It conditions minds against quiet, concerted study, against imagination unassisted by visuals, against linear, sequential analysis of texts, against an idle afternoon with a detective story and nothing else. This explains why teenagers and 20-year-olds appear at the same time so mentally agile and culturally ignorant."

The market trend of consumers resorting to audiovisuals for entertainment rather than books is evidenced in the large chain bookstores where shelf space for books is losing out to DVDs and audiotapes. A 2004 report from the National Endowment for the Arts showed a significant decline in book reading from previous generations. Amazing as it may seem to older citizens, there are those among the younger generations who take pride in the fact that they have never read a book.

"Today's rising generation thinks . . . highly of its lesser traits," wrote Bauerlein. "It wears anti-intellectualism on its sleeve, pronouncing book-reading an old-fashioned custom, and it snaps at people who rebuke them for it." After noting a number of surveys on knowledge before an audience of students in 2004, Bauerlein stated, "You are six times more likely to know who the latest American Idol is than you are to know who the Speaker of the House is." His taunting remark prompted a cry from the audience, "American Idol is more important!"

"She was right," acknowledged Bauerlein. "In her world, stars count more than the most powerful world leaders. Knowing the names and ranks of politicians gets her nowhere in her social set, and reading a book about the Roman Empire earns nothing but teasing. More than just dull and nerdish, reading is counterproductive. . . . The middle school hallways can be as competitive and pitiless as a Wall Street trading floor or an episode of *Survivor*. To know a little more about popular music and malls, to sport the right fashions and host a teen blog, is a matter of survival."

DANGEROUS TEACHING

THE CURRENT EDUCATION SYSTEM seems to have forgotten about developing students' critical thinking. John Taylor Gatto, who taught school in New York City for more than two decades, summed up this fact of modern life in his 1992 book *Dumbing Us Down: The Hidden Curriculum of Compulsory Schooling*. After teaching for some years, Gatto grew to understand that the education system does not exist to increase students' knowledge and power, but to diminish it. "Bit by bit, I began to devise guerrilla exercises to allow the kids I taught—as many as I was able— the raw material people have always used to educate themselves: privacy, choice, freedom from surveillance, and as broad a range of situations and human associations as my limited power and resources could manage."

"What we are seeing . . . is the psychologization of American education," stated an article in the April 1993 edition of *Atlantic Monthly*. "A growing proportion of many school budgets is devoted to counseling and other psychological services. The curriculum is becoming more therapeutic: children are taking courses in self-esteem, conflict resolution, and aggression management. Parental advisory groups are conscientiously debating alternative approaches to traditional school discipline, ranging from teacher training in mediation to the introduction of metal detectors and security guards in the schools. Schools are increasingly becoming emergency rooms of the emotions, devoted . . . to repairing hearts."

According to Gatto, real teaching can be dangerous. Government monopoly of schools has evolved in such a way that the premise of teaching students to think for themselves jeopardizes the total institution should it spread. The occasional teacher who attempts to instill critical thinking is merely an annoyance to the chain of command.

However, should what Gatto considers the central but false assumptions underlying modern education—such as the idea that it is difficult to learn to read, or that kids resist learning, and many more—be exposed, the ramifications could be extreme. "[T]he very stability of our economy is threatened by any form of education that might change the nature of the human product schools turn out; the economy schoolchildren currently expect to live under and serve would not survive a generation of young people trained, for example, to think critically," Gatto predicted.

"Over the years, I have come to see that whatever I thought I was doing as a teacher, most of which I actually was doing was teaching an invisible curriculum that reinforced the myths of the school institution and those of an economy based on caste," he added.

An overview of the current educational system provoked the questions: Do younger people just wake up one morning and decide they are not interested in history, politics, or world events? Or does popular culture draw them away from classical education and critical thinking? Also, what is the "invisible curriculum" referred to by Gatto and where did it come from?

Perhaps a quick review of education history in America can provide the answer.

WORKERS NOT THINKERS

SERIOUS ATTENTION TO EDUCATION as a means of social control began in Europe and with the same minds whose philosophies led to Communist and Nazi totalitarianism.

"Education should aim at destroying free will so that after pupils are thus schooled they will be incapable throughout the rest of their lives of thinking or acting otherwise than as their school masters would have wished," proclaimed Johann Gottlieb Fichte in 1810. Fichte, a teacher of philosophy and psychology at Prussian University in Berlin, was a great influence on Georg Hegel and other thinkers of the period. "When the technique has been perfected, every government that has been in charge of education for more than one generation will be able to control its subjects securely without the need of armies or policemen."

One major influence on both Adolf Hitler and Karl Marx, as well as the modern globalists, was Georg W. F. Hegel, whose words and works have often been appropriated to justify the means of the powerful. Hegel once wrote, "The State is the absolute reality and the individual himself has objective existence, truth and morality only in his capacity as a member of the State." Hegel is also most noted for his "Hegelian Dialect"—thesis, antithesis, synthesis, also known as problem, reaction, solution. The globalists, however, have bastardized Hegel's mere philosophical diagram of human interaction. Rather than wait for a problem to deal with, they create the problem, then offer a draconian solution. After compromise and negotiation, they still have advanced their agenda without the opposition realizing their design.

So, could it be the case that Rockefeller's contributions to education were really part of a secret agenda to create solutions for problems that didn't exist? Any serious study may find that the American education establishment has been created and guided for many years by the same Hegel-inspired globalist elite who created both Russian communism and German national socialism. (See Jim Marrs's *Rule by Secrecy* for further details.) The oil magnate John D. Rockefeller Sr., whose dominant oil empire was initially funded by the Rothschild-controlled National Bank of Cleveland, created the General Education Board (GEB) in 1903 to dispense Rockefeller donations to education. By 1960, it had ceased operating as a separate entity, and its programs were rolled into the Rockefeller Foundation. In 1917, the GEB made a $6 million grant to Columbia University to create the New Lincoln School, a private experimental coeducational school in New York City. According to school literature, the facility's "predecessor was founded as Lincoln School in 1917 by the Rockefeller-funded General Education Board as 'a pioneer experimental school for newer educational methods,' under the aegis of Columbia University's Teachers College."

According to the late Eustace Mullins, the authorized biographer of poet Ezra Pound, who in 1948 encouraged Mullins to research globalist control in finances, health, and education: "From this school descended the national network of progressive educators and social scientists, whose pernicious influence closely paralleled the goals of the Communist Party, another favorite recipient of the Rockefeller millions. From its outset, the Lincoln School was described frankly as a revolutionary school for the primary and secondary schools of the entire United States. It immediately discarded all theories of education which were based on formal and well-established disciplines, that is the McGuffey Reader type of education which worked by teaching such subjects as Latin and algebra, thus teaching children to think logically about problems."

Another institution of higher learning long funded by the Rockefellers is the University of Chicago and closely connected to it is the English world's most accepted authority on everything—Encyclopaedia Britannica, Inc. In 1943, advertising executive William Benton purchased the encyclopedia and operated it as a charity for the University of Chicago, eventually contributing more than $125 million to the school. From 1945 to 1953, Benton served as a senator from Connecticut after defeating Prescott Bush (the grandfather of former president George W. Bush) and was active in global affairs. Given Benton's power and influence, it is no stretch to think that his globalist beliefs could have easily permeated those things that received his benefaction.

Today, the William Benton Foundation also owns Compton's Encyclopedia and Merriam-Webster Inc., one of the world's leading publishers of dictionaries and thesauri. Reflecting the rise of the Internet coupled with a general decrease in reading today, Britannica's encyclopedia sales have precipitously dropped in recent years, yet it is still regarded as one of the most credible sources of information in the Western world.

The linchpin of Rockefeller's attempt to shape American education was his formation of the General Education Board and his continuing support of the University of Chicago. "The creation and funding of the University of Chicago had done much to enhance Rockefeller's public relations profile among Baptists and educators. . . . The only difficulty was that education, on the whole, wasn't in bad shape," explained Paolo Lionni, author of *The Leipzig Connection*. Lionni's 1993 book traced the deleterious effects of experimental psychology on the education system back to German professor of philosophy Wilhelm Maximilian Wundt, the founder of experimental psychology. "The indigenous American educational system was deeply rooted in the beliefs and practices of the Puritan Fathers, the Quakers, the early American patriots and philosophers. Jefferson had maintained that in order to preserve liberty in the new nation, it was essential that its citizenry be educated, whatever their income. Throughout the country, schools were established almost immediately after the colonization of new areas."

Lionni noted, "Educational results far exceeded those of modern schools. One has only to read old debates in the *Congressional Record* or scan the books published in the 1800's to realize that our ancestors of a century ago commanded a use of the language far superior to our own. Students learned how to read not comic books, but the essays of Burke, Webster, Lincoln, Horace, Cicero. Their difficulties with grammar were overcome long before they graduated from school, and any review of a typical elementary school arithmetic textbook printed before 1910 shows dramatically that students were learning mathematical skills that few of our current high school graduates know anything about. The high school graduate of 1900 was an educated person, fluent in his language, history, and culture, possessing the skills he needed in order to succeed."

According to author William H. Watkins, John D. Rockefeller Sr. was more concerned with shaping a new industrial social order than providing a useful education. "The Rockefeller group demonstrated how gift giving could shape education and public policy," commented Watkins. Rockefeller's agenda for dumbing down the population through new education led by his GEB shows itself in a letter written by Frederick T. Gates, Rockefeller's choice to head the board. Gates wrote, "In our dreams, we have limitless resources and the people yield themselves with perfect docility to our molding hands. The present education conventions fade from their minds, and unhampered by tradition, we work our own good will upon a grateful and responsive rural folk. We shall not try to make these people or any of their children into philosophers or men of learning, or men of science. We have not to raise up from among them authors, editors, poets or men of letters. We shall not search for embryo great artists, painters, musicians nor lawyers, doctors, preachers, politicians, statesmen, of whom we have an ample supply.

"The task we set before ourselves is very simple as well as a very beautiful one, to train these people as we find them to a perfectly ideal life just where they are. So we will organize our children and teach them to do in a perfect way the things their fathers and mothers are doing in an imperfect way, in the homes, in the shops and on the farm."

As recent as 1973, psychiatrist Dr. Chester M. Pierce, speaking at a Childhood International Education Seminar, echoed Gates's condescending arrogance but went even further by proclaiming, "Every child in America entering school at the age of five is insane because he comes to school with certain allegiances to our founding fathers, toward our elected officials, toward his parents, toward a belief in a supernatural being, toward the sovereignty of this nation as a separate entity. . . . It's up to you teachers to make all these sick children well by creating the international children of the future."

Of Gates's letter, Lionni stated that while "it would be false to say John D. Rockefeller was a mastermind of international intrigue and deception, it would not be false to say that Rockefeller money has been used in various ways to forward social and global control through economics, foundations, the United Nations, universities, banking, industry, medicine, and of course, education, psychology and psychiatry.

"That's a tremendous amount of control and involvement for one group!" noted Lionni, who then asked, "What if the theories and practices they funded and continue to fund are fundamentally flawed and don't lead to the best possible situations in the various fields mentioned? Well, the views in most of those areas *are* fundamentally flawed and they *don't* lead to the best solutions in 'mental health', education, medicine, sanity and happiness [original emphasis]. But, most likely, despite all 'humanitarian' posturing, they were never intended to."

Other Rockefeller-connected entities that still shape society in the United States include the Brookings Institution, the National Bureau of Economic Research, the Public Administration Clearing House, the Council of State Governments, and the Institute of Pacific Relations. Paul Volcker, a former Rockefeller assistant, was named chairman of the U.S. central bank, the Federal Reserve System, during the Carter administration and served there until 1987.

Norman Dodd (now deceased), who was the director of research in 1953–54 for the House Select Committee to Investigate Foundations and Comparable Organizations, reported that in 1952, the president of the Ford Foundation—part of the globalist syndicate working for financial, educational, and political control—told him bluntly that "operating under directive from the White House" his foundation was to "use our grant-making power so as to alter our life in the United States that we can be comfortably merged with the Soviet Union." Now, with the collapse of communism and the advent of the United Nations, NATO, and other economic treaties, it seems like this globalist goal is close to becoming realized.

Dodd also stated that the congressional investigation found that the Guggenheim, Ford, and Rockefeller foundations and the Carnegie Endowment were "working in harmony to control education in the United States," adding that these entities had been subverted from the original goals of their creators by subsequent directors, either working for or indoctrinated by the globalists. This is yet another example of wealth taking control of existing organizations.

Some of the past and current organizations and foundations that have had an impact on American education and that are linked by membership or funding to the globalist plutocracy include: the Agency of International Development; American Civil Liberties Union; American Council of Race Relations; American Press Institute; Anti-Defamation League; Arab Bureau; Aspen Institute; Association of Humanistic Psychology; Battelle Memorial Institute; Center for Advanced Studies in the Behavioral Sciences; Center for Constitutional Rights; Center for Cuban Studies; Center for Democratic Institutions; Christian Socialist League; Communist League; Environmental Fund; Fabian Society; Ford Foundation; Foundation for National Progress; German Marshall Fund; Hudson Institute; Institute for Pacific Relations; Institute on Drugs, Crime and Justice; International Institute for Strategic Studies; Mellon Institute; Metaphysical Society; Milner Group; Mont Pelerin Society; National Association for the Advancement of Colored People; National Council of Churches; New World Foundation; Ayn Rand Institute; Stanford Research Institute; Tavistock Institute of Human Relations; Union of Concerned Scientists; International Red Cross; and the YMCA.

According to Beverly Eakman, a former educator, government speechwriter, and author of *Walking Targets: How Our Psychologized Classrooms Are Producing a Nation of Sitting Ducks,* foundation-subsidized educators like G. Stanley Hall, Abraham Flexner, John Gardiner, Theodore Sizer, Ronald Havelock, John Goodlad, Benjamin Bloom, and Ralph Tyler brought to the classroom the psychology principles of the World Federation of Mental Health: "For openers, they worked to ensure that school curriculum and testing ditched the traditional focus on excellence and academics to concentrate on a subjective socialization (i.e., socialist) agenda that targeted the child's 'belief system.' To illustrate the radical nature of this step, one need only quote from the 'father of modern education,' John Dewey. In his acclaimed book *School and Society* he wrote: 'There is no obvious social motive for the acquiring of learning [and] . . . no clear social gain at success thereat.' Fast-forward to 1981 and to the 'father of outcome-based education,' Benjamin Bloom. In *All Our Children Learning,* Bloom averred that 'the purpose of education is to change the thoughts, feelings and actions of students . . . by [challenging] the student's fixed beliefs.'"

Eakman pointed to one example of how perceptions can be changed. " . . . [R]ugged individualism is an expression nobody hears much anymore, but folks used to hear with regularity," she noted. "Rugged individualism encompassed a range of characteristics—independence, self-sufficiency, thinking for oneself. In the 1970s, the axe was laid to all three. Negative terminologies like 'loner' and 'misfit' redefined the individualist. 'Independence' was scrapped for *interdependency,* self-sufficiency for *redistribution,* and 'thinking for oneself' was equated with *intolerance.* Today, any close reading of the newspaper reminds us daily that the 'loner' requires psychiatric intervention, and maybe drugs as well. . . .

"By 1989, the much-ballyhooed 'paradigm shift,' as it was dubbed by behaviorist educrats, occurred in American schools, and the free world was hurled into 'free fall': clandestine censorship counselors in university dorms, encounter-style techniques masquerading as 'class discussions' in high schools, massive invasions of privacy under the cover of 'academic testing,' 'value-neutral' courses in ethics, and world history that bestowed upon even the most heinous regimes the moral equivalence of Jeffersonian democracy. Little wonder that by the 1990s battalions of psychiatrists were being dispatched to every school district to help contain the new brand of war games: a tsunami of school shootings and mass murders perpetrated by kids raised on a diet of behavior modification and psychiatric drugs."

The changing of a student's beliefs, or "behavior modification," is a technique long studied by the CIA and other agencies seeking methods of mind control. It should be obvious that to modify anyone's behavior, first one must find out what people—preferably children—are thinking and then set about changing any "offending" attitudes.

It has been well documented in a number of books and articles that the U.S. intelligence community has heavily influenced the American education system to propagate its views and philosophies. David N. Gibbs, an associate professor of political science at the University of Arizona, believes that influence is always supported by the distribution of money. He wrote, "While pundits never tire of the cliché that American universities are dominated by leftist faculty, who are hostile toward the objectives of established foreign policies, the reality is altogether different: The CIA has become 'a growing force on campus,' according to a recent article in the *Wall Street Journal*. The 'Agency finds it needs experts from academia, and colleges pressed for cash like the revenue.' Longstanding academic inhibitions about being publicly associated with the CIA have largely disappeared: In 2002, former CIA Director Robert Gates became president of Texas A & M University, while the new president of Arizona State University, Michael Crow, was vice-chairman of the Agency's venture capital arm, In-Q-Tel Inc. . . . The CIA has created a special scholarship program, for graduate students able and willing to obtain security clearances. According to the London *Guardian,* 'the primary purpose of the program is to promote disciplines that would be of use to intelligence agencies.' And throughout the country, academics in several disciplines are undertaking research (often secret) for the CIA."

The Constitution never states that the federal government should control education. Education should never be the responsibility of the federal government but that of parents and local educators. Many people see government as a means to control education by selecting what to teach and what alternative theories to suppress. Many parents fear brainwashing in public or private schools. They look to schools to teach their children to be open-minded, to be able to read and write, and to fully understand the Constitution.

Yet schools have often fallen short of what parents want and, rather, have seemed to embrace what John D. Rockefeller Sr. wanted; he is often quoted as saying, "I don't want a nation of thinkers, I want a nation of workers." A 2006 report by the Federation of American Scientists seemed to echo Rockefeller's request for workers over thinkers by arguing for increased use of video games in the classroom. The report stated, "Workforce globalization is rapidly expanding. . . . The United States cannot compete in this highly-connected system of global commerce on the basis of low wages, commodity products and standardized services. It must compete by taking the lead in the next generation of knowledge creation, technologies, products and services, business models, and dynamic management systems. . . . When individuals play modern video and computer games, they experience environments in which they often must master the kinds of higher-order thinking and decision-making skills employers seek today." Others, such as author Beverly Eakman, contest the idea that such games can truly prepare young persons for the workplace.

Given the men behind America's education history and the mind-numbing curriculum they produced that is now used by teachers, it becomes understandable why our entire educational system merely churns out young people prepared for either wage slavery or to become teachers.

The late author and media critic Neil Postman wrote, "In order to understand what kind of behaviors classrooms promote, one must become accustomed to observing what, in fact, students actually *do* in them. What students do in a classroom is what they learn (as [John] Dewey would say), and what they learn to do is the classroom's message (as [media commentator Marshall] McLuhan would say). Now, what is it that students *do* in the classroom? Well, mostly they sit and listen to the teacher. Mostly, they are required to believe in authorities, or at least pretend to such belief when they take tests. Mostly they are required to *remember* [original emphasis]. They are almost never required to make observations, formulate definitions, or perform any intellectual operations that go beyond repeating what someone else says is true. They are rarely encouraged to ask substantive questions, although they are permitted to ask about administrative and technical details. (How long should the paper be? Does spelling count? When is the assignment due?) It is practically unheard of for students to play any role in determining what problems are worth studying or what procedures of inquiry ought to be used. Examine the types of questions teachers ask in classrooms, and you will find that most of them are what might technically be called 'convergent questions,' but what might more simply be called 'Guess what I am thinking' questions."

Postman and his coauthor Charles Weingartner concluded in their book *Teaching as a Subversive Activity* that contemporary curriculums are designed as a distraction to prevent students from knowing themselves and the world about them.

And the deficiencies of a weakened education system are passed along to future teachers. "It starts almost immediately," noted the two authors, "because the [teachers] have been victims—in this case for almost 16 years—of the kind of schooling we have described . . . as producing intellectual paraplegics. The college students [future teachers] we are now talking about are the ones who were most 'successful' in conventional school terms. That is, they are the ones who learned best what they were required to do: to sit quietly, to accept without question whatever nonsense was inflicted on them, to ventriloquize on demand with a high degree of fidelity, to go down only on the down staircase, to speak only on signal from the teacher and so on. All during these 16 years, they learned not to think, not to ask questions, not to figure things out for themselves. They learned to become totally dependent on teacher authority, and they learned it with dedication."

TWIXTERS

BUT IS TIME CONSUMED with DVD movies and video games or merely regurgitating facts back to a teacher truly preparing youth for gainful employment? Not if you pay attention to those who are called "Twixters," a new word for single, middle-class twenty-to-thirty-plus somethings who work in low-paying jobs (usually service), engage in serial dating, maintain old school friendships, and generally live with their parents or room with other Twixters.

Bob Schoeni, a professor of economics and public policy at the University of Michigan, has reported that the percentage of twenty-six-year-olds living with their parents has nearly doubled from 11 percent to 20 percent since 1970. According to Schoeni, youngsters between the ages twenty-five and twenty-six garner an average of $2,323 a year in financial support from their parents.

Laziness and a lack of initiative cannot be totally blamed for this phenomenon. Around 1980, most financial aid for college came in the form of grants. Today, lending is the common way to gain money for education. According to a study reported in 2005 by the Center for Economic and Policy Research, college graduates in 2005 owed 85 percent more in student loans than in the 1990s. A *Time* magazine poll showed 66 percent of student respondents owed more than $10,000 upon graduation and 5 percent owed a crippling $100,000 or more.

Such numbers fail to reflect burgeoning debt for students who abuse the credit cards often sent unsolicited by the giant credit companies. According to the public policy group Demos, credit-card debt for Americans aged eighteen to twenty-four more than doubled from 1992 to 2001. With such a debt load hanging over them, it is small wonder that young people, including the Twixters, can't seem to gain the financial independence to move out of their parents' house. Given the rise of Twixterization of the nation's young adults, the widespread use of video games, computer networking sites (Twitter, MySpace, Facebook, etc.), and the popular mass media, there seems to be too much competing for the attention of today's student. Add this to an overburdened and inadequate educational system and you have a recipe for intellectual disaster.

The consensus of thoughtful experts is that a dumbed-down education system produces dumbed-down teachers who produce dumbed-down students. The result is a dumbed-down population, the exact situation desired by old man Rockefeller and the elite globalists. The correlation is uncanny. It begs the question: Is this sheer happenstance or a conscious agenda?

PART III

HOW TO CONTROL ZOMBIES

The only sure bulwark of continuing liberty is a government strong enough to protect the interests of the people, and a people strong enough and well enough informed to maintain its sovereign control over the government.

—FRANKLIN D. ROOSEVELT

ONCE A NATION OF zombies has been created, the population must be kept docile and under control. This can be done through legislation and regulations, increasing police powers, and drugging the food and water supplies. But many commentators have written about how so many Americans become zombielike while sitting mesmerized before their TVs for more than eight hours a day. Between September 2007 and September 2008, the average household watched TV for more than 8 hours a day, a record high since the 1950s when TV viewer polls first began. In the third quarter of 2008, Americans watched more than 142 hours of TV a month, up 5 hours from the same period in 2007.

What is most essential to control is that the zombies are unaware they are being controlled. This, of course, would require controlling the mass media. Could this be happening in the United States, home of the First Amendment, and with a proud heritage of a free press?

MEDIA CONTROL AND FEARMONGERING

THE INTERNET HAS DONE a marvelous job of bringing alternative news and information to the people, but it has only done that for those who own and can use a computer. Everyone else is at the mercy of the corporate-controlled mass media, whether it be broadcast, cable, or satellite. America's mass media is currently in the hands of only five major multinational corporations: AOL-Time Warner, the Walt Disney Company, Viacom, Vivendi Universal, and News Corporation.

Media mogul Ted Turner once observed, "The media is too concentrated, too few people own too much. There's really five companies that control 90 percent of what we read, see and hear. It's not healthy." Not to mention Bertlesmann AG, which has become the largest English-language print publisher in the world and has roots in Nazi Germany.

The face of the media has changed considerably since 1975, when cable TV served less than 15 percent of the viewing population and satellite TV and the Internet did not even exist as we know them today. More than thirty years later, less than 15 percent of American homes don't have either satellite or cable TV, and one-third of the population receives its news through the Internet.

GOVERNMENT-DICTATED NEWS

FAR TOO OFTEN THE relationship between the government and the media corporations shapes what the news covers. "As technology blurs the distinction between print and electronics, the success of media businesses depends increasingly on the decisions of government, embodied in regulations, legislation, and judicial rulings," explained Leo Bogart (who died in 2005), a former Media Studies Fellow and general manager of the Newspaper Advertising Bureau. "This must make the people who run them more sensitive to the political effects of their news coverage. As political advertising has become a considerable component of television revenues, politicians have found it increasingly necessary and expedient to court the media, creating a new source of pressure on journalists."

Media reformist Robert McChesney agreed with Bogart, writing, "Professional journalism is now about currying close relations to the powerful so you have access to their news. When the powerful are entirely in agreement on an issue, for example, whether or not the U.S. has the right to invade another country (taken as a given by many people in power), the journalists don't ask questions. They reproduce the elite consensus, take it as a given. In fact, if a journalist were to question the right of the U.S. to invade a country, they would be regarded by the professional news community as un-professional. They would be seen as someone who was bringing their ideological agenda or axe to grind to the discussion. When a journalist dares to question the motives of those in power, they are framed as bringing their own personal political bias into news reporting. But when a journalist just reports and repeats what people in power say and doesn't try to weigh in with critical observations, they are regarded as professional, 'fair and balanced.'"

Editors, particularly those in publicly held corporations whose executives are cautious about reactions on Wall Street, do not have to be ordered to kill stories or slant the coverage. They intuitively understand the views and interests of their bosses and act accordingly. This capacity to anticipate the owners' desires is why they are made editors.

FEARMONGERING

WITH THEIR MASTERS CRACKING the whip, the "watchdog" media have turned into lapdogs for their corporate (and political) owners, which in turn has allowed the government to manipulate the public through national fearmongering.

One of the best examples of fearmongering came in early 2006, when President Bush—under fire for the unresolved wars in Iraq and Afghanistan, the torture of terrorist suspects, and unconstitutional spying on Americans—declared: "We cannot let the fact that America hasn't been attacked in four and a half years since September 11 lull us into the illusion that the threats to our nation have disappeared."

Bush then went on to describe a thwarted terrorist attack on Los Angeles in 2002, revealing that the attack in California was planned by a man named Hambali, reportedly a key lieutenant of Khalid Sheik Mohammed, the alleged mastermind of the 9/11 attacks. Both Hambali and Mohammed were reportedly captured in 2003.

According to Bush, al Qaeda leaders Hambali and Mohammed recruited Asian men who were supposed to use shoe bombs to blow open the cockpit door of a commercial airliner and then crash the plane into the U.S. Bank Tower in Los Angeles. Bush mistakenly referred to this building as "Liberty Tower," but was quickly corrected that its original name had been "Library Tower." Bush said the plot was foiled when a key Asian al Qaeda member was arrested. Bush declined to name the suspect or his nationality.

Soon, this story filled the mass media airwaves as some stations aired scenes from the Hollywood alien invasion film *Independence Day* as graphic representation of the destruction of the U.S. Bank Tower. But even before Americans could let out their collective sigh of relief at being spared further carnage, serious questions arose over Bush's statement. Many thoughtfully wondered why Bush had not called attention to saving the Los Angeles building early in 2003, soon after the attacks were thwarted and the criminals behind the attacks were captured. If Bush had delivered this news in 2003, he might have helped calm or prevent the large and numerous antiwar demonstrations conducted prior to the invasion of Iraq.

Public skepticism increased when Los Angeles mayor Antonio Villaraigosa told newsmen he knew nothing of the attempted attack and felt "blindsided" by Bush's announcement. Prior communication with the White House had been "nonexistent," despite the fact that the mayor had requested to meet with Bush at least two times over security issues. "I'm amazed that the president would make this [announcement] on national TV and not inform us of these details through the appropriate channels," Villaraigosa told newsmen. "I don't expect a call from the president—but somebody."

Others were even less considerate when characterizing Bush's breaking news. Doug Thompson, a writer for the Internet's oldest political news site, Capitol Hill Blue, said he was contacted by members of the U.S. intelligence community who disputed Bush's claim. Thompson said he was able to confirm the credentials of at least four of the persons who contacted him. All of those who contacted him asked not to be identified for fear of reprisals. "The president has cheapened the entire intelligence community by dragging us into his fantasy world," Thompson quoted a longtime CIA operative as saying. "He is basing this absurd claim [regarding the Los Angeles attack] on the same discredited informant who told us al Qaeda would attack selected financial institutions in New York and

Washington." Suspiciously enough, during the heat of the presidential election in August 2004, the Bush White House tried to increase the terror alert level by claiming attacks were imminent on major financial institutions. This alert was later withdrawn after administration officials admitted it was based on old information from a discredited source.

It has not always been the case that American leaders with a strong siege mentality broadcasted warnings of imminent attack to the public. In a prophetic testimony before joint hearings of the Senate Armed Services Appropriations and Intelligence committees in the spring of 2001, Colin Powell, then secretary of state, explained why Americans should not give up their freedoms in search of security. "If we adopted this hunkered-down attitude, behind our concrete and our barbed wire, the terrorists would have achieved a kind of victory," he declared.

This type of reasoned rhetoric changed completely after constitution-ally questionable laws and regulations were put into effect after the terror-ist attacks in New York City and at the Pentagon later that year. Within days of the 9/11 attacks, President Bush declared a "war on terrorism."

PATRIOT ACT

SINCE 9/11 THE GOVERNMENT has used nationalism as cover for implementing measures to control the population. Secret evidence, closed trials, false imprisonment, warrantless searches, involuntary drugging, the seizing of private property—all seem like something from a 1930s to-talitarian regime, but fear has pushed many Americans into a zombielike passiveness to authority.

A shining example of fear-based legislation is a dreadful piece of leg-islation entitled the Uniting and Strengthening America by Providing Appropriate Tools Required to Intercept and Obstruct Terrorism Act of 2001, more commonly referred to as the PATRIOT Act. The name is reminiscent of Hitler's 1933 "Enabling Act" legislation passed hurriedly following the burning of Berlin's Reichstag (its Parliament building) in 1933. This law, which founded Hitler's Third Reich, was called "The Law to Remove the Distress of the People and State."

Similarly, the PATRIOT Act was 342 pages long and made many changes to more than fifteen different U.S. statutes, most of them enacted in the wake of previous misuse of surveillance powers by the FBI and CIA. It was hurriedly signed into law by President George W. Bush on October 26, 2001, a little more than one month following the 9/11 attacks.

According to some congressmen, many lawmakers had not even read the entire document when it was passed. The ACLU also reported that some members of Congress had less than one hour to read the extensive changes of law contained within the act. The speed with which this legislation was presented to Congress caused some observers to believe that it had long been prepared and simply needed some provocation to put it into effect. Civil libertarians felt those two facts alone should be cause for wholesale dismissals of the obliging members of Congress.

Representative Ron Paul, who ran for president in 2008, confirmed rumors that the bill was not read by most members of the House prior to their vote. "It's my understanding the bill wasn't printed before the vote— at least I couldn't get it. They played all kinds of games, kept the House in session all night, and it was a very complicated bill. Maybe a handful of staffers actually read it, but the bill definitely was not available to members before the vote." Paul added he objected to how opponents were stigmatized by the name alone. "The insult is to call this a 'patriot bill' and suggest I'm not patriotic because I insisted upon finding out what was in it and voting no. I thought it was undermining the Constitution, so I didn't vote for it—therefore I'm somehow not a patriot. That's insulting."

Provisions of the original PATRIOT Act that most concerned civil libertarians were the following:

- The federal government may now monitor religious and political institutions without suspecting criminal activity to assist terrorism investigations (a violation of the First Amendment right of association).

- The feds now can close to the public once-open immigration hearings and secretly detain hundreds of people without charge while encouraging bureaucrats to resist Freedom of Information requests (a violation of Amendments 5 and 6 guaranteeing due process, speedy trials, and freedom of information).

- The government may prosecute librarians or other keepers of records if they tell anyone that the government subpoenaed information related to a terrorism investigation (a violation of the First Amendment right of free speech).

- The government now may monitor conversations between federal prisoners and their attorneys and may even deny access to lawyers to Americans accused of crimes (a violation of the Sixth Amendment right to have legal representation).

- The government now may search and seize individual and business papers and effects without probable cause to assist an antiterrorism investigation (a violation of the Fourth Amendment right against unreasonable searches and seizures).

- The government now may jail Americans indefinitely without a trial or charges (a violation of the Sixth Amendment right to a speedy trial and individuals' right to be informed of the charges against them).

After later reviewing the act further, Representative Ron Paul said, "The worst part of this so-called antiterrorism bill is the increased ability of the federal government to commit surveillance on all of us without proper search warrants." This section of the PATRIOT Act, entitled "Authority for Delaying Notice of the Execution of a Warrant," is commonly referred to as the "sneak-and-peek" provision. It allows authorities to search personal property without warning.

Congressman Paul pointed out that the act's supporters were flawed in thinking that the government would act in a restrained and responsible manner. "I don't like the sneak-and-peek provision because you have to ask yourself what happens if the person is home, doesn't know that law enforcement is coming to search his home, hasn't a clue as to who's coming in unannounced . . . and he shoots them. This law clearly authorizes illegal search and seizure, and anyone who thinks of this as antiterrorism needs to consider its application to every American citizen."

Since the ratification of the PATRIOT Act, many critics have argued that the surveillance portions are unconstitutional. The Fourth Amendment to the U.S. Constitution states: "The right of the people to be secure in their persons, houses, papers, and effects, against unreasonable searches and seizures, shall not be violated; and no warrants shall issue, but upon probable cause, supported by oath or affirmation, and particularly describing the place to be searched, and the persons or things to be seized."

Award-winning investigative reporter Kelly O'Meara spent sixteen years working as a congressional staffer to four members of Congress prior to working as an investigative journalist. She holds a BS in political science from the University of Maryland and makes her home in Alexandria, Virginia. O'Meara wrote, "With one vote by Congress and the sweep of the president's pen, say critics, the right of every American fully to be protected under the Fourth Amendment against unreasonable searches and seizures was abrogated."

Such perversion of the Constitution was aggravated on March 4, 2010, with the introduction of a bill called the Enemy Belligerent Interrogation, Detention, and Prosecution Act of 2010. This legislation expanded the Bush-era term "enemy combatant" to "enemy belligerent," defined as any individual, including American citizens, suspected of any affiliation with terrorism or supporting "hostilities against the United States or its coalition partners." Such suspects, under this law, must be turned over to military authorities and can be detained without charge, denied the Miranda warning of self-incrimination and legal representation, and held for "the duration of the hostilities." Despite Obama's earlier voiced opposition to Bush's Military Commissions Act, he was expected to support this bill, which was introduced by Democratic senator Joe Lieberman (CFR) and Republican senator John McCain. Constitutional attorney and author of the *New York Times* bestseller *How Would a Patriot Act?* Glenn Greenwald called the Enemy Belligerent Act "probably the single most extremist, tyrannical and dangerous bill introduced in the Senate in the last several decades, far beyond the horrendous habeas-abolishing Military Commissions Act."

LASER OR TASER

UNDER STATUES WITHIN THE PATRIOT Act, David Banach of Parsippany, New Jersey, was accused in 2005 of using a laser beam to temporarily blind the pilot and copilot of a jet plane passing over his house on December 29, 2004. Banach denied any evil intent and said he was simply using the laser to point out stars for his seven-year-old daughter.

Though the airplane landed safely and without incident, and though the FBI found no terrorist connection and acknowledged Banach's actions were simply "foolhardy and negligent," Banach faced a twenty-five-year prison sentence and a $500,000 fine. Banach was eventually released from jail after posting $100,000 bail. But then in early 2006, Banach was found guilty of violating a portion of the PATRIOT Act having to do with interfering with pilots of commercial aircraft. He was given a two-year probated jail sentence.

This probation sentence may be understandable since it was learned that the government was testing a laser system in the same area as Banach when he was arrested. In early 2005, Transportation Secretary Norman Mineta announced that the U.S. military would activate the ground-based lasers to warn off pilots whenever unauthorized or unresponsive aircraft entered restricted zones in the Northeast. During testing of this system, pilots began to report incidents of lasers being shone into their cockpits. Mineta's announcement came on January 12, 2005, the same day that technical testing of the laser warning system was completed. This means the U.S. government made the public believe that terrorists were testing laser beams to bring down aircraft when, in fact, it was the government testing lasers.

Another energy device being widely used by the government is the Taser, an electroshock weapon that disrupts voluntary muscle control.

Police too often use electric Tasers in situations that don't require it, even though some 129 people nationwide have died in connection to the device. The medical examiner in Tarrant County, Texas, Dr. Nizam Peerwani, told the media he believes that Tasers are safe. But following three deaths in Fort Worth, he said he would like to see more studies done on the Taser being used on people who are high on drugs, agitated, or suffering from heart problems. Peerwani said that in at least one case, the death of seventeen-year-old Kevin Omas, who died after being Tasered three times by police, he believes the use of the Taser was a contributory factor in the death.

POLICE TACTICS AND FEMA

NATIONAL FEARMONGERING MAY HAVE something to do with the rise of a police state in certain locations. Most local police no longer wear the traditional blue uniforms, and the slogan "To Serve and Protect" has largely been eliminated from their vehicles. Today, many officers, particularly in the large urban centers, wear black and, in serious situations, don body armor and helmets based on the German World War II design.

Beginning in 2008, public fears were further heightened when stories appeared on the Internet concerning stockpiles of cheap plastic sealable coffins discovered in the country. The stories were documented with photos. One such place, reportedly containing some half a million coffins, was in middle Georgia near the town of Madison, just east of Atlanta, home of the Centers for Disease Control.

Then rumors spread about plans for roadblocks, mandatory vaccinations, and quarantine holding centers for those who resisted relocation. Apparently, some rumors were based on information about the changing role of law enforcement. Greg Evensen, a former Kansas state trooper. recalled, "Our nation's police forces prior to . . . Richard Nixon were centered on community policing. Most of their time was spent on looking for, identifying, and monitoring criminals, and responding to unusual or dangerous events that were beyond the control of ordinary folks.

"As government began its sickening expansion, policing became a meaner and nastier job. It was made that way by badge-wearing thugs who didn't hesitate to do whatever they were told by the S.A.C. (Special Agent in Charge) of the FBI, BATF(E), U.S. Marshal's Office, right down to armed poultry inspectors—yes, they have them and they are really tough on criminal chickens. The 'us against them' mentality and the 'mission essential' attitude justified SWAT teams, 'dynamic entries,' and later use of Mace, Tasers, flashbang grenades, and 'routine' use of submachine guns—all in the name of 'taking down' the accused—no matter the charge. . . . Now we have become eaves-dropping, roadblock-setting, door-crashing, face-grinding, arm-breaking, pursuit-driven bastards that have sold their asses to the government masters, hell-bent on establishing the true reincarnation of the dreaded SS. That is no overstatement. . . .

"There are significant numbers of officers at all levels that simply detest the forced training at FEMA centers, the requirements to stop Patriots and others simply because they 'look' dangerous, and are exercising free speech statements on their vehicles," Evensen added.

Evensen referenced stories from other officers who turned whistle-blowers and warned, "Have you been made aware of the massive roadblock plans to stop all travelers for a vaccine bracelet (stainless steel band with a micro-chip on board) that will force you to take [a vaccine] shot? Refuse it? You will be placed on a prison bus and taken to a quarantine camp. What will you do when your children are not allowed into school without the shot? What will you do when you are not allowed into the workplace without the vaccine paperwork? Buy groceries? Go to the bank? Shop anywhere? Get on a plane, bus or train? Use the toilet in the mall? Nope. Police officers will become loathed, feared, despised and remembered for their 'official' duties."

Certainly, most Americans, lulled by the corporate mass media, must assume Evensen's predictions are paranoid delusions. Yet these Americans should examine the evidence around them.

Though many claim fearmongering is a tactic reserved only for the Bush administration, President Obama did nothing to stop the fearmongering following his election. In fact, in March 2009, he announced America's new regional strategy in the "Afpak [Afghanistan-Pakistan] theater." Mimicking Bush administration rhetoric, Obama declared, "Multiple intelligence estimates have warned that al Qaeda is actively planning attacks on the U.S. homeland from its safe havens in Pakistan." He vowed to send an additional four thousand troops to train recruits for the Afghan National Army, saying, "I want the American people to understand that we have a clear and focused goal: to disrupt, dismantle and defeat al Qaeda in Pakistan and Afghanistan." This hawkish rhetoric was backed up in late 2009 when Obama increased troops levels in Afghanistan by thirty thousand while pledging to begin the withdrawal of U.S. forces in 2011. Little media notice was given to the fact that while attention was focused on Afghanistan and Iraq, Obama quietly was returning an estimated one million U.S. troops home to reinforce the new Northern Command (USNORTHCOM), formed under President Bush "to provide command and control of Department of Defense homeland defense efforts and to coordinate defense support of civil authorities." Was Obama anticipating civil unrest?

DESIGNATED TERRORISTS

CONCERNS OVER POLARIZING THE population were raised again in March 2009 after the release of an unclassified "reference aid" from the Department of Homeland Security's Strategic Analysis Group and the Extremism and Radicalization Branch of the DHS Environment Threat Analysis Division. This "aid" was aimed not only at department offices but also "to assist federal, state, local, and tribal homeland security and law enforcement officials in conducting analytic activities." Apparently, this means aiding lawmen in determining who in Homeland Security's opinion might be considered a terrorist. This document, entitled a "Domestic Extremism Lexicon," lists, along with animal rights and environmental extremists, Aryan prison gangs, black nationalists and neo-Nazis, Cubans "who do not recognize the legitimacy of the Communist Cuban Government," lone terrorists, Jewish extremists, the patriot and tax resistance movements, and even "alternative media," defined as "various information sources that provide a forum for interpretations of events and issues that differ radically from those presented in mass media products and outlets." Following a public outcry, the extremist list reportedly was withdrawn, although for how long remains a question. Meanwhile, copies remain on the Internet, and some law enforcement officers still recall its words.

"The federal government is training its enforcers that people who don't believe everything they see on Fox News, CNN or read in the *New York Times* are to be treated as a 'threat' and a potential violent domestic terrorist," railed Internet commentator and author Paul Joseph Watson.

A pamphlet prepared by the Texas Department of Public Safety in 2004 and entitled "Terrorism: What the Public Needs to Know" was a recipe for paranoia and witch-hunting. It includes these pointers on how to spot a terrorist:

- Will employ a variety of vehicles and communicate predominately by cell phone, e-mail, or text messaging services
- Well prepared to spend years in "sleeper" mode until it is time to attack
- In many cases, will try to fit in and not draw attention to themselves
- May appear "normal" in their appearance and behavior while portraying themselves as a tourist, student, or businessperson
- May be found traveling in a mixed group of men, women, and children of varying ages, who are unaware of their purpose
- Trained to avoid confrontations with law enforcement and therefore can be expected to project a "nice-guy" image
- Known to use disguises or undergo plastic surgery, especially when featured on police wanted posters

Another example of the emergence of a police state is the quiet but sudden appearance over the past few years of steel cables attached to metal posts in the medians on freeways in and out of major cities. When these concrete strips first appeared, many people thought they were bicycle or jogging paths. But the steel cables revealed their true purpose—a barricade to prevent anyone from making a U-turn. Such impediments have joined the thousands of concrete barriers already in place on most freeways and interstate highways.

But was there a huge problem with U-turning traffic to begin with? None that anyone could recall. Then what was the purpose of spending millions of dollars on major highways when the economy was at a low ebb?

Conspiracy-minded individuals believe that these barriers are in preparation for future roadblocks to prevent city dwellers from leaving town. Anyone caught in line for a checkpoint, similar to those already in use in Los Angeles and other major cities, will find they will be unable to turn around. No other purposes for these barricades have been publicized. Because local police do not have the personnel to administer this level of police state activity, it may be up to the military to take charge, despite the Posse Comitatus Act prohibiting such action.

Very little national media coverage was given to the heavy-handed police reaction against protesters at the late September 2009 G-20 meetings in Pittsburgh, Pennsylvania.

Frustrated at not being able to approach the meeting place, some two thousand demonstrators (called "anticapitalists" by one news report) clashed with black-clad helmeted police armed with dogs, gas, rubber pellet–filled shotgun charges called "beanbags," and advanced technologies like long-range acoustic device (LRAD) sonic cannons. Allegations that violence at the G-20 was initiated by government agent provocateurs were supported by YouTube videos of a supposed black-clad "anarchist" posing for photos with grinning police officers. Officials later claimed the youth in the video was forced to pose by the men in riot gear.

During the same G-20 protests, two hundred people were arrested and dozens of bystanders, including passing students and journalists, were gassed near the Oakland Thomas Merton Center, a city center containing several universities, museums, and hospitals, as well as an abundance of stores and restaurants. Many complained that the crisis was akin to a military-style occupation.

"The police were beating people and gassing people who were wandering out of restaurants . . . wandering out of their dorms," said Nigel Parry, a journalist with Twin Cities Indymedia. Another journalist, Melissa Hall, said police erased her video footage and damaged her camera.

"This was unjust," complained twenty-three-year-old Nathan Lanzendorfer. "I was peaceful. I had done nothing wrong." Lanzendorfer showed newsmen large purple spots on his legs and one arm where he said police shot him at close range with beanbag rounds.

Elizabeth Pittinger, executive director of the city's Citizen Police Review Board, said her group had received fifty complaints about the police and that the board would conduct a comprehensive investigation of the police response. She was "very disturbed" over the arrest of journalists, including *Pittsburgh Post-Gazette* reporter Sadie Gurman.

The country's police-state mentality has trickled down to those who do not even hold status as actual police officers and at times even jeopardizes the public's basic rights as guaranteed by the Constitution. For example, a video taken at an August 2009 town hall meeting in Reston, Virginia, and placed on the Internet shows an unnamed man being ordered to lower a sign depicting President Barack Obama with a clown face, presumably to make him look like the Joker, a character from the popular Batman franchise. Wesley Cheeks Jr., a school security officer, ordered the man to lower his sign. When the demonstrator argued that he was only exercising his constitutional rights, Cheeks threatened him with arrest.

"This used to be America," the man groused.

Typifying the change in the attitude of police, who once considered themselves servants of the people, the officer responded, "Well, it ain't no more, okay!"

OBAMA'S SCHOOL TALK

When kindergartners in B. Bernice Young School sang a medley of two short songs praising the president in February 2009, alarm bells went off among many who are concerned about the globalists' control of media over what they perceived as undue worship of a public leader. Then, in October of that year, President Obama addressed the nation's schoolchildren directly. Many school districts declined to broadcast President Obama's remarks to America's youth because they felt the president intended to use his office to politicize public school classrooms with "training materials."

The training materials produced by the U.S. Department of Education's Teaching Ambassador Fellows and handed out to prekindergarteners through twelfth graders stated:

"Before the Speech: Teachers can build background knowledge about the President of the United States and his speech by reading books about presidents and Barack Obama and motivate students by asking the following questions:

Who is the President of the United States?
What do you think it takes to be President?
To whom do you think the President is going to be speaking?
Why do you think he wants to speak to you?
What do you think he will say to you?"

Other topics of discussion in these materials included: "Why is it important that we listen to the President and other elected officials, like the mayor, senators, members of congress, or the governor? Why is what they say important?"

Republicans argued over what they saw as political propaganda in the talk's preparatory materials, such as the brochure's suggestion for students to "Write letters to themselves about what they can do to help the president." In this case, wording was changed to "Write letters to themselves about how they can achieve their short-term and long-term education goals."

"We changed it to clarify the language so the intent is clear," explained White House spokesman Tommy Vietor.

Nevertheless, a number of parents objected to a president having access to all schoolchildren. Regine Gordon of Tampa, Florida, and mother of a six-year-old student, told newsmen, "It's a form of indoctrination, and I think, really, it's indicative of the culture that the Obama administration is trying to create. It's very socialistic. . . . It's kind of like going through the children to get to their parents. Children are very vulnerable and excited. I mean, this is the president. I think it's an underhanded tactic and indicative of the way things are being done."

One Texas school district declined to make listening to Obama's school talk mandatory. The superintendent sent a memo to all teachers explaining:

"The decision not to require all students Pre-K to grade 12 to watch the speech together in school was based in the following concerns:

1. The Federal and State School Accountability systems are so demanding that it is difficult to defend stopping instruction for any reason. School districts, campuses, and now teachers are being compared based on student performance. The changing system and increasing demand for results requires everyone to focus on the instructional mission at all times.

2. It is difficult to comprehend that anyone could make a single presentation that was equally meaning for to Pre-K students and Seniors in High School.

3. The timing of a 'first day' speech two and a half weeks into the school year is less than ideal.

4. The time of day the speech was scheduled creates a number of potential scheduling issues for each campus. These challenges would have ultimately cost the district a full day of instruction.

5. School districts have not stopped instruction at all grade levels to watch a speech in the past.

6. Communication about the speech and lead time provided to school districts was less than ideal.

7. It is difficult to show anything to a classroom of students that you have not previewed prior to the demonstration. A one time speech is clearly different than a ceremonial event like an Inauguration.

8. The speech has been presented as an optional and not required event."

But other school districts were not as lenient as the one in Texas. The superintendent of School District No. 3 in Tempe, Arizona, Dr. Arthur Tate Jr., stated parents would not be permitted to pull their children out of class during Obama's speech. "I have directed principals to have students and teachers view the president's message on Tuesday," stated Tate. "In some cases, where technology will not permit access to the White House Web site, DVDs will be provided to classes on subsequent days. I am not permitting parents to opt out students from viewing the president's message, since this is a purely educational event."

The fact that a president, and one who was surrounded by so much controversy, would address all of the nation's schoolchildren renewed the concerns of many Americans over the security of American values. For example, many saw Obama's nationwide talk to children and his call for involuntary service as ominous signs of indoctrinating the youth. Many even compared all this to the Hitler Youth movement practiced in Nazi Germany.

The controversy over Obama's talk to schoolchildren renewed the earlier criticism over songs about the president being taught to the B. Bernice Young School kindergarten children. As news about the song gained nationwide attention, the New Jersey Department of Education issued a mild rebuke. Spokeswoman Beth Auerswald said the department desired "to ensure students can celebrate the achievements of African Americans during Black History Month without inappropriate partisan politics in the classroom." The teacher involved retired.

LEADER CONTROL

You cannot help the poor by destroying the rich. You cannot strengthen the weak by weakening the strong. You cannot bring about prosperity by discouraging thrift. You cannot lift the wage earner up by pulling the wage payer down. You cannot further the brotherhood of man by inciting class hatred. You cannot build character and courage by taking away people's initiative and independence. You cannot help people permanently by doing for them, what they could and should do for themselves.

—WILLIAM J. H. BOETCKER, Presbyterian minister

(often erroneously attributed to Abraham Lincoln)

EARLY IN HIS PRESIDENCY, Barack Obama found himself caught in a tangle of misstatements and reversals of promises. His promise for a "change" in politics rang hollow, and many of his former supporters felt betrayed.

Obama was caught in one falsehood after another, backing down on his campaign promise to dismantle the U.S. missile defense system. Then, in late 2009, he reversed his promised reversal, putting the system back on track.

As mentioned earlier in "Manufactured AIDs," in a statement revealing the ongoing reach of the globalists, Obama's national security adviser, General James L. Jones, told attendees at the Munich Security Conference, "As the most recent national security adviser of the United States, I take my daily orders from Dr. Kissinger, filtered down through General Brent Scowcroft and Sandy Berger, who is also here. We have a chain of command in the National Security Council that exists today." Kissinger is one man widely viewed as the architect of a U.S. foreign policy that has turned foreign extremists into implacable enemies.

Strangely enough, Obama was selected to receive the Nobel Peace Prize in October 2009. The corporate mass media, which thirsted for good news during a bad economy, publicized the announcement widely.

But as Nancy Gibbs, writing for *Time* magazine observed, "The last thing Barack Obama needed at this moment in his presidency and our politics is a prize for a promise." She noted that "when reality bites, it chomps down hard." Because none of Obama's political goals had been accomplished a year after his election, Gibbs said "a prize for even dreaming them can feed the illusion that they have."

Gibbs compared Obama's failed promises with the accomplishments of another Nobel Prize candidate, Greg Mortenson. The son of a missionary, Mortenson is a former Montana mountaineer and was nominated by the U.S. Congress for his humanitarian work in building 130 schools for girls in Muslim countries hostile to education for women. "Sometimes the words come first. Sometimes, it's better to let actions speak for themselves," mused Gibbs.

The prize especially angered those who saw Barack Obama as a "peace candidate" when he only increased spending in Iraq after being elected. In late 2009, the Obama administration announced plans to send an additional thirty thousand troops to Afghanistan. In fact, Obama's total military expenditures have grown at a greater rate than those under George W. Bush, almost universally considered a "hawk" president.

Obama's reversal of his campaign pledge to dismantle the U.S. missile defense system meant that the United States would continue to aim nuclear missiles at Moscow from Poland, Ukraine, and perhaps Georgia. It should be noted that the Obama administration waffled several times over the controversial missile system. In 2008, the Obama campaign stated its opposition to a program that many saw as merely a continuation of the old "Star Wars" program of the Reagan years. But in early 2009, Vice President Joe Biden told a European audience the United States would continue the missile defense system after consulting with NATO countries and assuring Russia the weapons are only meant for Iran. Russia has long perceived the program as a threat to its security. Later that year, both Secretary of State Hillary Clinton and President Obama backpedaled in an attempt to use the missile defense system as an incentive for Iran to discontinue its nuclear program. They suggested that the missile system might be shelved depending on Iran's actions.

But by the fall of 2009, it seemed as if the missile defense system was doomed. Despite news that a stockpile of nuclear weapons had been discovered in Iran, on September 17, the White House announced it was scrapping the strategic missile defense system in favor of smaller SM-3 interceptors capable of intercepting Iranian missiles. According to the *New York Times,* the decision was "one of the biggest national security reversals of [Obama's] young presidency."

The reversal created both confusion among eastern European allies and anger from conservatives who accused Obama of caving in to objections from the Russians.

Typically, Obama appeared to be trying to placate both sides, saying, "President Bush was right that Iran's ballistic missile program poses a significant threat. This new approach will provide capabilities sooner, build on proven systems and offer greater defenses against the threat of missile attack than the 2007 European missile defense program."

The brouhaha over the missile defense system presents a small glimpse into the power struggles taking place between globalists eager for higher defense budgets, and hence more defense profits, and those who are trying to push their one-world socialist agenda by merging the nations.

A COUNCIL CABINET

TO ANALYZE THE TRUTH about the Obama administration, it is instructive to look behind his rhetoric at the individuals and groups that shape his administration. Many believe that Obama's cabinet members have ties to secretive societies that may have orchestrated the present economic recession.

People in Obama's cabinet who are members of the Council on Foreign Relations include: Defense Secretary Robert Gates, Homeland Security Secretary Janet Napolitano, Commerce Secretary Bill Richardson, UN ambassador Susan Rice, national security adviser General James L. Jones, Treasury Secretary Timothy Geithner, economic adviser Paul Volcker, and director of the National Economic Council L. H. Summers.

Many of these same names plus others appear on the roster of the Trilateral Commission, cofounded in 1973 by Zbigniew Brzezinski and David Rockefeller Sr. Alan Greenspan and Paul Volcker, both of whom went on to head the Federal Reserve System, also were founding members of the commission. Brzezinski, a former chairman of the commission, was Obama's principal foreign policy adviser during the 2008 campaign, and it was generally accepted that Trilateral members groomed Obama for office. The Trilateral Commission is generally considered a spin-off organization of the Council on Foreign Relations that was designed to include Asian nations.

Additional Trilateralists in the Obama administration include: Tim Geithner; Susan Rice; the deputy national security adviser, Thomas Donilon; the director of national intelligence, Admiral Dennis C. Blair; the assistant secretary of state for Asia and the Pacific, Kurt M. Campbell; the deputy secretary of state, James Steinberg; and State Department special envoys Richard Haass, Dennis Ross, and Richard Holbrooke.

"According to official Trilateral Commission membership lists, there are only 87 members from the United States (the other 337 members are from other countries). Thus, within two weeks of his inauguration, Obama's appointments encompassed more than 12 percent of [the] Commission's entire U.S. membership," noted researchers at Project Censored.

Patrick Wood, author of the Project Censored paper "Obama: Trilateral Commission Endgame," stated, "The concept of 'undue influence' comes to mind when considering the number of Trilateral Commission members in the Obama administration. They control the areas of our most urgent national needs: financial and economic crisis, national security, and foreign policy. The conflict of interest is glaring. With 75 percent of the Trilateral membership consisting of non-U.S. individuals, what influence does this super-majority have on the remaining 25 percent? For example, when Chrysler entered bankruptcy under the oversight and control of the Obama administration, it was quickly decided that the Italian carmaker Fiat would take over Chrysler. The deal's point man, Treasury Secretary Timothy Geithner, is a member of the Trilateral Commission. Would you be surprised to know that the chairman of Fiat, Luca di Montezemolo, is also a fellow member? Congress should have halted this deal the moment it was suggested."

In his book *With No Apologies,* the late senator Barry Goldwater described the commission with distrust: "In my view, the Trilateral Commission represents a skillful, coordinated effort to seize control and consolidate the four centers of power: political, monetary, intellectual, and ecclesiastical. All this is to be done in the interest of creating a more peaceful, more productive world community. What the Trilateralists truly intend is the creation of a worldwide economic power superior to the political governments of the nation-states involved. They believe the abundant materialism they propose to create will overwhelm existing differences. As managers and creators of the system they will rule the future." Trilateralists and others within globalist societies have displayed very little concern for the United States as a sovereign nation. Their policies often run counter to the best interests of the United States and, in fact, appear to support the allegation that they seek a one-world government.

A COMFORTABLE STAFF

IT MAY BE THAT President Obama is connected to too many of the nation's elite globalists to affect policy in any one direction without appearing hypocritical. He also came under fire for perceived conflicts of interest, such as the time he golfed with Robert Wolf, president of UBS Investment Bank and the chairman and CEO of UBS Group Americas. Wolf had been an early supporter of Obama and had raised $250,000 for his campaign back in 2006. Moreover, according to ProPublica reporter Sharona Coutts and Stephen Kohn, executive director of the National Whistleblowers Center, their golf game took place only three days after UBS had reached an agreement with the IRS to give up the names of forty-five hundred American clients suspected of hiding billions in secret Swiss bank accounts. This was done in an effort to avoid criminal charges. UBS already had admitted to aiding the clients in tax evasion and agreed to pay a $780 million fine. In February 2009, Wolf was appointed to the President's Economic Recovery Advisory Board, tasked with trying to fix America's new depression.

What is perhaps both ironic and tragic about the UBS situation is that the man who had been the whistle-blower on UBS's shady dealings was sentenced to a forty-month prison sentence. As a UBS banker involved with the Swiss accounts, Bradley Birkenfeld had initiated the investigation. It was Birkenfeld's disclosures and cooperation with authorities that provided inside information into the bank's conduct and allowed the government to gain $780 billion in fines and stop a massive tax-evasion scheme. Yet, for his trouble, Birkenfeld was given a harsher sentence by Fort Lauderdale federal judge William Zloch than was asked for by the case's prosecutors.

Stephen Kohn pointed out that while the government announced it would investigate the forty-five hundred names handed over by UBS, Birkenfeld had actually turned in fifty-two thousand account identifications. "But the whole case is puzzling," Kohn added, "because if the United States wants to crack tax fraud, if they want to crack money laundering or stop these practices in secret banks, why are they putting into prison for forty months the whistleblower? The billionaire to whom he was serving got probation. How are they ever going to get another banker to step forward and cooperate with an investigation?"

But it was not just Obama's corporate buddies that raised the ire of many citizens. His wife took flak as more outrage resulted from the news of First Lady Michelle Obama's well-paid staff.

"No, Michele Obama does not get paid to serve as the First Lady and she doesn't perform any official duties. But this hasn't deterred her from hiring an unprecedented number of staffers to cater to her every whim and to satisfy her every request in the midst of the Great Recession," wrote Dr. Paul L. Williams in the *Canada Free Press*. "Just think Mary Lincoln was taken to task for purchasing china for the White House during the Civil War. And Mamie Eisenhower had to shell out the salary for her personal secretary."

Williams quoted Mrs. Obama's statement from the Democratic National Convention—"In my own life, in my own small way, I have tried to give back to this country that has given me so much. See, that's why I left a job at a big law firm for a career in public service." Paul Williams then lamented, "If you're one of the tens of millions of Americans facing certain destitution, earning less than subsistence wages stocking the shelves at Wal-Mart or serving up McDonald cheeseburgers, prepare to scream and then come to realize that the benefit package for these servants of Miz Michelle are the same as members of the national security and defense departments and the bill for these assorted lackeys is paid by John Q. Public."

The listing of White House staffers assigned to the first lady and their salaries made ripples through the Internet. They included:

Susan Sher (Chief of Staff)—$172,200

Jocelyn C. Frye (Deputy Assistant to the President and Director of Policy and Projects for the First Lady)—$140,000

Desiree G. Rogers (Special Assistant to the President and White House Social Secretary)—$113,000

Camille Y. Johnston (Special Assistant to the President and Director of Communications for the First Lady)—$102,000

Melissa E. Winter (Special Assistant to the President and Deputy Chief of Staff to the First Lady)—$100,000

David S. Medina (Deputy Chief of Staff to the First Lady)—$90,000

Catherine M. Lelyveld (Director and Press Secretary to the First Lady)—$84,000

Frances M. Starkey (Director of Scheduling and Advance for the First Lady)—$75,000

Trooper Sanders (Deputy Director of Policy and Projects for the First Lady)—$70,000

Erinn J. Burnough (Deputy Director and Deputy Social Secretary)—$65,000

Joseph B. Reinstein (Deputy Director and Deputy Social Secretary)—$64,000

Jennifer R. Goodman (Deputy Director of Scheduling and Events Coordinator for the First Lady)—$62,000

Alan O. Fitts (Deputy Director of Advance and Trip Director for the First Lady)—$60,000

Dana M. Lewis (Special Assistant and Personal Aide to the First Lady)—$57,500

Semonti M. Mustaphi (Associate Director and Deputy Press Secretary to the First Lady)—$52,500

Kristen E. Jarvis (Special Assistant for Scheduling and Traveling Aide to the First Lady)—$50,000

Tyler A. Lechtenberg (Associate Director of Correspondence for the First Lady)—$45,000

Samantha Tubman (Deputy Associate Director, Social Office)— $43,000

Joseph J. Boswell (Executive Assistant to the Chief of Staff to the First Lady)—$40,000

Sally M. Armbruster (Staff Assistant to the Social Secretary)— $36,000

Natalie Bookey (Staff Assistant)—$35,000

Deilia A. Jackson (Deputy Associate Director of Correspondence for the First Lady)—$35,000

Williams noted this list did not include makeup artist Ingrid Grimes-Miles, age forty-nine, and "First Hairstylist" Johnny Wright, thirty-one, both of whom traveled to Europe with the Obamas aboard Air Force One.

While to many the first lady's staff seems excessively large, "Michelle Obama's staff is not 'unprecedented,' but rather on a par with her predecessor's," according to the Internet fact-checking site, snopes.com. Though Michelle Obama has more than twenty staffers working for her, the 2008 White House staff list included sixteen names working for First Lady Laura Bush.

HELPING HAMAS TERRORISTS

MANY BELIEVED OBAMA'S POLICY of unrestrained illegal immigration would destroy the last vestiges of national cohesion, especially when he issued Presidential Determination No. 2009-15, entitled "Unexpected Urgent Refugee and Migration Needs Related to Gaza." In early 2009, alternative news outlets reported that Obama actually was aiding immigration through organizations formerly linked to terrorism, such as the Islamic Resistance Movement (Hamas).

In Presidential Determination No. 2009-15, sent to Secretary of State Hillary Clinton in January 2009, Obama claimed authority under the Migration and Refugee Assistance Act of 1962 to spend up to $20.3 million from the United States Emergency Refugee and Migration Assistance Fund for assistance in relocating Palestinian refugees and conflict victims in Gaza to the United States. Though Obama's action received little notice in the corporate mass media, it nevertheless stirred up both the anti-Obama crowd as well as the pro-Israel lobby.

Such actions play into the hands of those who decry Obama as pro-Muslim. They remember the words of Libya's Mu'ammar Gadhafi, who, in 2008, said, "There are elections in America now. Along came a black citizen of Kenyan African origins, a Muslim, who had studied in an Islamic school in Indonesia. His name is Obama. All the people in the Arab and Islamic world and in Africa applauded this man. They welcomed him and prayed for him and for his success, and they may have been involved in legitimate contribution campaigns to enable him to win the American presidency."

OBAMA'S "CIVILIAN ARMY"

ANY VIABLE SOCIALIST SYSTEM has needed to indoctrinate its citizenry and use government snitches to gather intelligence. Under President Obama, the United States is no different. President Obama and his chief of staff, Rahm Emanuel, have tried to create a seven-million-person "civilian army" reportable to only the president. In July 2008, Obama stated, "We cannot continue to rely only on our military in order to achieve the national security objectives that we've set. We've got to have a civilian national security force that's just as powerful, just as strong, just as well-funded."

In 2008, Obama's official campaign website, Change.gov, announced that President-elect Obama would expand national service programs like AmeriCorps and the Peace Corps and create a new Classroom Corps to help teachers in low-income schools. Additionally, Obama planned to create a new Health Corps, Clean Energy Corps, and Veterans Corps for his civilian army.

Originally, Change.gov stated, "Obama will call on citizens of all ages to serve America by developing a plan to *require* [emphasis added] 50 hours of community service in middle school and high school and 100 hours of community service in college every year."

Following a barrage of criticism over drafting children into Obama's "national security force," the website's wording was softened to read, "Obama will call on citizens of all ages to serve America by setting a goal that all middle school and high school students do 50 hours of community service a year and be developing a plan so that all college students who conduct 100 hours of community service receive a universal and fully refundable tax credit ensuring that the first $4,000 of their college education is completely free."

Some critics, such as those writing for WorldNetDaily.com and *Modern Conservative,* found Obama's "civilian security force" reminiscent of Hitler's brownshirts, the Hitler Youth, and youth brigades in Russia and other Communist countries. Scarier words came from Obama's chief of staff, Rahm Emanuel. In his 2006 book *The Plan: Big Ideas for America,* Emanuel wrote, "It's time for a real Patriot Act that brings out the patriot in all of us. We propose universal civilian service for every young American. Under this plan, all Americans between the ages of eighteen and twenty-five will be asked to serve their country by going through three months of basic training, civil defense preparation, and community service."

Due to public resistance, such a plan for universal civilian service languished until passage of Obama's health-care plan, euphemistically called the Patient Protection Affordable Care Act, on March 23, 2010. The Senate version, adopted by the House, included a provision for establishing both a National Health Service Corps as well as a Ready Reserve appointed by the president. The purpose of the Ready Reserve was to meet a perceived need for "additional Commissioned Corps personnel available on short notice (similar to the uniformed service's reserve program) to assist regular Commissioned Corps personnel to meet both routine public health and emergency response missions." Is this Obama's "civilian army"?

ACORN AND SEIU

ZOMBIES LIKE TO FLOCK together in packs or large crowds under the control of a master. In the horror movies, one rarely saw a zombie roaming about alone. Today, American zombies prefer to gather in large groups and they too are likely to follow some master, whether it be a gang boss, guru, or political chieftain.

Political leaders do not work alone. Rather, they are surrounded by staffs, mentors, guides, advisers, and other string pullers, and they have minions working for them in the form of nonprofit organizations. Liberals during the 2000 presidential election accused Bush's Republican supporters of Nazi-style Stormtrooper tactics when they surrounded and occupied a building conducting a recount vote in Palm Beach County, Florida.

But those who cause problems haven't always sat on the right side of the political fence. Another example of Stormtrooper tactics can be seen in an attack by a Service Employees International Union (SEIU) member on Kenneth Gladney, a young black man who was trying to distribute small flags with a Revolutionary War motto on them at a St. Louis Tea Party meeting. According to Gladney's attorney, David B. Brown, "Kenneth was approached by an SEIU representative as Kenneth was handing out 'Don't Tread on Me' flags to other conservatives. . . . The SEIU representative demanded to know why a black man was handing out these flags. The SEIU member used a racial slur against Kenneth, [and] then punched him in the face. Kenneth fell to the ground. Another SEIU member yelled racial epithets at Kenneth as he kicked him in the head and back. Kenneth was also brutally attacked by one other male SEIU member and an unidentified woman."

Also consider the recent activities of ACORN (the Association of Community Organizations for Reform Now), SEIU, and the political action group MoveOn. In 2009, the Census Bureau enlisted ACORN to promote the 2010 census, already under criticism for marking individual homes with global positioning system (GPS) satellite coordinates. Census director Robert Groves, in a letter to ACORN president Maude Hurd, stated that the community organization had become a distraction. ACORN national deputy director, Brian Kettenring, said, "We will continue to do what we've said we'll do, which is encourage people in communities to participate fully in the census." On September 11, 2009, the U.S. Census Bureau discontinued its partnership with ACORN in the 2010 census after hidden-camera videos recorded four ACORN employees giving tax advice on running a brothel.

Furor arose after a young activist named James O'Keefe posed as an aspiring politician and told ACORN workers he planned to use his girlfriend's prostitution income to fund future campaigns. He claimed to want advice on how to obtain a loan for a house from which to conduct her profession. In the videos, ACORN workers gave instructions for creating a company in order to prove they had enough income for a house that could serve as a brothel and home for teenage prostitutes.

After secretly filming at the ACORN's Baltimore office, and later in Washington, D.C., the activists posted the videos on YouTube. ACORN officials said the four offending workers were fired. Yet despite the terminations and assurances from the president and the executive director of ACORN Housing that ACORN Housing staff members are expected to behave ethically and comply with the law, a growing number of Republican politicians began calling for congressional hearings and IRS audits of ACORN.

After the ACORN videos were released on YouTube, the Senate voted 83–7 to block the Housing and Urban Development Department (HUD) from giving grants to ACORN, which ended housing and community funding to the organization. This meant that ACORN would not receive grants for programs such as counseling low-income people on how to get mortgages and for fair housing education and outreach.

Only three days after the Senate stopped ACORN funds, the House followed suit, voting 345–75 (the "nays" being all Democrats) to deny the organization all federal funds. "ACORN has violated serious federal laws, and today the House voted to ensure that taxpayer dollars would no longer be used to fund this corrupt organization," remarked Republican representative Eric Cantor of Virginia.

Most caring Americans agree that although it is noble to try to help the poorer segments of our society, it is counterproductive, even outrageous, to support groups with taxpayer money who commit corrupt and unlawful acts as well as preach hatred of the United States, causing further division within the nation. This division plays right into the hands of the globalist fascists who, through their control of the mass media and party politics, placed Obama in office and were using his supporters to advance their socialist agenda.

TIPS AND OTHER SNOOPS

OBAMA'S CIVILIAN ARMY SCHEME is only the latest in attempts to recruit Americans to spy on their fellow Americans. In midsummer 2002, a program called TIPS (Terrorism Information and Prevention System) was launched by President George W. Bush. The program was part of a larger program called the "Citizen Corps," which was a program first created by President Bush to mobilize the nation's citizenry against national security threats. On its website, TIPS describes itself as "a national system for concerned workers to report suspicious activity." In published material, TIPS advocates said the program was to be administered by the Justice Department, coordinated by FEMA, and operated under the Homeland Security Department. It would involve "millions of American workers who, in the daily course of their work, are in a unique position to see potentially unusual or suspicious activity in public places." This, of course, referred to postal carriers, meter readers, repair personnel, or anyone who might have an ax to grind against their neighbors. The program was quickly dropped, however, following public outrage, and after the U.S. Postal Service stated that it would not participate in the snitch program.

The U.S. Postal Service stated it had "been approached by Homeland Security regarding Operation TIPS; however, it was decided that the Postal Service and its letter carriers would not be participating in the program at this time." Nothing was mentioned about whether or not individual carriers could join on their own or if "at this time" left open the possibility that the postal service may participate in TIPS in the future. Despite the postal service's public reticence, some researchers believe that postal service employees may still be reporting suspicious behavior. It is just not done officially.

Other critics immediately compared the TIPS plan to the Nazi Gestapo, the former East German secret police service, and to Fidel Castro's Committee for the Defense of the Revolution (CDR), an organization established by Castro on September 28, 1960. With the CDR, Cubans are encouraged to spy on and report any "counterrevolutionary" behavior by their neighbors. An estimated eight million Cubans belong to more than 121,000 committees in the CDR system.

The American Civil Liberties Union (ACLU) and other public watchdog organizations reacted negatively to Operation TIPS, saying it would create an atmosphere in which Americans would be spying on one another. "The administration apparently wants to implement a program that will turn local cable or gas or electrical technicians into government-sanctioned Peeping Toms," declared ACLU legislative counsel Rachel King. Of TIPS, Rutherford Institute executive director John Whitehead said, "This is George Orwell's '1984.' It is an absolutely horrible and very dangerous idea. It's making Americans into government snoops. President Bush wants the average American to do what the FBI should be doing. In the end, though, nothing is going to prevent terrorists from crashing airplanes into buildings."

Even former Homeland Security chief Tom Ridge was forced to backpedal over the TIPS organization, saying, "The last thing we want is Americans spying on Americans."

Although Ridge still vouched for the TIPS program, the Citizen Corps softened both its language and details about the program after it began to make a national stir.

In July 2002, the website stated Operation TIPS "will be a nationwide program giving millions of American truckers, letter carriers, train conductors, ship captains, utility employees and others a formal way to report suspicious terrorist activity. Operation TIPS, a project of the U.S. Department of Justice, will begin as a pilot program in 10 cities.... Operation TIPS, involving 1 million workers in the pilot stage, will be a national reporting system that allows workers, whose routines make them well-positioned to recognize unusual events, to report suspicious activity. . . . Everywhere in America, a concerned worker can call a toll-free number and be connected directly to a hotline routing calls to the proper law enforcement agency or other responder organizations when appropriate."

In an Orwellian act of word changing, by early August 2002, the list of occupations that would participate in TIPS was dropped and the words "suspicious terrorist activity" and "unusual events . . . suspicious activity" were changed to "suspicious and potentially terrorist-related activity" and "Potentially unusual or suspicious activity in public places."

The TIPS program was merely an official way for Americans to snoop, and further what author Jim Redden called modern society's "snitch culture." From the schoolkid Drug Abuse Resistance Education (DARE) program to professional finger pointers such as the Southern Poverty Law Center (SPLC), more and more Americans were being encouraged to spy and report on one another. It's one thing to keep an eye out for strangers in the neighborhood and quite another to constantly snoop on the activities of neighbors.

Many people believe that neighborhood snooping went out with Bush-era fearmongering. However, these people should know that the Major Cities Chiefs Association, which includes police chiefs from sixty-three of the largest departments in the United States and Canada, endorsed a program called iWATCH during an annual conference in Denver on October 3, 2009. Los Angeles police chief William Bratton, whose department developed the iWATCH program, called it "the 21st century version of Neighborhood Watch." The program's watchword is "If you see something, say something."

As a policy counsel for the ACLU and a former FBI agent who worked terrorism cases, Mike German was unenthusiastic about iWATCH despite assurances that the program would not infringe on individual liberties. German told the Associated Press he suspects people will fall back on personal biases and stereotypes of what they think a terrorist should look like when deciding to report someone to the police. He said, "That just plays into the negative elements of society and doesn't really help the situation."

There have been many cases where innocent people have had their lives unsettled, ruined, or even lost due to egregious snitching. Although these stories are usually not played up in the corporate mass media, the purchase of "snitch" information continues to be a mainstay of federal law enforcement. In 1994, the DEA spent $31.7 million and Customs spent $16.5 million to pay thousands of informants.

Although accurate numbers are hard to come by, former Miami police supervisor and DEA special agent Dennis G. Fitzgerald's book *Informants and Undercover Investigations* reported that a 2005 inspector general's report revealed the DEA has about four thousand "confidential sources" at hand on any given day. They may be paid up to $100,000 a year for their information, although their paycheck must be approved by DEA headquarters.

The FBI can pay up to $25,000 to informants for information on serious crimes. Under a program called "Rewards for Justice," both the U.S. State and Treasury departments can offer money to informants for information leading to the arrest and conviction of any terrorist or terrorist group. By September 2005, more than $50 million had been paid out from this fund. One can only imagine how the lure of $100,000 to a million dollars simply to find some sort of terrorist activity could highly induce a greedy person to make false claims.

ASSET FORFEITURE FUND

MUCH OF THE MONEY paid to government informants comes from the Asset Forfeiture Fund (AFF), a controversial program that confiscates real assets, such as property, homes, cars, aircraft, boats, jewelry, financial instruments, and even whole businesses, from those convicted of a crime. In 2005 alone, $614.5 million worth of assets were deposited into the AFF.

Under the AFF, there are two types of forfeiture—criminal and civil. In a criminal forfeiture case, the defendant is always innocent until proven guilty, and it is the responsibility of the prosecution to prove that the defendant should forfeit his or her property. In civil forfeiture proceedings, the court presumes the defendant is guilty and the property owner has the burden of proving the property was not involved in any wrongdoing. Critics have complained that the AFF can seize an innocent person's property even if it is used by someone else to commit a crime without the owner's permission or knowledge. Forfeiture laws have now been used by local law enforcement in connection with local issues, including unsafe housing, prostitution, and even drunk driving.

The AFF's property seizures prompted chairman of the House Judiciary Committee Henry Hyde to state, "They don't have to convict you. They don't even have to charge you with a crime. But they have your property." Bob Barr, a Libertarian Party presidential nominee in 2008, noted, "In many jurisdictions, it has become a monetary tail wagging the law enforcement dog." The practice of confiscating property has prompted protests from civil libertarians and attorneys, but in today's fearful society, the protests have not worked.

CHEMTRAILS

IN APRIL 2009, PRESIDENT Obama's science adviser John P. Holdren stated publicly that the federal government was going to increase geoengineering, perhaps to include spraying "pollutants" into the upper atmosphere to retard global warming. Some suspicious researchers saw this as the first public acknowledgment of what many people believe is a controversial aerial spraying program known as Chemtrails. Holdren gave no details on what the "pollutant" particles might be, nor did he explain how they would be dispersed into the sky. Some scientists, including the late Dr. Edward Teller, the "father of the hydrogen bomb," a founder of the "Star Wars" missile defense system, and the inspiration for the character Dr. Strangelove in Stanley Kubrick's 1964 film of that name, have proposed using balloons or military aircraft to seed the sky with millions of tons of sulfur or heavy metals to create a cloud cover to deflect sun rays and prevent further heating of the Earth. Some scientists warned such a program would turn blue skies milky white and perhaps cause droughts and further ozone depletion.

Holdren conceded the possibility of "grave side effects," but said, "We might get desperate enough to want to use it."

Many researchers and bloggers believed that Holdren's proposed program has been secretly under way since about 1997. "Reports of chemtrails, jet plumes emitted from planes that hang in the air for hours and do not dissipate like condensation trails, often blanketing the sky in criss-cross patterns, have increased dramatically over the last 10 years. Many have speculated that they are part of a government program to alter climate, inoculate humans against certain pathogens, or even to toxify humans as part of a population reduction agenda," stated one such article entitled "Weather Wars and Chemtrails." "The project is closely tied to an idea by Nobel Prize winner Paul Crutzen, who 'proposed sending aircraft 747s to dump huge quantities of sulfur particles into the far-reaches of the stratosphere to cool down the atmosphere.' Such programs merely scratch the surface of what is likely to be a gargantuan and overarching black-budget funded project to geo-engineer the planet, with little or no care for the unknown environmental consequences this could engender."

Obama augmented this potential public hazard by appointing people to the Department of Agriculture who were fully aligned with genetic engineering, the use of fluoride, and the irradiation of food. As usual, this was done in the name of the public health.

GLOBAL SWARMING

ADDING TO CONTROVERSIES OVER the environment was the fact that the scientific community hasn't even reached a consensus about whether or not the planet is warming. As it stands, scientists may be arguing forever. Contradicting reports further complicated the debate. In the spring of 2009, news headlines proclaimed, "Antarctic Ice Melting Faster Than Expected." Only one week later, there were headlines reading, "Antarctic Ice Spreading." Both events were blamed on an increase in man-made gases. Clearly something is happening to the planet, because data continue to indicate that something is quickly changing the planet's Arctic regions. The latest satellite observations of sea ice in the Artic show the ice cover appears to be shrinking and the ice cap is getting smaller, and thinner as well. The ice has been receding more in the summers and not growing back to its previous size and thickness during the winters.

But what if it's the case that politics, and not solid science, is behind the endless global warming debate? The debate intensified in 2006 with the release of former vice president Al Gore's documentary *An Inconvenient Truth.* Since its release Gore has traveled the world forcefully arguing that human-generated carbon emissions from automobiles, aircrafts, and factories are the cause of the earth's warming trend.

But Gore's critics have pointed out that huge profits can be made from being a global warming alarmist and that Gore heads a firm that is in line to reap the benefits of Gore's eco-conservatism. According to online reports, Gore helped found Generation Investment Management, which invests in solar, wind, and other projects that reduce energy consumption around the globe. Critics claim that as chairman of the firm, Gore stands to profit handsomely from his global environmental crusade.

With great irony, the Tennessee Center for Policy Research (TCPR) criticized Gore by pointing out that Gore's Tennessee home used "Twenty times as much electricity as the average household nationwide." Gore supporters pointed out that Gore's "home" of twenty rooms needed special security measures and that the Gore family bought energy produced from renewable sources, such as wind and solar. However, the TCPR reported that, according to its analysis, the Gores still consumed 10 percent more yearly energy than before their home was equipped with eco-friendly energy-saving devices. Despite news reports on his home as an "energy hog" by Fox News and *BusinessWeek,* a Gore spokeswoman defended his lifestyle by countering that his investments in renewable energy compensated for his power consumption.

When Texas Republican representative Joe Barton tried to question Al Gore's carbon emissions statistics during testimony before the House Energy and Commerce Committee, the former vice president compared scientists who question global warming to indicted stock swindler Bernie Madoff.

"It is important to look at sources of science you rely on," Gore told Barton. "With all due respect, I believe you have relied on people you have trusted who have given you bad information. I don't blame the investors who trusted Bernie Madoff but he gave them bad information."

Confusion over the global warming issue's legitimacy continued with the news that, in August 2009, the University of East Anglia's Climate Research Unit (CRU) admitted it had destroyed the raw data for its global surface temperature research because of an alleged lack of storage space. CRU data served as the foundation for several major international studies claiming that the earth's global warming crisis is a real issue. These studies were used to gain support for the cap-and-trade legislation passed by Congress in mid-2009, officially called the American Clean Air and Security Act. "CRU's destruction of data, however, severely undercuts the credibility of those studies," stated a news release from the Competitive Enterprise Institute (CEI), a Washington-based public interest group dedicated to free enterprise and limited government.

In October 2009, the advocacy group petitioned the Environmental Protection Agency (EPA) to review its policies based on the CRU research. CEI general counsel Sam Kazman said, "EPA is resting its case on international studies that in turn relied on CRU data. But CRU's suspicious destruction of its original data, disclosed at this late date, makes that information totally unreliable. If EPA doesn't reexamine the implications of this, it's stumbling blindly into the most important regulatory issue we face."

Some critics claim that Gore is operating on bad information as well and note that the current warming trend may include our entire solar system—polar caps on Mars are shrinking, ice is melting on the moons of Jupiter, and the outer planets appear more luminescent, which is an indication that they too are warming. Speculation over why the solar system is warming systemwide ranges from lack of sunspot activity to the approach of some celestial object.

If the entire solar system is warming, then this theory calls into question the idea that global warming on Earth is only induced by humans. On May 19, 2009, record low spring temperatures were recorded in twenty-eight states. "If there had been record warmth in 28 states, you would have seen 'we're-causing-global-warming' headlines plastered across the front page of almost every newspaper in the country, and TV hosts would have gleefully announced the dire news.... But had you even heard about this?" asked author and attorney Alfred Lambremont Webre who, along with former Fairchild Industries corporate manager Dr. Carol Rosin, founded the Institute for Cooperation in Space.

Proponents on both sides of issues such as the global warming argument accuse each other of using fearmongering tactics to impose a police state on the public.

The federal EPA pointed to the combustion of fossil fuels in vehicles, gas and coal plants, and industry as the world's largest source of carbon dioxide (CO_2), or greenhouse gas, emissions. This conclusion remains controversial. In December 2009, the EPA went so far as to declare carbon dioxide—a product of normal human expiration (we all breath out CO_2)—a health hazard, paving the way for more regulation of emissions. Adding to the argument that the government is using fearmongering to achieve control over a dumbed-down public, Richard S. Lindzen, a professor of atmospheric science at MIT, pointed out that CO_2 is not a pollutant. "[I]t's a product of every living creature's breathing, it's the product of all plant respiration, it is essential for plant life and photosynthesis, it's a product of all industrial burning, it's a product of driving—I mean, if you ever wanted a leverage point to control everything from exhalation to driving, this would be a dream. So it has a kind of fundamental attractiveness to bureaucratic mentality," he said.

Fears over global warming resulted in the controversial American Clean Energy and Security Act of 2009, legislation that requires a limit, or cap, of how much CO_2 a company may emit into the air. Companies that reduce CO_2 more rapidly than others may conduct emission trading, selling or trading their emission credits to firms that overproduce. Opponents of cap and trade voiced fears over loss of jobs and federal control not only over businesses but private homes. Despite a narrow House victory (219 to 212), Speaker Nancy Pelosi pronounced, "We passed transformational legislation which takes us into the future." She took congratulatory phone calls from Obama, Al Gore, and Senate majority leader Harry Reid.

Obama had joined the global warming chorus in April 2007, with a plan for a national low-carbon fuel standard (NLCFS). Speaking to students at the University of New Hampshire's Durham campus, then-senator Obama proclaimed, "This is our generation's moment to save future generations from global catastrophe by creating a market for clean-burning fuels that can stop the dangerous transformation of our climate. In states like New Hampshire and California, people are taking the lead on producing fuels that use less carbon. It's time we made this a national commitment to reduce our dependence on foreign oil and take the equivalent of 32 million cars' worth of pollution out of the atmosphere."

His fervent support for reducing vehicle pollution indicated that Obama either did not truly understand what is causing solar-system-wide warming or has fallen in line with the globalist attempt to control the lives of individuals through the scare tactics of global warming.

A POLICE STATE

A state of war only serves as an excuse for domestic tyranny.

—ALEKSANDR SOLZHENITSYN,

Nobel Prize–winning novelist and survivor of the Communist gulag system

FEARMONGERING HAS ALWAYS BEEN a favored tool of despots and tyrants trying to limit public freedom. Only dazed zombies accept increasing social control by allowing threats of terrorism and depression to be held over their heads.

MODEL STATE EMERGENCY HEALTH POWERS ACT

IMAGINE A PIMPLE-FACED EIGHTEEN-YEAR-OLD dressed in camouflage and armed with a fully loaded M-16 at your door informing you that you must leave your home because the authorities fear a pandemic in your city. If you protest and say you'll stay and take your chances, you are in violation of the law and subject to arrest, fine, and imprisonment. After seeing the boy's armed companions, you decide to join your neighbors in a military truck destined for a "relocation camp" many miles from your home. At the camp, you are instructed to stand in line for a vaccination against flu, smallpox, anthrax, or whatever the latest threat might be. You recall how, in past years, so many vaccines were proven to be tainted. Yet if you refuse the inoculation, you are again subject to fine and jail.

If this sounds like a paranoid view of an Orwellian nightmare, it should be noted that since 2002 laws authorizing such action had already been passed in thirty-eight states and the District of Columbia. Under this act, named the Model State Emergency Health Powers Act, authorities would be able to federalize all medical personnel, from EMTs to physicians, and enforce quarantines. They would have the right to vaccinate the public, with or without its consent, seize and destroy private property without compensation, and ration medical supplies, food, fuel, and water in a declared emergency.

The legislation was drawn up as a model law for the federal Centers for Disease Control and Prevention following the anthrax attacks that occurred in the Capitol after 9/11. Not much has been mentioned in recent years about these attacks. What many don't know is that investigations showed that the anthrax pathogens from the attacks were military grade and unavailable outside the U.S. Armed Forces. To date, one suspect was exonerated after several years, and the next man named in those attacks died while in federal custody.

After being passed, the Model State Emergency Health Powers Act was then sent to each state legislature with a federal endorsement. Federal officials claimed the laws were needed to provide local authorities the legal right to make quick decisions in an emergency involving contagious or deadly pathogens.

Though many states modified or outright rejected this legislation, no one should relax just yet.

GOVERNMENT CAMPS

THE CREATION OF INTERNMENT or relocation camps in times of crisis is nothing new.

During the War Between the States, President Abraham Lincoln suspended the writ of habeas corpus, a critical mainstay of American justice that allows defendants the right to face their accusers. This action was later overturned by the U.S. Supreme Court because Lincoln did not previously obtain approval from Congress to suspend habeas corpus.

But then during World War I, the Supreme Court upheld the right of the president to seize the property of enemy aliens without a hearing. The court stated, "National security might not be able to afford the luxuries of litigation and the long delays which preliminary hearings traditionally have entailed." These words led to the Supreme Court consent for the rounding up, incarceration, and property seizure of Japanese Americans following the attack on Pearl Harbor.

Even former attorney general Janet Reno, who headed the U.S. Department of Justice during the tragedies at Ruby Ridge and Waco, expressed concern over the creation of a martial state. "I have trouble with a war that has no endgame and I have trouble with a war that generates so many concerns about individual liberties," she told an audience at Old Dominion University in 2002.

Reno asked Americans to remember the lessons learned from the unjust imprisonment of Japanese Americans during World War II and added that the government would be hard-pressed to find a legal basis for prosecuting many of the Taliban and al Qaeda prisoners detained at Guantanamo Bay Naval Base in Cuba.

Initially, most Americans thought little of jailing terrorism suspects in Cuba. It was only after the prosecutions that followed the horrors at Abu Ghraib Prison in Iraq that the average American began to question the American military's methods.

Democratic senator Richard J. Durbin, a member of the Senate Judiciary Committee, tried for more than two years to conduct hearings on the treatment of the Guantanamo prisoners. Durbin had spent six and a half years as a prisoner in North Vietnam. "This is not a new question," he told fellow Congress members in 2005. "We are not writing on a blank slate. We have entered into treaties over the years, saying this is how we will treat wartime detainees. The United States has ratified these treaties. They are the law of the land as much as any statute we passed. They have served our country well in past wars. We have held ourselves to be a civilized country, willing to play by the rules, even in time of war. Unfortunately, without even consulting Congress, the Bush administration unilaterally decided to set aside these treaties and create their own rules about the treatment of prisoners."

Durbin pointed out that President Bush and his appointees had unilaterally created a new detention policy based on the belief that prisoners in the war on terrorism have no legal rights—no right to a lawyer, no right to see the evidence against them, no right to challenge their detention. In fact, the U.S. government has claimed detainees have no right to challenge their detention, even if they claim they were being tortured.

"For example," he explained, "they [the Bush administration] have even argued in court they have the right to indefinitely detain an elderly lady from Switzerland who writes checks to what she thinks is a charity that helps orphans but actually is a front that finances terrorism."

Senator Durbin shocked his colleagues and angered Bush supporters when he cited an FBI account of how Guantanamo prisoners had been chained to cells in extreme temperatures and deprived of food and water. Durbin stated, "If I read this to you and did not tell you that it was an FBI agent describing what Americans had done to prisoners in their control, you would most certainly believe this must have been done by Nazis, Soviets in their gulags, or some mad regime—Pol Pot or others—that had no concern for human beings. Sadly, that is not the case. This was the action of Americans in the treatment of their prisoners."

It is disturbing to note that some days later, Durbin was forced to issue an apology. A tearful Durbin told his fellows, "Some may believe that my remarks crossed the line. To them, I extend my heartfelt apologies." Is this what happens to a person who brings reason and compassion to the table amid fearmongering?

There are those who claim that the quarantine centers such as the one at Guantanamo are simply unsubstantiated theory. These people should think again.

One of former attorney general John Ashcroft's visions was to have the power to strip American citizens of their constitutional rights, including access to the court system. Ashcroft wanted to indefinitely imprison those citizens considered "enemy combatants" in internment camps.

"The proposed camp plan should trigger immediate congressional hearings and reconsideration of Ashcroft's fitness for this important office," declared Jonathan Turley, a professor of constitutional law at George Washington University Law School. Turley, who previously actively supported Ashcroft during his contentious nomination hearing, reflected, "Whereas al Qaeda is a threat to the lives of our citizens, Ashcroft has become a clear and present threat to our liberties."

Apparently, Ashcroft's plan to build camps was continued after he left office in February 2005. In January 2006, the engineering and construction subsidiary of Halliburton Co., KBR, announced it was awarded a contingency contract from the Department of Homeland Security to construct emergency facilities in support of its Immigration and Customs Enforcement (ICE) unit. The maximum total value of the contract is $385 million and consists of a one-year base period with four one-year options. KBR had been working under similar grants since 2002.

This call for contractors to build holding camps only increased the anxiety of conspiracy theorists. Such facilities, whether in the form of empty but maintained military bases or newly built accommodations, do exist in the United States. They only await inmates.

CAMP FEMA

THE FEDERAL EMERGENCY MANAGEMENT AGENCY (FEMA) is designated as the lead agency concerned with detention centers under the Department of Homeland Security. This organization has plans in its files to evacuate cities and use sprawling temporary camps to house evacuees.

Under the pretext of planning for a war on terrorism, FEMA has dusted off and augmented contingency plans to counter the effects of nuclear, biological, and chemical attacks. In mid-2002, FEMA asked its vendors, contractors, and consultants to envision the logistics of millions of Americans being displaced in the event that American cities come under attack. The firms were given a deadline of January 2003 to be ready to establish displaced-persons camps. FEMA made it known that it already had ordered significant numbers of tents and trailers to be used for housing.

The Internet is full of sites detailing a string of concentration camps across America, many based on ill-informed rumor and speculation. But others are quite real, ready and waiting for lines of detainees or dissidents to be herded inside. While some may cast this off as paranoia, these camps do exist. Most of them are situated in military bases that are reported closed or maintained by skeleton crews. Others are operated by FEMA. Some facilities began as World War II camps for Axis prisoners.

During the Clinton administration, the Base Re-Alignment and Closing (BRAC) program closed several bases but kept them maintained for "future use" in case of "national emergency," be it war, pandemic, insurrection, or natural disaster. All bases have barracks, dining facilities, latrines, and showers, and can house thousands of people on short notice.

Military installations that have been designated as such centers include Fort Chaffee, Arkansas, an active National Guard training base with hundreds of buildings that include barracks, warehouses, mess halls, motor transport facilities, and a railroad track with loading docks. This facility was converted into a concentration camp within seventy-two hours to house Mariel boatlift refugees from Cuba in 1980 and Vietnamese refugees following the fall of Saigon. Warehouses at Fort Chaffee currently hold massive amounts of barbed wire, fencing material, mattresses, blankets, and other supplies. Eglin Air Force Base, Florida, has also been used to hold various refugees in the past, primarily Cubans. Other military bases designated as potential detention centers include McAlester Ammunition Depot and Fort Sill in Oklahoma; Fort Drum, New York; Fort Irwin and Twenty-nine Palms Marine Corps Base, California; Fort Lewis, Washington; Fort Bliss, Texas; Fort McCoy, Wisconsin; Camp Grayling, Michigan; Fort Riley, Kansas; and Minot Air Force Base, North Dakota.

Author and retired U.S. Army Reserve lieutenant colonel Craig Roberts said he chased down a number of rumors and thirdhand accounts of American concentration camps and found many untrue or outdated. But he also found some sites fully ready to operate as detention facilities. Roberts explained, "In actuality, there are two true sets of camps. First is the military, which can use any base at will to house detainees. . . . Fort Chaffee, for instance, has already been used twice for this purpose—first with the Cuban 'boat people,' and second for Vietnamese refugees. It continues to have warehouses full of mattresses, bunks, barbed wire rolls, fence posts, etc. All of these were to be used around empty barracks to provide for a detention facility if needed. . . . Some of the 'closed' military bases have been designated as 'emergency holding facilities' and already have barracks, mess halls, compounds and latrines in place. All they need are guards, administrators, logistics people and they're in business. All this can be accomplished in 72 hours. . . . Operational plans are in existence for the 'handling of civilian prisoners and laborers on military installations, both male and female.'

"The second category is FEMA. We know they have let a contract for 1,000 'emergency relocation camps' in case of widespread terrorism, biological or chemical attacks on the cities. Again, this can be speedy. The President can declare a national emergency, evoke [Executive Order] 11490, and take over the country without deferring to Congress or the Constitution. Bingo! New World Order in a couple of days."

Roberts said the worst of these camps is in Alaska. "There is a million acre 'facility' near Elmendorf AFB in Alaska that is called the 'Alaska Mental Health Facility,'" said Roberts. "An exercise by the AF was conducted in the mid-90s where they tested how many cargo aircraft could fly in and out of Elmendorf from all over the country in a three-day period, operating 24/7. To me, there is only one 'cargo' that they would be taking from all over the country to Alaska, right next to the mental health facility and that's people. You can look it up yourself. I think it's the Alaska Mental Health Act, dated 1955 or '56. The place has been federalized and still exists. It is our version of Siberia and the gulag."

On January 22, 2009, Democratic representative Alcee Hastings of Florida moved to legalize relocation camps by introducing H.R. 645, the National Emergency Centers Establishment Act. The purpose of the bill was to direct the secretary of Homeland Security to establish national emergency centers on military installations. The bill was referred to the Committee on Transportation and Infrastructure, in addition to the Committee on Armed Services, then later to the Subcommittee on Terrorism, Unconventional Threats and Capabilities, where it remained as of early 2010.

With members of the military being called upon to combat terrorism and illegal drugs, many fear military control over the civilian population, even in times of "national emergency," will lead to draconian measures such as the establishment of large concentration camps.

The original creation and maintenance of such camps has been lost in a bewildering maze of executive orders (EOs) dating back to World War II. Other relocation EOs can be traced to the Kennedy presidency and were issued under the duress of the cold war and the Cuban missile crisis.

A brief search of the FEMA website shows that even now the government has many plans to evacuate major cities—whether it be because of tornadoes, hurricanes, flooding, or a nuclear or biological strike. Where are those people to go? Who will feed them? How will they live? The answers to these questions remain elusive. And in the meantime, dozens of large military installations sit, mutely awaiting future inhabitants.

Alongside many former military bases that are technically closed but still being maintained, a growing number of civilian locations are being prepared for any "emergency." One such facility is the State Fair of Virginia's new home in Caroline County, designated as an emergency shelter in any emergency requiring a mass evacuation. The fair opened in late September 2009 on a new 360-acre site. Formerly known as Meadow Farm, the fair was a result of a 2007 deal between the State Fair of Virginia Inc. and the Virginia Department of Emergency Management. The twenty-year agreement was reached after the state agreed to appropriate about $2.4 million to help build the main exhibition hall, which officials said could house almost fifteen hundred people in the event of an emergency.

Under the agreement, the facility is required to have a kitchen capable of preparing up to sixty-five hundred meals a day, restrooms that can accommodate more than twenty-two hundred people, backup generator power, ten acres of parking lots suitable for emergency-response operations, and two twenty-thousand-square-feet paved pads that could be used for tents or other temporary structures to include sheltering household pets.

In case doubts remain that the government is building and maintaining internment camps, the U.S. Army advertised jobs in 2009 for "internment/resettlement specialists." These "specialists," according to an army description, "are primarily responsible for day-to-day operations in a military confinement/correctional facility or detention/internment facility. I/R Specialists provide rehabilitative, health, welfare, and security to U.S. military prisoners within a confinement or correctional facility; conduct inspections; prepare written reports; and coordinate activities of prisoners/internees and staff personnel."

Why are all these specialists needed? Clearly, there is no pressing need for additional personnel to intern or resettle the mere handful of al Qaeda members caught to date. Citizens should ask themselves who the quarantine centers are being prepared for and what sort of resettlement program these specialists will administer.

AMERICAN POLICE FORCE

ADDING TO CONCERNS OVER a police state presided over by FEMA or the military came word of what appeared to be the growth of private security contractors, similar to Blackwater (now Xe Services, LLC) and DynCorp.

In September 2009, representatives from the American Police Force (APF) held a news conference to announce plans to create a $27 million high-security prison and police training facility in Hardin, Montana. A spokesperson declined to name the force's parent company. "Confusion and secrecy about American Police Force has grown during the last few weeks," noted reporters Nick Lough and Katie Ussin of KURL8 TV News. "While they gave details for the site, other questions went unanswered. Where will the prisoners come from? What experience does APF have in prisoners and training police officers?"

APF spokesperson Becky Shay denied there was any secrecy involved. "APF has been here for 10 months but it has never been stealth," said Shay, who had only days before taken the job of APF public relations director after covering the detention facility story for the *Billings Gazette*. Shay assured the media that the private police group would not house terror suspects from Guantanamo Bay, Cuba, a rumor that persisted in the area after the Hardin City Council approved the plan.

Associated Press reporter Matthew Brown wrote that the Two Rivers Detention Center was promoted "as the largest economic development project in decades in the small town of Hardin when the jail was built two years ago. But it has been vacant ever since." Brown noted that the bonds used to finance the facility fell into default and that questions had arisen over the legitimacy of the APF.

"Government contract databases show no record of the company," wrote Brown. "Security industry representatives and federal officials said they had never heard of it. On its Web site, the company lists as its headquarters a building in Washington near the White House that holds 'virtual offices.' A spokeswoman for the building said American Police Force never completed its application to use the address."

On its website, the APF claimed to sell assault rifles, weapons, and military supplies internationally while providing security, investigative work, and other services to clients "in all 50 states and most countries." APF literature also boasted "rapid response units awaiting our orders worldwide" capable of fielding a battalion-size team of special forces soldiers within seventy-two hours.

Maziar Mafi, a personal injury and medical malpractice attorney who was hired by APF, said APF was a new spin-off of a major security firm founded in 1984. He declined to name the parent firm.

Oddly, the APF's logo is the double-headed eagle with fleur-de-lis emblem of the Republic of Serbia. Today, Serbia is a member of the United Nations, the Organization for Security and Cooperation in Europe, and the Council of Europe and is a candidate for membership in the European Union. Could there be a connection between the APF and the globalists?

Writing anonymously due to fear of retaliation, one Hardin resident posted this on the Internet: "We have found out that our little town of Hardin is the 'test town' for President Obama's new law to privatize the police force of local communities. Last night [September 4, 2009], the city council voted to disband our sheriff's department and to bring in a private security company to police the town. Interestingly, earlier in the day, the mayor when asked in an interview about the privatization of our police department completely denied it and said that would not be done without a council meeting. Then that evening, a council meeting was held in regards to that very thing. At the beginning of this month, our local prison signed an agreement with the American Police Force which is a subsidiary of a larger private security force that the U.S. used in the Iraq war and Hurricane Katrina.

"Yesterday, a convoy of twelve 'blacked out' Mercedes-Benz SUV's were brought into town. They were already painted with Hardin's colors and 'Hardin Police Force' was already painted on them! Hardin's sheriff's department will no longer be in operation after the month of October. During October, the Sheriff's Department is to train this new security force in all the logistics of running the town of Hardin. If you go on the American Police Force site, you might notice that the logo they use is actually a Russian [Serbian] logo. I have been told that the man who came with this new security force as the captain has a thick Russian accent."

The story of Hardin and the APF reached the national media and prompted Montana attorney general Steve Bullock to launch an investigation after it was learned that the man representing the APF, "Captain" Michael Hilton, was a Serbian immigrant with a long list of aliases. Apparently, Hilton had served time in U.S. prisons for fraud and had more than $1 million in judgments against him.

While the corporate mass media only publicized the jobs that would be created in Hardin and speculated that the APF could be a con, the alternative media noted that the APF claimed to run the U.S. Training Center in Moyock, North Carolina, the same center connected to Blackwater, the controversial recipient of large U.S. government contracts in Iraq and Afghanistan. Officials at Xe/Blackwater tried to distance the company from the APF, stating the force was using its name illegally. Xe spokeswoman Stacy DeLuke said, "It's bizarre. They have nothing to do with us. We have nothing to do with them." However, the contact address in Moyock for both the U.S. Training Center (formerly the Blackwater Training Center) and the American Police Force was the same.

Paul Joseph Watson, a researcher and columnist for talk-show host Alex Jones's web page, PrisonPlanet.com, wrote, "The fact that APF's training center plans to recruit foreign assets who could then be patrolling the streets of America bossing U.S. citizens is obviously a frightening prospect, completely unconstitutional, and another reason why APF needs to abandon its plans to act as a private police force completely."

William N. Grigg, writing on the Internet blog Pro Libertate, observed there was "something utterly surreal about the Hardin case; it's as if some kind of martial law melodrama were being played out as an enhanced 'reality' program—something like *Red Dawn* meets *Jericho* with a touch of the Orson Welles' 'War of the Worlds' broadcast added for good measure."

THE POSSE COMITATUS ACT

LESS THAN A MONTH after the 9/11 attacks, former governor of Pennsylvania Tom Ridge arrived in his new office steps away from the

Oval Office of President Bush, the man who created his job. Ridge's new job was to head up the recently created Office of Homeland Security, which would coordinate forty-six different federal government agencies in an effort to protect the American people from terrorists. From its inception, the position was designed to become a permanent government department.

By 2006, Ridge had used his power as secretary of the Department of Homeland Security to create a vast network of small suburban and rural counties that had their own Homeland Security departments and that were all answerable to the national department. This concentration of so much power into the DHS under a leader with a controversial past has been cause for concern by many Americans. During the Vietnam War, Tom Ridge participated in the infamous Phoenix Program, a "pacification" program responsible for the assassination of forty-five thousand Vietnamese and the torture and abuse of thousands.

Douglas Valentine, author of *The Phoenix Program,* wrote in a posting on the website Disinfo.com, "During the Vietnam War, under the CIA's Phoenix program—which is the model for the Homeland Security Office—a terrorist suspect was anyone accused by one anonymous source. Just one. The suspect was then arrested, indefinitely detained in a CIA interrogation center, tortured until he or she (in some cases children as young as twelve) confessed, informed on others, died, or was brought before a military tribunal (such as Bush is proposing) for disposition.

"In thousands of cases, innocent people were imprisoned and tortured based on the word of an anonymous informer who had a personal grudge or was actually a Viet Cong double agent feeding the names of loyal citizens into the Phoenix blacklist. At no point in the process did suspects have access to due process or lawyers, and thus, in 1971, four U.S. Congresspersons stated their belief that the Phoenix Program violated that part of the Geneva Conventions guaranteed to protect civilians in time of war. . . ."

When Ridge was appointed, the White House announced that he would work in conjunction with Bush's deputy national security adviser, U.S. Army general Wayne Downing. This announcement indicated to the public that the military would play a prominent role in counterterrorism activities. Few thought to ask if this was a violation of the Posse Comitatus Act (PCA), the law that prohibits the U.S. military from conducting law enforcement duties against the American public.

The PCA has never really been challenged in this nation's history because it addresses a grievance that was used to proclaim the American Revolution when early colonists were forced to feed and quarter King George's troops while submitting to the troops' authority. The act embodies the traditional principle of separation of military and civilian authority, one of the fundamental precepts of a democratic government and a cornerstone of American liberty. Posse Comitatus, Latin for a support group of citizens for law enforcement (i.e., a posse), wasn't passed until 1878. The act was a direct result of the outrage over Southern states being at the mercy of inept or corrupt military authorities during Reconstruction. The Posse Comitatus Act specifically prohibits most members of the armed services (the Coast Guard is exempted for its coastal protection duties) from exercising police powers on nonfederal property within the United States.

Posse Comitatus has been slowly shredded since 1981, when Congress allowed an exception to be made for the war on drugs. The military was allowed to be used for drug interdiction along the nation's borders. This small, and what appeared to be sensible, action at the time soon grew out of proportion. Congress, still unable to come to grips with the true social causes of drug abuse, in 1989 designated the Department of Defense as the lead agency in drug interdiction.

In the tragedy at Waco on April 19, 1993, military snipers were on hand and General Wesley Clark used tanks from Fort Hood to bulldoze the burning Branch Davidian church. Clark's command was authorized because federal officials used the pretext that the Davidians were involved with drugs. No evidence of drugs was ever found.

On April 19, 1995, the Murrah Federal Building in Oklahoma City was bombed, and President Bill Clinton proposed yet another circumvention of the PCA to allow the military to help civilian investigators look for weapons of mass destruction. Around the same time, Congress considered legislation to allow troops to enforce customs and immigration laws at the borders. This legislation didn't pass.

Only one year later, Bob Dole promised on the presidential campaign trail to heighten the military's role in the war on drugs. Another primary contender, Lamar Alexander, suggested that a new branch of the military be created to substitute for the INS and Border Patrol.

By 2010, the Posse Comitatus Act was about finished after it was announced that for the first time an active U.S. Army unit—the Third Infantry Division's First Brigade Combat Team—was to be redeployed inside the United States under the Northern Command (NORTHCOM). After spending sixty months in Iraq quelling insurgents, the team was available when called upon to "help with civil unrest and crowd control or to deal with potentially horrific scenarios such as massive poisoning and chaos in response to a chemical, biological, radiological, nuclear or high-yield explosive (CBRNE) attack . . . subdue unruly or dangerous individuals." They carry equipment to construct roadbloacks and install spike strips for slowing and stopping or controlling traffic, as well as shields, batons, and beanbag bullets for nonlethal crowd control.

"The need for reaffirmation of the PCA's principle is increasing," wrote legal scholar Matthew Hammond in the *Washington University Law Quarterly,* "because in recent years, Congress and the public have seen the military as a panacea for domestic problems."

He added, "Major and minor exceptions to the PCA, which allow the use of the military in law enforcement roles, blur the line between military and civilian roles, undermine civilian control of the military, damage military readiness, and inefficiently solve the problems they supposedly address. Additionally, increasing the role of the military would strengthen the federal law enforcement apparatus that is currently under close scrutiny for overreaching its authority."

In the weeks following 9/11 and before the creation of the Transportation Security Administration (TSA), military troops patrolled airports and the streets of Washington and New York without protest. Such scenes were a brief glimpse of life under martial law. In 2005, President Bush announced that he would use military troops in the event of a national pandemic. In 2009, the military was an integral part of the swine flu general vaccination process. Why is it that the military's role in daily life has kept steadily increasing when the nation hasn't been attacked again? Could it be that martial law was planned years ago?

Those who doubt the veracity of this should just ask the residents of Kingsville, Texas.

On the night of February 8, 1999, a series of mock battles using live ammunition erupted around the town of twenty-five thousand. As part of a military operation named Operation Last Dance, eight black helicopters roared over the town. Ferried by the choppers, soldiers of the elite 160th Special Operations Aviation Regiment, known as the Night Stalkers, staged an attack on two empty buildings using real explosives and live ammunition. One of the helicopters nearly crashed when it hit the top of a telephone pole and started a fire near a home. Additionally, an abandoned police station was accidentally set on fire and a gas station was badly damaged when one or more helicopters landed on its roof.

Citizens of Kingsville were terrified during the drill. Police chief Felipe Garza and Mayor Phil Esquivel were the only ones notified of the attack in advance. Both men refused to give any details of the operation, insisting they had been sworn to secrecy by the military. Only Arthur Rogers, the assistant police chief, would admit to what happened. "The United States Army Special Operations Command was conducting a training exercise in our area," he said. He refused to provide any details.

The local emergency management coordinator for FEMA, Tomas Sanchez, was not happy with the attack and with the lack of information and warning. When asked what the attack was all about, Sanchez, a decorated Vietnam veteran with thirty years' service in Naval Intelligence, replied that, based on his background and knowledge, the attack was an operational exercise. The scenario of the exercise was that "Martial law has been declared through the Presidential Powers and War Powers Act, and some citizens have refused to give up their weapons. They have taken over two of the buildings in Kingsville. The police cannot handle it. So you call these guys in. They show up and they zap everybody, take all the weapons and let the local PD clean it up." Sanchez and other military experts told World

Net Daily that the night attack indicated the use of Presidential Decision Directive (PDD) 25, a top-secret document that apparently authorizes military participation in domestic police situations. Some speculated that PDD 25 may have surreptitiously superseded the 1878 Posse Comitatus Act.

Asked for comment, George W. Bush, then Texas governor, said it was not his job to get involved in the concerns over the Night Stalkers and the use of live ammunition in a civilian area of his state.

Just in case one might be tempted to think that the events in Kingsville were simply some aberration from the distant past, a similar military exercise took place in 2009. Soldiers from Fort Campbell, including the 101st Airborne Division (Air Assault) and other infantry brigades, performed a training air assault in Troy, Tennessee, on September 29–30. It was called Operation Diomedes, after the ancient Greek warrior who wounded Aphrodite, the goddess of love.

After being helicoptered from Fort Campbell, soldiers were dropped into multiple locations throughout the town. Once on the ground, the troops were to clear predetermined buildings in four different objective areas based on a combat scenario.

Military spokesmen said this air assault was the first time that soldiers from the 101st Airborne had conducted such training in the area. The purpose of the exercise was to provide the troops with "pertinent realistic training in unfamiliar terrain to prepare them for possible contingency operations around the world." Some saw this exercise as practice for the military capture of small towns in the United States.

Exercises similar to those in Troy and Kingsville may have occurred as early as 1971, when plans were drawn up to merge the military with police and the National Guard. In that year, Senator Sam Ervin's Subcommittee on Constitutional Rights discovered that military intelligence had established an intricate surveillance system to spy on hundreds of thousands of American citizens. Most of the citizens were antiwar protesters. This plan to merge the police with the military included exercises that were code-named Garden Plot and Cable Splicer. Britt Snider, who was lead researcher on military intelligence for Ervin's subcommittee, said the plans seemed too vague to get excited about. "We could never find any kind of unifying purpose behind it all," he told a reporter. "It looked like an aimless kind of thing."

Yet four years later the plans began to come into sharper focus. In the *New Times* magazine, Ron Ridenhour and Arthur Lubow reported that "[C]ode named Cable Splicer cover[s] California, Oregon, Washington and Arizona, [and is] under the command of the Sixth Army. . . . [It] is a plan that outlines extraordinary military procedures to stamp out unrest in this country. . . . Developed in a series of California meetings from 1968 to 1972, Cable Splicer is a war plan that was adapted for domestic use procedures used by the US Army in Vietnam. Although many facts still remain behind Pentagon smoke screens, Cable Splicer [documents] reveal the shape of the monster that the Ervin committee was tracking down."

During the time when Cable Splicer was being carried out, several full-scale war games were conducted with local officials and police working side by side with military officers in civilian clothing. Many policemen were taught military urban pacification techniques. Afterward, they returned to their departments and helped create the early SWAT (Special Weapons and Tactics) teams.

Representative Clair Burgener of California, a staunch Reagan Republican who had attended the Cable Splicer II kickoff conference, was flabbergasted when shown Cable Splicer documents. "This is what I call subversive," he said. Subcommittee chief counsel Doug Lee read through the documents and blurted out, "Unbelievable. These guys are crazy! We're the enemy! This is civil war they're talking about here. Half the country has been designated as the enemy." Britt Snider agreed, stating, "If there ever was a model for a takeover, this is it."

The war on terrorism and the more recent flu alarms have provided the pretext for the activation of plans like Cable Splicer, which is a clear violation of the Posse Comitatus Act.

Diana Reynolds, formerly an assistant professor of politics at Bradford College and a lecturer at Northeastern University, wrote a paper entitled "The Rise of the National Security State: FEMA and the NSC in 1990." In the paper, Reynolds argued:

The Rex-84 Alpha Explan (Readiness Exercise 1984, Exercise Plan) indicates that FEMA in association with 34 other federal civil departments and agencies conducted a civil readiness exercise during April 5–13, 1984. It was conducted in coordination and simultaneously with a Joint Chiefs exercise, Night Train 84, a worldwide military command post exercise (including Continental U.S. Forces or CONUS) based on multi-emergency scenarios operating both abroad and at home. In the combined exercise, Rex-84 Bravo, FEMA and DOD led the other federal agencies and departments, including the Central Intelligence Agency, the Secret Service, the Treasury, the Federal Bureau of Investigation, and the Veterans Administration through a gaming exercise to test military assistance in civil defense.

The exercise anticipated civil disturbances, major demonstrations and strikes that would affect continuity of government and/or resource mobilization. To fight subversive activities, there was authorization for the military to implement government ordered movements of civilian populations at state and regional levels, the arrest of certain unidentified segments of the population, and the imposition of martial rule.

In 1984, the military's involvement with civilian authorities inspired then attorney general William French Smith to write a critical letter stating, " . . . In short I believe that the role assigned to the Federal Emergency Management Agency (FEMA) on the revised Executive Order exceeds its proper function as a coordinating agency for emergency preparedness."

In January 2005, fears that secretive, overreaching agencies with military connections might violate the Posse Comitatus Act were substantiated. During this time, news outlets reported that since 2002 the Pentagon's Defense Intelligence Agency (DIA) had operated an intelligence-gathering and support unit called the Strategic Support Branch (SSB) with authority to operate clandestinely anywhere in the world to support antiterrorism and counterterrorism missions. The SSB previously had been unknown, operating under an undisclosed name.

Military involvement in daily life has even reached the halls of education. In 2002, the principal of Mount Anthony Union High School in Bennington, Vermont, was shocked to receive a letter from military recruiters demanding a list of all students, including names, addresses, and telephone numbers. Because the school's privacy policy prevented the disclosure of such individual information, the principal told the recruiters no. However, the principal was soon shocked to learn that buried deep within President Bush's new No Child Left Behind Act, public schools must provide their students' personal information to military recruiters or face a cutoff of federal funds.

Republican representative David Vitter of Louisiana, who sponsored the recruitment requirement in the education bill, noted that in 1999, more than nineteen thousand U.S. schools denied military recruiters access to their records. Vitter said such schools "demonstrated an anti-military attitude that I thought was offensive." What could be offensive about wanting to protect the privacy of the nation's students?

"I think the privacy implications of this law are profound," commented Jill Wynns, president of the San Francisco Board of Education. "For the federal government to ignore or discount the concerns of the privacy rights of millions of high school students is not a good thing, and it's something we should be concerned about."

Not only are high school students being bothered by Homeland Security, but also mere kindergarten students who only want to play outside. In May 2002, Scott and Cassandra Garrick sued the Sayreville School District in New Jersey after their six-year-old child and three classmates were disciplined for playing cops and robbers. Apparently, other students saw the youngsters playing on the school yard pretending to use their fingers as guns. The other students told a teacher and the kindergartners were suspended from school.

U.S. District Judge Katharine S. Hayden dismissed the parents' civil suit, claiming school authorities have the right to restrict violent or disruptive games. Yet the parents' attorney, Steven H. Aden, remarked, "They have the right to be children. The school and the courts shouldn't censor their play [even if] it's politically incorrect."

Incidents like this are rarely covered in the corporate mass media. They are never distributed to a large audience, but they often worry thoughtful people.

"I'm terrified," said Ellen Schrecker, author of *Many Are the Crimes: McCarthyism in America.* "What concerns me is we're not seeing an enormous outcry against this whole structure of repression that's being rushed into place...."

ACLU president Nadine Strossen also voiced concern. "I've been talking a lot about the parallels between what we're going through now and McCarthyism. The term 'terrorism' is taking on the same kind of characteristics as the term 'communism' did in the 1950s. It stops people in their tracks and they're willing to give up their freedoms. People are too quickly panicked. They are too willing to give up their rights and to scapegoat people, especially immigrants and people who criticize the war." Paul Proctor, a columnist for NewsWithViews.com and periodicals across America, added, "Besides being unconstitutional and un-American, snooping on innocent people in a free society is cowardly, divisive and just plan evil.... But, you see—terrorists don't want your freedom—they want your life. It is tyrants and dictators that want your freedom."

DEFICIENT BORDER PATROL

SOME PEOPLE DESIRE TO alter or even abolish the Posse Comitatus Act, particularly with regard to border security. In an October 2001 letter to Donald Rumsfeld, Republican senator John Warner wrote, "Should this law [PCA] now be changed to enable our active-duty military to more fully join other domestic assets in this war against terrorism?" Warner's question continued into late 2002 when then representative Tom Tancredo, members of the Immigration Reform Caucus, and families of victims slain in the course of conflicts between immigration officers and illegal aliens petitioned Congress demanding that military troops patrol the U.S. borders. Jumping on the antiterrorist bandwagon, Tancredo stated, "As long as our borders remain undefended, we cannot claim that we are doing everything possible to protect the nation from terrorism.... It's time to authorize the deployment of military assets on our borders."

During this time, a FEMA official named John Brinkerhoff wrote a paper arguing that "President Bush and Congress should initiate action to enact a new law that would set forth in clear terms a statement of the rules for using military forces for homeland security and for enforcing the laws of the United States. Things have changed a lot since 1878, and the Posse Comitatus Act is not only irrelevant but also downright dangerous to the proper and effective use of military forces for domestic duties." This paper remained on the Homeland Security website well into 2006.

Critics have questioned how the massive restructuring of the U.S. government under the Homeland Security Act has helped protect the nation, especially when the national borders are left dangerously open to both the nation's north and south. To many thoughtful Americans it seems like a commonsense first step in the war on terrorism to tighten security on the U.S. borders. Yet this does not appear to have happened. Despite an increasing clamor for tighter security, a flood of illegal immigrants continues unabated on the southern border. Though politicians spouted rhetoric about building an effective fence, only a few motion sensors and cameras ended up being deployed.

In late 2009, Border Patrol director of media relations Lloyd Easterling confirmed that the number of agents on the U.S.-Mexico border would be cut by 384. This would bring the total of agents on the Mexican border to 17,015 and agents on the Canadian border to 2,212.

Terence P. Jeffrey, editor in chief of CNSNews.com, reported that in a review Homeland Security officials said the Border Patrol's goal for fiscal 2009 was to have 815 of the 8,607 miles of border under "effective control." Jeffey noted, "The review also said the Border Patrol's goal for fiscal 2010 was to again have 815 miles of border under 'effective control,' meaning DHS was not planning to secure a single additional mile of border in the coming year." U.S. Customs and Border Protection defines "effective control" as the Border Patrol's ability to detect and apprehend an illegal immigrant crossing a particular area of the border.

OFFICE OF STRATEGIC INFLUENCE

THE DEFENSE DEPARTMENT'S RECENT actions have done little to inspire confidence that traditional American liberties will be respected if the Posse Comitatus Act is abolished.

Following the 9/11 attacks, the Pentagon announced that it was creating the Office of Strategic Influence (OSI), which was designed to present a more favorable image of the U.S. military to foreign news media. The new organization provoked an immediate controversy when it was learned it planned to influence international opinion by planting false stories in the foreign media. Considering the closeness of the world today, thanks to the Internet, such false stories could easily find their way back into the domestic media. This was nothing new. The CIA had used similar tactics for decades, but this was too blatant. Even the major media, including the *New York Times,* were stirred to action. "Mingling the more surreptitious activities [such as Pentagon covert operations like computer network attacks, psychological activities, and deception], with the work of traditional public affairs would undermine the Pentagon's credibility with the media, the public and governments around the world," noted the *Times.*

In a rare step backward, in early 2002 the government announced the OSI would be closed. Though Donald Rumsfeld, then secretary of defense, argued that criticism of the OSI was "off the mark," he nevertheless admitted that "the office has been so damaged that . . . it's pretty clear to me that it cannot function."

Rumsfeld refused to let the matter lie. At a November 18, 2002, press briefing, the defense secretary defiantly stated, "And then there was the Office of Strategic Influence. You may recall that. And 'oh my goodness gracious isn't that terrible, Henny Penny, the sky is going to fall.' I went down that next day and said fine, if you want to savage this thing? Fine, I'll give you the corpse. There's the name. You can have the name, but I'm gonna keep doing every single thing that needs to be done and I have." True to his word, only the name of the office was abolished, and OSI activities were merely shifted to another agency. Yet this time, Rumsfeld's vow to continue his program of disinformation was not repeated in the corporate mass media. With this type of propaganda and disinformation program continuing, small wonder many concerned Americans are suspicious about the accuracy of the corporate mass media news content.

OATH KEEPERS

"I, _____, DO SOLEMNLY swear (or affirm) that I will support and defend the Constitution of the United States against all enemies, foreign and domestic; that I will bear true faith and allegiance to the same; that I take this obligation freely, without any mental reservation or purpose of evasion; and that I will well and faithfully discharge the duties of the office on which I am about to enter. So help me God."

This is the U.S. Uniformed Services Oath of Office administered to the military and law enforcement officers. Nowadays, those who try to stay true to this oath are viewed as zealous or jingoistic. Even the word "patriot" is now looked at as un-American.

One group of government employees that have been adamant about staying true to their oaths is the Oath Keepers. According to the Oath Keepers website, "Oath Keepers is a non-partisan association of currently serving military, reserves, National Guard, veterans, Peace Officers, and Fire Fighters who will fulfill the Oath we swore, with the support of like minded citizens who take an Oath to stand with us, to support and defend the Constitution against all enemies, foreign and domestic, so help us God. Our Oath is to the Constitution. Our motto is 'Not on our watch!'" Some domestic police officials and soldiers serving in Iraq and Afghanistan have begun wearing Oath Keeper patches on their uniforms.

The Oath Keepers' patriotism was soon attacked by commentators such as Bob Hanafin, a staff writer for the military and foreign affairs journal *Veterans Today* and a retired air force officer. In a "special report" headlined "Are Right Wing Extremists Trying to Recruit Our Troops?"

Hanafin wrote, "Taking oaths to disobey orders of any Chain of Command is not only illegal under the UCMJ [Uniform Code of Military Justice], but it is also illegal under the Hatch Act which most of our troops, including Junior Officers don't know the meaning of." He added, "We will leave it to the Department of Homeland Security (unless they have been too intimidated by the right wing), Department of Defense, and our readers to decide if what Oath Keepers is doing to attract our troops to take an Oath to potentially disobey orders from their Chain of Command is legitimate or if the Department of Defense and Homeland Security needs to closely monitor such recruitment."

Hanafin's article caused reactions from a number of Oath Keepers, such as Patrick M. Fahey, who responded on his blog by writing, "The Hatch Act of 1939['s] main provision is to prohibit federal employees from engaging in partisan political activity (that is why during basic training, military recruits are told you can't protest in uniform). . . . Now, on to the Uniform Code of Military Justice. The UCMJ is the governing body of law for the military. It lists out what you can be charged for, the regulations for pre-trial confinement, non-judicial punishment, etc. Mr. Hanafin says that under the UCMJ, it is unlawful to disobey an order, which would be correct in most circumstances."

Fahey dissected eight separate articles of the UCMJ and concluded that the Oath Keepers would not violate the requirements of these articles. Oath Keepers have pledged to disobey only orders that fall into the following categories:

1. We will NOT obey orders to disarm the American people.
2. We will NOT obey orders to conduct warrantless searches of the American people.
3. We will NOT obey orders to detain American citizens as "unlawful enemy combatants" or to subject them to military tribunal.
4. We will NOT obey orders to impose martial law or a "state of emergency" on a state.
5. We will NOT obey orders to invade and subjugate any state that asserts its sovereignty.

6. We will NOT obey any order to blockade American cities, thus turning them into giant concentration camps.
7. We will NOT obey any order to force American citizens into any form of detention camps under any pretext.
8. We will NOT obey orders to assist or support the use of any foreign troops on U.S. soil against the American people to "keep the peace" or to "maintain control."
9. We will NOT obey any orders to confiscate the property of the American people, including food and other essential supplies.
10. We will NOT obey any orders which infringe on the right of the people to free speech, to peaceably assemble, and to petition their government for a redress of grievances.

Oath Keepers founder Stewart Rhodes noted that his group only wants to have military and law enforcement members think carefully about the orders they receive. "For example, if a police officer feels he is being asked to do an illegal search of a home or vehicle, he should stand down," Rhodes said.

Yet despite the fact that the Oath Keeper pledge was seemingly written with the general good in mind, the organization was attacked by the Alabama-based Southern Poverty Law Center (SPLC), which described Oath Keepers as "a particularly worrisome example of the Patriot revival."

"I'm not accusing [Oath Keepers founder] Stewart Rhodes or any member of his group of being Timothy McVeigh or a future Timothy McVeigh," said SPLC spokesman Mark Potok. "What's troubling about Oath Keepers is the idea that men and women armed and ordered to protect the public in this country are clearly being drawn into a world of false conspiracy theory."

HATE CRIMES

IT SHOULD BE NOTED that the SPLC has come under its share of criticism, despite well-funded promotion and close ties to Democratic politics. Founded in 1971 by Morris Dees and Joseph J. Levin Jr., the nonprofit center has fought to enact laws against hate crimes ranging from hate-inspired murder to outbursts of speech. The SPLC often invokes the name of Oklahoma City bomber McVeigh, thought to be connected to white supremacist groups, as an example of the results of hate crime.

The SPLC regularly conducts highly profitable fund-raising activities. According to Ken Silverstein of *Harper's* magazine, one example occurred about ten years ago when "The Center earned $44 million … $27 million from fund-raising and $17 million from stocks and other investments—but spent only $13 million on civil rights programs, making it one of the most profitable charities in the country." By 2005, the SPLC reported an endowment fund of more than $152 million.

Armed with millions of dollars, the SPLC has displayed a hatred for haters. The center disparaged the Minuteman Project, a group devoted to preventing illegal border crossings into the United States from Mexico. On the SPLC website, it was reported, "major elements of the Minuteman anti-immigration movement are broadening their agenda to become part of a resurgent antigovernment 'Patriot' movement." Internet commentator Judy Andreas, who conducted an in-depth investigation of the SPLC, stated, "The Southern Poverty Law Center may, at one time, have been a force dedicated to preserving the American Values of freedom and constitutional rights through American law, but they have strayed far afield of their initially stated goals. Today they are wildly flailing accusations at anyone who, in their estimation, is guilty of a 'hate' crime. And as their brush continues to broaden, they are busily applying sweeping strokes to the word 'racism' and hungrily scanning the landscape for anyone who dares to oppose the actions of any non-white individual, group or nation. Whether the non-white individual has committed an act of criminal conduct appears to be irrelevant to the SPLC."

Despite trying to obtain donations while pressing for more "hate crime" legislation, the SPLC's own website, quoting crime experts, admitted that the whole hate crime reporting system is plagued with errors. Regardless, the SPLC's lobbying efforts proved successful in late October 2009, when Congress passed (68–29 in the Senate; 237–180 in the House) the Matthew Shepard Hate Crimes Prevention Act, named after a gay Wyoming college student murdered in 1998. This bill broadened the formerly narrow range of actions (e.g., discrimination in admittance to school or voting) that makes it acceptable for the federal government to intervene in cases where a state is unwilling or unable to prosecute an alleged hate crime.

Joe Solmonese, president of the Human Rights Campaign, one of the nation's largest gay rights group, was pleased with the bill, describing it as "our nation's first major piece of civil rights legislation for lesbian, gay, bisexual and transgender people. Too many in our community have been devastated by hate violence."

Yet there was still controversy with the bill. After Iowa Republican representative Steve King failed in his effort to add an amendment specifying that pedophiles could not use the law as protection of their conduct, the bill was derisively nicknamed "The Pedophile Protection Act." South Carolina senator Jim DeMint believed the bill was a "dangerous step" toward thought crimes and serves "as a warning to people not to speak out too loudly about their religious views." How could legislation that has prompted so much controversy and debate sail through Congress?

The answer is simple. It was attached to the $680 billion defense appropriations bill. Congress had the choice of voting for expanding the dubious hate crime laws or voting against funding the nation's military personnel.

"The inclusion of the controversial language of the hate crimes legislation, which is unrelated to our national defense, is deeply troubling," said Alabama senator Jeff Sessions, one of the Republicans who voted against the bill.

DARPA

EVADING CONFINEMENT IN A military-run FEMA camp or hoping for assistance from an Oath Keeper may be the least of a modern American's worries. The government is now intruding into the public's lives on a wide scale. In any police state, surveillance of the population is paramount.

In late 2002, it was revealed that the army's Intelligence and Security Command (INSCOM) in Fort Belvoir, Virginia, planned to use high-powered computers to secretly search the citizenry's e-mail messages, credit card purchases, telephone records, and bank statements on the chance that one might be associated, or sympathetic to, terrorists. Meanwhile, the Pentagon's new Office of Information Awareness (OIA) was to create a "vast centralized database" filled with information on the minutest details of citizens' private lives.

Opponents of this plan were incensed at the appointment of John Poindexter, a former national security adviser and vice admiral, as the head of the OIA. Poindexter had previously been involved in the Iran-Contra scandal during the Reagan administration, which involved the illegal sale of weapons to Iran and using the profits to fund the CIA-backed Contra army fighting in Nicaragua, all done in defiance of Congress. Poindexter lost his national security adviser job in 1990 after being convicted of lying to Congress, defrauding the government, and destroying evidence. But from 1996 to 2003, as vice president of Syntek Technologies, Poindexter worked with the Defense Advanced Research Projects Agency (DARPA) to develop Genoa, a powerful search engine and information-harvesting program. In 2002 and 2003, he served as director of the DARPA Information Awareness Office (IAO), tasked with creating all-encompassing electronic and computer surveillance—total information awareness—in the war on terrorism. But after a public outcry over the possibility that such a mass surveillance program could be turned on honest Americans, Congress stopped IAO funding and Poindexter lost his job. Researchers believe that some of the IAO programs continue to operate today under different names and other funding.

Christopher H. Pyle, a teacher of constitutional law and civil liberties at Mount Holyoke College, wrote, "That law enforcement agencies would search for terrorists makes sense. Terrorists are criminals. But why the Army? It is a criminal offense for Army personnel to become directly involved in civilian law enforcement [the Posse Comitatus Act]. Are they seeking to identify anti-war demonstrators whom they harassed in the 1960s? Are they getting ready to round up more civilians for detention without trial, as they did to Japanese Americans during World War II? Is counterterrorism becoming the sort of investigative obsession that anti-Communism was in the 1950s and 1960s, with all the bureaucratic excesses and abuses that entailed? This isn't the first time that the military has slipped the bounds of law to spy on civilians. In the late 1960s, it secretly gathered personal information on more than a million law-abiding Americans in a misguided effort to quell anti-war demonstrations, predict riots and discredit protesters. I know because in 1970, as a former captain in Army intelligence, I disclosed the existence of that program."

While writing two book-length reports on army spying for Senator Sam Ervin's Subcommittee on Constitutional Rights, Pyle was struck by the harm that could be done to the nation if the government ever gained untraceable access to the financial records and private correspondences of its critics. "Army intelligence was nowhere near as bad as the FBI [with its infamous COINTELPRO], but it responded to my criticisms by putting me on Nixon's 'enemies list,' which meant a punitive tax audit. It also tried to monitor my mail and prevent me from testifying before Congress by spreading false stories that I had fathered illegitimate children. I often wondered what the intelligence community could do to people like me if it really became efficient."

Today, national security programs are gaining that efficiency, thanks to the proliferation of computer technology. In his book *The Puzzle Palace,* author James Bamford wrote about two new locations for National Security Agency (NSA) databases—one in Utah and the other in Texas. Apparently lacking space at Fort Meade, Maryland, NSA headquarters needed new space to house "trillions of phone calls, e-mail messages, and data trails: Web searches, parking receipts, bookstore visits, and other digital 'pocket litter.'

"Unlike the British government, which, to its great credit, allowed public debate on the idea of a central data bank, the NSA obtained the full cooperation of much of the American telecom industry in utmost secrecy after September 11. For example, the agency built secret rooms in AT&T's major switching facilities where duplicate copies of all data are diverted, screened for key names and words by computers, and then transmitted on to the agency for analysis. Thus, these new centers in Utah, Texas, and possibly elsewhere will likely become the centralized repositories for the data intercepted by the NSA in America's version of [a] 'big brother database,'" wrote Bamford.

Bamford was skeptical as to whether the NSA's surveillance would even benefit the country in any ways. "Based on the NSA's history of often being on the wrong end of a surprise and a tendency to mistakenly get the country into, rather than out of, wars, it seems to have a rather disastrous cost-benefit ratio. Were it a corporation, it would likely have gone belly-up years ago," said Bamford. "The September 11 attacks are a case in point. For more than a year and a half the NSA was eavesdropping on two of the lead hijackers, knowing they had been sent by bin Laden, while they were in the US preparing for the attacks. The terrorists even chose as their command center a motel in Laurel, Maryland, almost within eyesight of the [NSA] director's office. Yet the agency never once sought an easy-to-obtain FISA warrant to pinpoint their locations, or even informed the CIA or FBI of their presence."

Although more thoughtful and aware citizens are concerned enough with the potential misuse of the nationwide electronic surveillance systems increasingly coming online, most are blissfully unaware that these systems, whether operated by military or civilian intelligence agencies, are commanded at the highest levels of the federal government, which have been demonstrated in this work to be filled with globalists seeking to control the world and its population.

ELECTRONIC SURVEILLANCE

THE POTENTIAL FOR "BIG BROTHER" surveillance has been part of American life for decades. As far back as 1975, Senator Frank Church performed a study of the National Security Agency (NSA) and warned Congress, "That [the NSA] capability at any time could be turned around on the American people and no American would have any privacy left, such is the capability to monitor everything: telephone conversations, telegrams, it doesn't matter. There would be no place to hide." It is ironic that Bush bypassed the 1978 Foreign Intelligence Surveillance Act (FISA), which was passed after President Nixon used the NSA to spy domestically on political enemies.

The PATRIOT Act, which clearly abridges many American rights, was built upon the little-known FISA bill, which cracked the door open to secret government searches. FISA was passed in the contingencies of the cold war and in the wake of revelations of abuse in surveillance by the FBI and CIA.

The FISA law created the secret federal Foreign Intelligence Surveillance Court (FISC), which meets in total secrecy to routinely approve covert surveillances on non-Americans by intelligence agencies. All applications to the court must be approved by the attorney general. Either federal prosecutors are extremely efficient and effective in their work or the federal judges (originally numbering seven but expanded to eleven by the PATRIOT Act) who make up this secret court are not picky about the Constitution because out of the some twelve thousand requests for secret surveillances and physical searches made during the first twenty-three years of the FISC, not one application was denied until four in 2003.

Why the sudden scrutiny of surveillance requests to the FISC? This was due to provisions of the PATRIOT Act, drafted by the Bush administration and secretly fine-tuned by the House and Senate leadership, that expanded FISA to include Americans.

Although FISA legislation was meant to impose limits and a review process upon warrantless surveillance and searches conducted for "national security" purposes, the current use of the FISA process, now expanded through the PATRIOT Act and its revisions, has resulted in the erosion of numerous constitutional rights and basic legal procedures traced back to the Magna Carta.

Church's warning that the NSA's surveillance ability could be turned on Americans became reality thirty years later when President George W. Bush ordered the NSA to monitor Americans without seeking warrants from the special intelligence court (FISC) or any other court. It was also revealed in 2006 that the NSA had already been secretly collecting phone call records of millions of Americans using data from AT&T, Verizon, and BellSouth, the three largest companies in the United States. General Michael Hayden, appointed director of the CIA in 2006, oversaw the program during his tenure heading the NSA.

During the Bush years, wiretapping and surveillance became highly politicized after the *New York Times* disclosed to the public news of a secret electronic monitoring program that had swept up information on American citizens for years without court approval. Controversy grew in early 2006 when it was reported that President Bush had instructed the NSA to electronically monitor Americans for signs of terrorism.

It is interesting to note that elements of the PATRIOT Act existed even before George W. Bush came into office. One feature of the PATRIOT Act that was approved in 2001 had actually been introduced (yet failed to pass) in 1998.

During the Clinton administration, there was a brief furor over proposed new federal banking regulations that would require all banks to report to the government any large deposits, withdrawals, or unusual activity from the banking public. Euphemistically called the "Know Your Customer" program, it heralded a new era where law-abiding citizens might have to defend their financial matters before government agents.

Under the program, banks would be required to create a profile of each customer and report any deviation from the profile to the feds. For example, if a person sold an unneeded car and then deposited the cash into his bank account, the banks would report this to the government. The bank computer would flag the transaction because the money from the car sale was an unusually large deposit based on the person's previous deposit re-cord. Federal authorities would be notified and soon agents would be sent to interrogate the customer on the chance he or she might be a drug dealer or terrorist.

In 1998, Representative Ron Paul planned to introduce legislation to stop this intrusive program, but an irate citizenry saved him the trouble. The schemers behind the proposal, the Federal Deposit Insurance Corp, the IRS, and other agencies, quickly backed off. Paul said quite propheti-cally, "Somehow, though, I imagine such action will not stop them, only slow them down."

Paul was right. Almost all of the provisions of the Know Your Cus-tomer program can be found in the PATRIOT Act.

This legislation can undermine the general public's ability to carry out their daily lives. It can red-flag your bank account if you deposit or with-draw a certain amount of money and this amount keeps changing. Once the government was notified if there was more than $10,000 involved. By 2010, this amount had dropped to $5,000. In early 2006, Rhode Island retired schoolteacher Walter Soehnge and his wife tried to pay down an excessive credit card bill with a JC Penney MasterCard. They sent in a check for $6,500 to pay down their debt. When the Soehnges found the money had not been credited to their account, they began to make inquiries. They were told that when a payment is much larger than usual, Homeland Security must be notified and that the money is held until a threat assessment is made.

The couple's money was eventually freed, although they never found out how making a large credit card payment posed a threat to national security.

"If it can happen to me, it can happen to others," Soehnge said.

Even after Congress revised the PATRIOT Act in 2006, portions of the act still concerned both libertarians and some congressmen. Represen-tative C. L. "Butch" Otter of Idaho was one of the three Republicans who found the entire act potentially unconstitutional from the onset. One sec-tion made it illegal for one citizen to tell another that the authorities were conducting searches of his property or business. "Section 215 authorizes the FBI to acquire any business records whatsoever by order of a secret US Court. The recipient of such a search order is forbidden from telling any person that he has received such a request. This is a violation of the First Amendment right to free speech and the Fourth Amendment protection of private property," commented Otter. "[S]ome of these provisions place more power in the hands of law enforcement than our Founding Fathers could have dreamt and severely compromises the civil liberties of law-abiding Americans. This bill, while crafted with good intentions, is rife with constitutional infringements I could not support."

The issue of leveling penalties against persons who reveal how the government intrudes on private life was a central point of controversy when Congress renewed the PATRIOT Act in late 2005. The PATRIOT Act's Section 215 contains a "gag order" clause that was retained by Congress only after legislators reached a compromise on the wording. The gag order clause makes permissible the following scenario: if someone's small business is searched by the FBI, that person is gagged from telling anyone that the feds were there. The compromise made gag orders effective only for a year after a secret search was conducted of a person's property. But even then, one year seems too long for Americans to wait to learn that their government is spying on them.

It is no surprise that this compromise wasn't sufficient for journalists covering the PATRIOT Act. According to a *New York Times* editorialist, "The compromise also fails to address another problem with Section 215: it lets the government go on fishing expeditions, spying on Americans with no connection to terrorism or foreign powers. The act should require the government, in order to get a subpoena, to show that there is a connection between the information it is seeking and a terrorist or a spy."

Given that the United States had been attacked on 9/11 and was fighting in Iraq, Attorney General Alberto Gonzales publicly argued that spying on the American public was within the legal rights of a wartime president. Yet when members of the Senate Judiciary Committee asked to see why Bush was within legal rights to spy on the public, the White House denied requests for classified legal documents that were behind Gonzales's defense.

Opponents to the NSA's warrantless spying claimed it not only was intrusive and a violation of constitutional safeguards on privacy, but also ineffective because it overloaded law enforcement agencies with bad leads. They also saw the surveillance program as a serious step to consolidating power in the executive branch. "The history of power teaches us one thing," said former Reagan administration attorney Bruce Fein, "if it's unchecked, it will be abused."

Nadine Strossen, a professor of law at New York University and president of the American Civil Liberties Union, was also appalled after reading the provisions of the PATRIOT Act. According to Strossen, many of the act's provisions have little or nothing to do with fighting terrorism.

"There is no connection between the September 11 attacks and what is in this legislation," Strossen argued. "Most of the provisions related not just to terrorist crimes but to criminal activity generally. This happened too, with the 1996 antiterrorism legislation where most of the surveillance laws have been used for drug enforcement, gambling and prostitution."

Strossen was right. By 2005, the PATRIOT Act provisions were often being used for cases other than terrorism. According to *New Jersey Star-Ledger* writer Mark Mueller, "While the Justice Department says it does not uniformly track the PATRIOT Act's use in such cases, a reading of government reports and congressional testimony shows it has been used hundreds of times against the likes of drug dealers, computer hackers, child pornographers, armed robbers and kidnappers. In Washington State, investigators invoked the law to surreptitiously bug a tunnel that had been bored beneath the US-Canadian border by drug runners. In Las Vegas, prosecutors used it to seize the financial records of a strip-club owner suspected of bribing local government officials."

Investigative reporter Kelly O'Meara noted that a similar antiterrorist act in England allows government investigators to obtain information from Internet service providers about their subscribers without a warrant. The British law is now being applied to minor crimes, tax collection, and public health measures so it probably won't be long before this insidious bypassing of former rights takes place in America.

Obama's opinions on electronic surveillance have been a lightning rod since he shifted positions during the 2008 presidential campaign. Early in his campaign, Obama had opposed granting immunity to telecommunications companies from lawsuits by Americans who believed their privacy had been violated in government electronic data collection programs. Yet in July 2008, then senator Obama, over the objections of liberals, voted to support expanding the Foreign Intelligence Surveillance Act, which grants immunity to the telecommunications firms.

This flip-flopping by Obama prompted Tucker Bounds, a spokesman at the time for the John McCain campaign, to say, "He's willing to change positions, break campaign commitments and undermine his own words in his quest for higher office." It might be noted that McCain, campaigning in Pittsburgh, was absent for the vote.

Many liberals on the House Judiciary Committee had wanted the Bush administration's surveillance excesses corrected. Yet Deputy Assistant Attorney General Todd Hinnen told the committee that although the Obama administration was willing to negotiate stronger privacy protection for the public, it insisted on keeping in place current authority to track suspects and obtain records.

Committee chairman John Conyers, a Michigan Democrat, groused, "You sound like a lot of people who came over from DOJ [the Department of Justice] before." What Conyers did not clarify was that the people from the DOJ were from the Bush administration.

In early 2009, the Electronic Frontier Foundation (EFF), a U.S.-based international nonprofit advocacy and legal organization dedicated to defending civil rights in the digital world, challenged government electronic surveillance by suing the NSA over the eavesdropping on millions of ordinary Americans. In April that same year, the Obama administration filed a motion to dismiss the suit. Obama defended his decision by adopting the Bush administration's argument—the courts cannot judge the legality of the NSA's warrantless wiretapping program. Furthermore, a court case would disclose "state secrets."

"President Obama promised the American people a new era of transparency, accountability, and respect for civil liberties," commented EFF senior staff attorney Kevin Bankston. "But with the Obama Justice Department continuing the Bush administration's cover-up of the National Security Agency's dragnet surveillance of millions of Americans, and insisting that the much-publicized warrantless wiretapping program is still a 'secret' that cannot be reviewed by the courts, it feels like deja vu all over again."

When asked what the Founding Fathers might say about the use of the PATRIOT Act, Congressman Ron Paul laughed and said, "Our forefathers would think it's time for a revolution. This is why they revolted in the first place. They revolted against much more mild oppression."

MAGIC LANTERN, FLUENT, DTECTIVE, AND ENCASE

THE GLOBALISTS HAVE NOT limited themselves to merely subverting and neutralizing the flow of information on the Internet. Law enforcement possesses a device called a "key logger," which can be secretly installed into computers using a viruslike program. The device, code-named Magic Lantern, allows authorities to capture passwords by recording every key stroke on the computer. Authorities can then use the passwords to access encrypted data files. The FBI has acknowledged using such a device in a recent gambling investigation.

Again, this kind of law enforcement tactic is beneficial to the public good if it is used on criminals. However, William Newman, director of the ACLU in western Massachusetts, said Magic Lantern technology could easily be used to spy on all Americans. He pointed out that federal law enforcement agencies now are permitted "the same access to your Internet use and to your email use that they had to your telephone records." Agencies could easily overstep their authority. "The history of the FBI is that they will do exactly that."

In a 2007 *Wired* magazine article, former hacker and senior editor Kevin Poulsen wrote about how the FBI used technology similar to Magic Lantern to catch a teenager making bomb threats. According to Poulsen, court papers offered "the first public glimpse into the bureau's long-suspected spyware capability, in which the FBI adopts techniques more common to online criminals." In an affidavit to the U.S. District Court in the Western District of Washington, FBI agent Norman Sanders described the software as a "computer and internet protocol address verifier (CIPAV)."

Poulsen explained that this software "was sent to the owner of an anonymous MySpace profile linked to bomb threats against Timberline High School near Seattle. The code led the FBI to 15-year-old Josh Glazebrook, a student at the school, who . . . pleaded guilty to making bomb threats, identity theft and felony harassment."

In July 2007, the U.S. Court of Appeals for the Ninth Circuit supported the Washington court's decision that this type of computer monitoring without a wiretap warrant is legally permissible because Internet users have no "reasonable expectation of privacy" when using the Internet.

Law enforcement and intelligence agencies are hard at work developing other types of technology that are purportedly necessary to fight terrorism. For example, the CIA is developing a program called "Fluent," which searches foreign websites for terrorist activities and displays an English translation back to Langley. Fluent may be used in conjunction with "Oasis," a technology that transcribes into English worldwide radio and TV broadcasts. The FBI and some police departments are now using a software program called "dTective" to record financial transactions with dramatically improved surveillance video feeds from banks and ATMs. The feds are even working on techniques for restoring videotapes and computer disks that have been destroyed, cut up, or tossed in water. One software program called "Encase" can recover deleted computer files and search for incriminating documents on any computer. This was used by the FBI to examine computers seized in the wake of the 9/11 attacks.

In 2010, testing was being done on a device that emits an electromagnetic pulse capable of disabling the engine of any vehicle. The developers hoped to have a portable model ready in the near future for use by police. They said it would signal the end of dangerous car chases.

All this surveillance technology could hypothetically lead to scary scenarios such as the one envisioned by *Village Voice* editor Russ Kick: "You just got a call that your sister is in critical condition in the hospital. So you jump in your car and hit the gas. Trouble is, the speed limit is 30 miles per hour and your car won't let you drive any faster. Or maybe you're lucky enough to have a vehicle that still lets you drive at the speed you choose. A cop pulls you over and demands a saliva sample, so he can instantly match your DNA to a data bank of criminals' genes. You refuse and are arrested. After booking you, the authorities force you to submit to 'brain fingerprinting,' a technology that can tell if memories of illegal events are in your mind.

"By this point, you're thinking this is a worst-case scenario, a science-fiction dysphoria. Well, wake up and smell the police state, because all this technology—and more—is already being implemented."

NATIONAL ID ACT

FOR YEARS, PUNDITS HAVE consistently brought up the idea of a national identification card while civil libertarians have consistently cooled the public's receptivity to such a concept—until now.

Even as the terror following 9/11 began to subside in 2002, Representative Jim Moran of Virginia cited increased concern over terrorism and introduced legislation in Congress titled the Driver's License Modernization Act of 2002 (H.R. 4633). The bill was styled as a law, which would set uniform standards for driver's licenses in all fifty states and the District of Columbia. But what was most disturbing was that it also included provisions to establish a national database and identification system. Moran's bill codified a plan developed by Congress that urged the Department of Transportation to develop electronic "smart" driver's licenses. The licenses would contain embedded programmable computer chips that could be read by law enforcement authorities across the nation.

"It's more of a national ID *system* [original emphasis], a linking of Department of Motor Vehicles—and the records they keep on you—across state lines, with some extra on-card security measures thrown in," wrote

Frank Pellegrini of Time.com. "The plan, Congress hopes, will be cheaper and easier to implement, and less likely to incur the talk-show ire of civil libertarians and states' rights purists (the same type who squawked in 1908 when the FBI was born). But the approach is mere stealth—50 different state ID cards all linked together is pretty much the same as one national ID card, just as all those new quarters are still worth 25 cents each, no matter which state is on the back."

The House bill also stated the new ID card must "conform to any other standards issued by the Secretary [of Transportation]," an open invitation for bureaucrat tinkering.

The Rearing and Empowering America for Longevity against acts of International Destruction (REAL ID) Act of 2005 was passed in an effort to set standards for all driver's licenses, making them acceptable for "official purposes" as defined by the secretary of Homeland Security. These purposes included entering any federal building and boarding any commercial airliner. But the states balked at the plan, not due to privacy and control concerns but because of the cost of implementing it, and by 2008, an extension was given to all states. As concerns over REAL ID grew, by October 2009 at least twenty-five states had passed resolutions or legislation withdrawing from REAL ID.

In April 2009, without acknowledging the rebellion of the states over the REAL ID Act, Homeland Security Secretary Janet Napolitano announced she was working with governors to repeal the REAL ID Act. Napolitano, a former governor of Arizona, said she wanted to substitute the federal law with "something else that pivots off of the driver's license but accomplishes some of the same goals." She added, "And we hope to be able to announce something on that fairly soon."

A CHIP IN YOUR SHOULDER

LOVERS OF LIBERTY REJOICED when Moran's bill failed to become law and the ensuing REAL ID Act bombed in the state houses. However, most states now issue driver's licenses with a magnetic strip capable of carrying computer-coded information.

Driver's licenses are not the only ID cards to contain computer-coded information. New York City became one of the first major cities to announce plans to try out microchipped identification cards for the city's 250,000 employees. Some 50,000 officers and workers for the NYPD were scheduled to receive ID cards. The state-of-the-art plastic cards contain microchips, holograms, and other security devices to prevent theft. On the front of this picture ID is the Statue of Liberty and two chips, one containing fingerprints and handprints and the other filled with personal information, including blood type and emergency telephone numbers. Police officials said that eventually the ID cards will be used in conjunction with "biometric" hand scanners to ensure the person bearing the card is the correct one. They also hoped to save money in computing paychecks by using the cards to keep track of employee hours.

Time's Frank Pellegrini has warned that the real fight for privacy will be over when and where citizens will have to show such IDs. "The average American's driver's license gets a pretty good workout these days," he said, "certainly far more than traffic laws themselves would seem to warrant— but you can only get arrested for *driving* without one. If the US domestic response starts to resemble Zimbabwe's, which passed a law in November [2001] making it compulsory to carry ID on pain of fine or imprisonment, well, that's something to worry about."

In 2002, author Steven Yates, a teaching fellow at the Ludwig von Mises Institute, warned, "The long and the short of it is, the Driver's License Modernization Act of 2002 would bring us closer than ever before to establishing a comprehensive national ID system. The present excuse is that extreme measures are necessary to 'protect us against terrorism.'

"It is a testimony to how much this country has changed since 9/11 that no one has visibly challenged H.R. 4633 as unconstitutional and incompatible with the principles of a free society. The 1990s gave us the obviously corrupt Clinton Regime and a significant opposition to federal power grabs. Now it's Bush the Younger, beloved of neocons [neoconservatives] who see him as one of their own and believe he can do no wrong. . . .

Clearly, the slow encirclement of law-abiding US citizens with national ID technology would advance such a cause [globalism or the New World Order] while doing little if anything to safeguard us against terrorism."

Yates predicted a chilling future where the feds could stifle dissent by "freezing" a dissident's assets by reprogramming his or her database information. Scanners would not recognize the dissenter and he or she would become officially invisible, unable to drive or work legally, have a bank account, buy anything on credit, or even see a doctor. "Do we want to trust *anyone* [original emphasis] with that kind of power?" he asked.

Already the practice of marking people for identification through computer systems is being played out in private industry. In late October 2002, Applied Digital Solutions, Inc., a high-tech development company headquartered in Palm Beach, Florida, announced a national promotion named "Get Chipped" for its new subdermal personal verification microchip. Applied Digital Solutions company literature states that its "VeriChip" is "an implantable, 12mm by 2.1mm radio frequency device . . . about the size of the point of a typical ballpoint pen. It contains a unique verification number. Utilizing an external scanner, radio frequency energy passes through the skin energizing the dormant VeriChip, which then emits a radio frequency signal containing the verification number. The number is displayed by the scanner and transmitted to a secure data storage site by authorized personnel via telephone or Internet."

The chip can be used to access nonpublic facilities such as government buildings and installations, nuclear power plants, national research laboratories, correctional institutions, and transportation hubs, either by itself or in conjunction with existing security technologies such as retinal scanners, thumbprint scanners, or face recognition devices. Applied Digital Solutions officials believe the chip will eventually be used in a wide range of consumer products, including PC and laptop computers, personal vehicles, cell phones, and homes and apartments. They said the implanted chip will help stop identity theft and aid in the war against terrorists.

In addition to "VeriChip Centers" in Arizona, Texas, and Florida, the firm also fields the "ChipMobile," a motorized marketing and "chipping" vehicle. The firm's Get Chipped campaign was launched just days after the Food and Drug Administration ruled that the chip is not a regulated medical device.

Tommy Thompson, a former Wisconsin governor and secretary of health and human services in the George W. Bush administration, subsequently joined the board of directors of VeriChip. He pledged to get chipped and encouraged Americans to do the same so their electronic medical records would be available in emergencies.

By early 2006, fears of mandatory chipping became reality when a Cincinnati video surveillance firm, CityWatcher.com, began to require employees who worked in its secure data center to implant the VeriChip device into their arm.

Many also feared that the microchips were being included in the swine flu vaccine. In September 2009, VeriChip Corp. announced that its stock shares had tripled after the company was granted an exclusive license to patents for "implantable virus detection systems in humans." The system used biosensors that can detect swine flu and other viruses and was intended to combine with VeriChip's implantable radio frequency identification devices (RFIDs) to develop virus triage detection systems, microchips in one's bloodstream broadcasting the body's information to whoever has a reader device.

The use of GPS devices is reminiscent of the 1987 film *The Running Man,* in which Arnold Schwarzenegger is equipped with a collar that will blow his head off if he leaves a certain area. So, if microchipping the population sounds like something from a science fiction movie, consider that the giant drug corporation Novartis has already tested a microchip that reminds a person to take his or her medicine by transmitting a signal to a receiver chip implanted in the patient's shoulder. The pill itself contains a tiny "harmless" microchip that signals the receiver chip each time a pill is taken. If the patient fails to take a pill within a prescribed time period, the receiver chip signals the patient or a caretaker to remind the person. Novartis's head of pharmaceuticals, Joe Jiminez, said testing of the "chip in the pill" to a shoulder receiver chip had been carried out on twenty patients by the close of 2009.

One shouldn't count on government watchdog organizations to always maintain privacy rights. In late 2002, the American Civil Liberties Union gave its stamp of approval to an electronic tracking system that uses GPS satellites to track suspects and criminals. Created by the Veridan company of Arlington, Virginia, this "VeriTracks" system not only keeps tabs on convicted criminals but also on suspects. It can even match a person's position to high-crime areas or crime scenes and suggest that the person may be involved in law breaking. Law enforcement agencies can create "electronic fences" around areas they deem off-limits to those wearing a cell-phone-size GPS receiver. The person who wears the module must tie it around his or her waist while an electronic bracelet worn on the ankle acts as an electronic tether to the GPS receiver that records the person's exact position. Should the wearer move outside the proscribed area, the authorities are signaled and a police unit is dispatched. At night, the wearer must place the module in a docking system to recharge batteries and upload its data to a central headquarters, which checks to see if the wearer has been at any crime scenes.

How do you get someone to agree to this monitoring system? Sheriff Don Eslinger of Seminole County, Florida, answered, "It's either wear the GPS device or go to jail. Most of them find this much more advantageous than sitting in a cold jail cell, and it also saves us between $45 and $55 a day." Eslinger said his county had equipped ten pretrial suspects with the GPS device as a condition of making bond. According to Eslinger, county officials hoped to expand the program to include nonviolent probationers and parolees.

For many, using GPS tracking devices to track criminals makes sense. Yet, disturbingly, surveillance technology has not been limited to felons and probationers. In Texas, some one thousand teenage drivers allowed an unnamed insurance company to place a transponder in their vehicles to keep track of their speed on the road.

Texas representative Larry Phillips introduced a bill in 2005 that would have required all state automobile inspection stickers to carry a built-in electronic transponder. The device would transmit information like the car's vehicle identification number (VIN), insurance policy number, and license plate number, and should the owner's insurance expire, the person would be mailed a $250 ticket. This bill was not passed.

The firm Digital Angel has developed a wristband that allows parents to log on to the Internet and instantly locate their children, who must wear the bracelet. Another company, eWorldtrack, is working on a child-tracking device that will fit inside athletic shoes. The German firm Siemens has tested a seven-ounce tracking device that allows constant communication between parents and their children.

Author and political critic Joe Queenan quipped, "Fusing digital mobile phone technology, a satellite-based global positioning system and good old-fashioned insanity, the device can pinpoint a child within several yards in a matter of seconds."

Support has grown in the American legal system for GPS surveillance technology. In spring 2002, the Nevada Supreme Court ruled it was okay for police to hide electronic monitoring devices on people's vehicles without a warrant for as long as they want. The court ruled that there is "no reasonable expectation of privacy" on the outside of one's vehicle and that attaching an electronic device to a man's car bumper did not constitute unreasonable search or seizure. In early 2004, a Louisiana court ruled it was permissible for police there to make warrantless searches of homes and businesses even without probable cause.

In September 2009, the Massachusetts Supreme Judicial Court ruled that the state constitution allows police to break into a suspect's car to secretly install a GPS tracking device, provided that authorities have a warrant before they do so. The unanimous ruling upheld the drug trafficking conviction of Everett H. Connolly, a Cape Cod man who was tracked by state police in 2004 after they installed a GPS device in his minivan. The court declared the GPS device an "investigative tool" and said it did not violate the ban on unreasonable search and seizure in the state's Declaration of Rights.

"We hold that warrants for GPS monitoring of a vehicle may be issued," Justice Judith Cowin stated in the court's opinion. "The Commonwealth must establish, before a magistrate ... that GPS monitoring of the vehicle will produce evidence that a crime has been committed or will be committed in the near future." Generally, search warrants expire after seven days, yet the court said GPS devices can be installed for up to fifteen days before police must prove that the devices need to remain in place.

In an attempt to provide protection against the widespread use of GPS devices by law enforcement, William Leahy, chief counsel for the Committee on Public Counsel Service, said the court's ruling means that police must persuade a judge they have probable cause before the GPS devices can be installed.

ECHELON AND TEMPEST

THOUGH GPS AND SURVEILLANCE systems are reasons for serious concern, the two greatest electronic threats to American privacy and individual freedom are Echelon and TEMPEST.

"The secret is out," wrote Jim Wilson in *Popular Mechanics.* "Two powerful intelligence gathering tools that the United States created to eavesdrop on Soviet leaders and to track KGB spies are now being used to monitor Americans." Echelon is a global eavesdropping satellite network and massive supercomputer system that operates from the National Security Agency's headquarters in Maryland. It intercepts and analyzes phone calls, faxes, and e-mail sent to and from the United States, both with or without encryption. Encrypted messages are first decrypted and then joined with clear messages. The NSA then checks all messages for "trigger words" with software known as "Dictionary." Terms like "nuclear bomb," "al Qaeda," "Hamas," "anthrax," and so on are then shuttled to appropriate agencies for analysis.

Although speculation and warnings about Echelon were circulating on the Internet for a number of years, it was not until 2001 that the U.S. government finally admitted the program's existence. This admission came after high-profile investigations in Europe discovered that Echelon had been used to spy on the two European companies Airbus Industries and Thomson-CSF.

Though the U.S. government revealed Echelon's use in 2001, the government had been using an early version of Echelon in the late 1960s and 1970s. During that time, Presidents Lyndon Johnson and Richard Nixon used NSA technology to gather files on thousands of American citizens and more than a thousand organizations opposed to the Vietnam War. In a program called Operation Shamrock, the NSA collected and monitored nearly every international telegram sent from New York.

Although paid for primarily by U.S. taxpayers, Echelon is now multinational and involves nations like the United Kingdom, Canada, Australia, New Zealand, and even Italy and Turkey. Most of the information that comes from Echelon goes to the CIA. According to *Popular Mechanics*'s Wilson, "Based on what is known about the location of Echelon bases and satellites, it is estimated that there is a ninety percent chance that NSA is listening when you pick up the phone to place or answer an overseas call. In theory, but obviously not in practice, Echelon's supercomputers are so fast, they could identify Saddam Hussein by the sound of his voice the moment he begins speaking on the phone."

Amazing as all this technology may sound, because the government now acknowledges its existence may mean that it is phasing the program out for another technology. The next system that the government uses may be a ground-based technology known as TEMPEST. To prevent your computer from causing static on your neighbor's TV, the Federal Communications Commission (FCC) certifies all electronic and electrical equipment. TEMPEST, or Telecommunications Electronics Material Protected from Emanating Spurious Transmissions, technology stemmed from simply shielding electronic equipment to prevent interference with nearby devices. But in the process of preventing unwanted electronic signals, researchers learned how to pick up signals at a distance. Advances in TEMPEST technology mean that somewhere out there, someone may be able to secretly read the displays on machines like personal computers, cash registers, television sets, and automated teller machines (ATMs) without the person using those machines knowing it.

Jim Wilson wrote that documents now available from foreign governments and older sources clearly show how these systems are used to invade our right to privacy. "We think you will agree it also creates a real and present threat to our freedom."

In September 2002, the Associated Press obtained U.S. government documents that showed that the Bush administration would create a fund that would combine tax dollars with funds from the technology industry to pay for "Internet security enhancements." Under the title "Executive Summary for the National Strategy to Secure Cyberspace," the documents discussed "sweeping new obligations on companies, universities, federal agencies and home users" to make the Internet more secure, presumably from terrorists.

This new Internet strategy was headed up by Richard Clarke, formerly a top counterterrorism expert in both the Bush and Clinton administrations, and Howard Schmidt, a former senior executive at Microsoft Corp. When released in 2003, the plan offered more than eighty recommendations for tightening Internet security.

THE CYBERSECURITY ACT OF 2009

ONE REASON THE GLOBALISTS want to shut down the free flow of information is that it interferes with their fearmongering and sociopolitical manipulation. With the introduction of Senate Bills No. 773 and 778, by Democratic senator Jay Rockefeller of West Virginia, legislators continued to put the power to shutter free speech into the hands of the executive branch. These bills are part of what is called the Cybersecurity Act of 2009, and they essentially give the president of the United States the power to shut down Internet sites he feels might compromise national security.

The bills put forth the idea of creating a new Office of the National Cybersecurity Advisor to protect the nation from cybercrime, espionage, and attack. The new cybersecurity adviser would report directly to the president. In the event of cyberattack, which is ill defined in the proposed laws, the president, through this national cybersecurity adviser, would have the authority to disconnect "critical infrastructure" from the Internet, which would include citizens' banking and health records. According to an early draft of the bill, the secretary of commerce would have access to all privately owned information networks deemed critical to the nation "without regard to any provision of law, regulation, rule or policy restricting such access."

In talks to Congress, Senator Rockefeller warned that "we must protect our critical infrastructure at all costs." And the bills' cosponsor, Maine Republican senator Olympia Snowe, said that failure to pass this law would risk a "cyber-Katrina." However, privacy advocates immediately attacked the legislation. Leslie Harris, president of the Center for Democracy & Technology, stated, "The cybersecurity threat is real, but such a drastic federal intervention in private communications technology and networks could harm both security and privacy."

Larry Seltzer, a technology writer for the Internet news source eWeek, agreed with Harris. "The whole thing smells bad to me. I don't like the chances of the government improving this situation by taking it over generally, and I definitely don't like the idea of politicizing this authority by putting it in the direct control of the president."

Jennifer Granick, civil liberties director at the Electronic Frontier Foundation, said that by concentrating Internet control in one individual, the Internet could actually become less safe. When one person can access all information on a network, "it makes it more vulnerable to intruders," argued Granick. "You've basically established a path for the bad guys to skip down." Granick added that the nonspecific scope of this legislation is "contrary to what the Constitution promises us." Should the Commerce Department decide to use information gained while accessing "critical infrastructure" on the Net against the user, privacy would be lost. According to Granick, this is a clear violation of the U.S. Constitution's Article IV, which states the "right of the people to be secure in their persons, houses, papers, and effects, against unreasonable searches and seizures, shall not be violated. . . ."

"Who's interested in this [legislation]?" asked Granick. "Law enforcement and people in the security industry who want to ensure more government dollars go to them."

TAKE A NUMBER

WITH HIGH-POWERED TECHNOLOGY AND the right legislation in place, America is coming closer and closer to the totalitarian surveillance society that George Orwell described in *1984*.

Consider how the government has effectively enumerated the American citizenry over fifty years:

- 1935—Social Security initiated.
- 1936—The current Social Security numbering system begins.
- 1962—The IRS starts requiring Social Security numbers on tax returns even though Social Security cards plainly state the number was "Not for Identification."
- 1970—All banks require a Social Security number.
- 1971—Military ID numbers are changed to Social Security numbers.
- 1982—Anyone receiving government largesse is required to obtain a Social Security number.
- 1984—Anyone being declared a dependent for IRS tax purposes needs a Social Security number. Within two years, even newborn babies were required to have a Social Security number under penalty of fine.

Free people are individuals. Enslaved serfs are numbered chattel. If Americans are to remain a truly free people, tight restrictions must be placed on the microchipping of the population as well as the frequency of times the State requires one to present a number for identification. At the rate things are going, George Orwell's *1984* vision of psychological and electronic tyranny is almost upon us.

HOMELAND SECURITY

IF SURVEILLANCE OF THE American public is being centralized, then it makes sense that the nation would need a more centralized law enforcement agency—in other words, a national police force.

During his nearly forty-year career as director of the FBI, J. Edgar Hoover continuously argued against the need for a national police force. This may have been due more to maintaining the independence of his bureau than any personal regard for civil liberties. Yet Hoover's objection struck a chord with the majority of Americans.

But with the hurried passage of a law creating the Department of Homeland Security (DHS) in November 2002, a nationalized police force was formed. This act was the greatest restructuring of the federal government since the National Security Act of 1947, yet this time it didn't include any of the previous act's deliberation and review. After 9/11, President Bush argued that this needed to be done rapidly, because the country faced "an urgent need, and [the government needed to move] quickly . . . before the end of the congressional session." Thus began the push to create the Department of Homeland Security with Tom Ridge holding a cabinet-level position controlling more than 170,000 federal employees and twenty-two federal agencies.

Despite congressional misgivings, the Homeland Security Act passed speedily through Congress with little or no revision. In the U.S. Senate, the proposal to create the department passed on a 98–1 vote (one senator did not vote). (As a side note: apparently, senators were so confident that they were about to do a genuine service for America that they voted themselves a pay raise for the fourth consecutive year.) The Homeland Security bill was signed into law by President Bush on November 25, 2002.

With the new office, Bush wanted to bring a myriad of government agencies under one central control. The agencies responsible for border, coastline, and transportation security were now under the command of the new office, and Bush remarked, "The continuing threat of terrorism, the threat of mass murder on our own soil, will be met with a unified, effective response." By 2006, Homeland Security encompassed more than eighty-seven thousand government jurisdictions at both the state and local level, with additional directorates carrying names, both familiar and not, such as Preparedness, Science & Technology, Management, Policy, FEMA, the TSA, Customs, Border Patrol, the INS, the Federal Law Enforcement Training Center, the U.S. Coast Guard, and the Secret Service.

Equally disturbing as Homeland Security's power was the news that the federal General Services Administration (GSA) had asked for $481.6 million in discretionary funds in its fiscal 2009 budget request, constituting a 103 percent increase from the $237.7 million in fiscal 2008. Legislators had earmarked the increase for the initial construction of a new DHS headquarters.

Recently, the DHS announced it will consolidate most of its sixty offices spread across the Washington, D.C., region into a single new headquarters building that should cost $3 billion. The move into the new HQ is scheduled to begin in 2011 with a 1.2-million-square-foot headquarters to be built in southeast Washington on the grounds of the closed St. Elizabeth's Hospital, ironically a former mental asylum.

The project continued to invite controversy, with opposition coming from numerous area organizations and think tanks. "Aside from the humorous nature of moving perhaps the most helter-skelter of all federal agencies onto the grounds of an old loony bin, this move is so shockingly idiotic that only the DHS could do it," wrote James Joyner, a former army officer, an editor for a nonpartisan group, and publisher of *Outside the Beltway,* an online journal of politics and foreign affairs analysis. "It was bad enough that the Powers That Be gave in to political pressure and headquartered DHS in D.C. proper rather than out in the much cheaper, more secure space in Chantilly, Virginia as originally planned. But now they're consolidating their critical functions into a single building?! . . . Dispersion would save the taxpayer billions in cost of living and inordinately improve the quality of life of most DHS employees." Joyner added that from a security standpoint, it would seem to be more advantageous to disperse Homeland Security components rather than concentrate them all in one location.

By 2010, the power of Homeland Security was being felt in even small police and sheriffs' departments across the nation. Tax money flowed into them providing everything from updated—and interlinked—computer systems to bulletproof vests, crowd-control devices, and armored cars and helicopters. More vocal critics were noting the similarities between Homeland Security and the German Gestapo, which by 1935 had brought all of Nazi Germany's law enforcement agencies under its control and considered any person, regardless of their position in society, suspect.

Could the consolidation of power within Homeland Security be a continuation of a plan to change America from a well-defended constitutional republic to a police state, a reincarnation of the Nazi Gestapo? No one knows for certain. Immediately upon taking office, President Bush ordered all records of former presidents, including Reagan and those of his father, sealed from the public. Under Bush's executive order, even if an ex-president wants to release his papers to the public, the sitting president has the right to prevent their release.

Although in April 2009, Obama did order some of Reagan's papers released from the National Archives, many remained kept from the public, including his own birth certificate and school records. He also fought to keep secret his White House visitor list. Due to this failure to act on his campaign pledge of "transparency" in government, by 2010, fears that a police state might be just around the corner were heightened.

None of this is really new. Plans to shift America into a police state date back to 1984, when Reagan's National Security Council (NSC) drafted a plan to impose martial law in the United States through FEMA. Marine lieutenant colonel Oliver North helped author the plan, which in 1987 was leaked to the media.

Arthur Liman, then chief counsel of the Senate Iran-Contra Committee, declared in a memo that North was at the center of what amounted to a "secret government-within-a-government." Oddly, this is a term similar to Bush's "shadow government." At the time, officials said North's involvement in the proposed plan to radically alter the American government by executive order was proof that he was involved in a wide range of secret activities, foreign and domestic, that went far beyond the Iran-Contra scandal.

North's shadow-government plan called for suspension of the Constitution and turning control of the government over to the little-known FEMA. Military commanders would be appointed to run state and local governments. In the event of a crisis such as "nuclear war, violent and widespread internal dissent or national opposition to a US military invasion abroad," the government would declare martial law. When he drafted these plans, North was the NSC's liaison to FEMA. Many people are bothered by the idea of being placed under martial law in the event of widespread internal dissent, especially when they look at the continuation of many of Bush's policies under Obama.

North's contingency plan was to be part of an executive order or legislative package that Reagan would sign but hold secretly within the NSC until such time as a crisis arose. It was never revealed whether Reagan had signed the plan.

When Bush took office, sealed Reagan's records from the public, and moved to create the Homeland Security Department, former Nixon counselor John Dean warned that America was sliding into a "constitutional dictatorship" and martial law. Further concerns were voiced by Timothy H. Edgar, legislative counsel for the American Civil Liberties Union. In testimony to various congressional committees, Edgar noted that the Homeland Security Department would have substantial powers and more armed federal agents with arrest authority than any other government agency. He questioned whether the new department would have structural and legal safeguards to keep it open and accountable to the public. "Unfortunately, legislation [to create the Homeland Security Department] not only fails to provide such safeguards, it eviscerates many of the safeguards that are available throughout the government and have worked well to safeguard the public interest," stated Edgar.

He then enumerated problem areas within the proposed Homeland Security Department, saying it undermines the Freedom of Information Act (FOIA) by allowing the various agencies to decide on their own which documents should be made public, and it limits citizen input by exempting advisory committees to Homeland Security from the Federal Advisory Committee Act (FACA). Passed in 1972, FACA was designed to ensure openness, accountability, and the balance of viewpoints in government advisory groups. Edgar also argued that, by allowing the secretary of Homeland Security to make his own personnel rules, the Homeland Security Department would silence whistle-blowers protected under the federal Whistleblower Protection Act (WPA). Last, the HSD might threaten personal privacy and constitutional freedoms because the vague wording in the Homeland Security Act does not provide sufficient guarantees.

The ACLU counsel was also hugely concerned by plans to combine the CIA and the FBI under Homeland Security. "The CIA and other agencies that gather foreign intelligence abroad operate in the largely lawless environment," noted Edgar. "To bring these agencies into the same organization as the FBI risks further damage to Americans' civil liberties."

Edgar added, "No one wants a repeat of the J. Edgar Hoover era, when the FBI [under the infamous COINTELPRO] was used to collect information about and disrupt the activities of civil rights leaders and others whose ideas Hoover disdained. Moreover, during the Clinton Administration, the 'Filegate' matter involving the improper transfer of sensitive information from FBI background checks of prominent Republicans to the White House generated enormous public concern that private security-related information was being used for political purposes. Congress should not provide a future Administration with the temptation to use information available in Homeland Security Department files to the detriment of its political enemies."

NO-FLY LIST

INFORMATION ABOUT HOMELAND SECURITY'S information databases became public when news spread about its "no-fly" lists of people suspected of terrorist connections. By early 2006, this list included 325,000 names, compiled from more than twenty-six terrorism-related databases from the intelligence and law enforcement communities.

"We have lists that are having baby lists at this point," commented

Timothy Sparapani, a former legislative attorney for privacy rights at the American Civil Liberties Union. "If we have over 300,000 known terrorists who want to do this country harm, we've got a bigger problem than deciding which names go on which list. But I highly doubt this is the case."

In early 2006, Alberto Gonzales, then attorney general, tried to assure members of the Senate Judiciary Committee that "information is collected, information is retained and information is disseminated in a way to protect the privacy interests of all Americans."

But Gonzales had a hard time convincing the late senator Edward Kennedy, a committee member who was prevented from flying five times in March 2004 because a "T. Kennedy" appeared on the DHS's no-fly list. In Washington, Boston, and other cities, airline employees refused to issue Kennedy a boarding pass because his name was on the no-fly list. Kennedy was delayed multiple times until supervisors were called and approved his travel. Even after supposedly clearing up the mistake in names, Kennedy was stopped again from flying in 2004, mostly at Boston's Logan International Airport. Banned from flying, even in his own hometown, prompted a personal telephone apology from Homeland Security Secretary Tom Ridge.

"That a clerical error could lend one of the most powerful people in Washington to the list—it makes one wonder just how many others who are not terrorists are on the list," commented senior ACLU counsel Reginald T. Shuford. "Someone of Senator Kennedy's stature can simply call a friend to have his name removed but a regular American citizen does not have that ability. He had to call three times himself."

Alarmingly, Timothy Edgar's fears that the lists will be used for political vengeance may already be realized. Arizona state treasurer Dean Martin told a local TV journalist that in 2009 his name suddenly appeared on the government's no-fly list after former Arizona governor Janet Napolitano became head of the DHS. Martin claims the blacklisting may be based on his past political rivalry with Napolitano. "My staff used to joke after my disagreements with the previous governor that I wouldn't be able to fly once she got back in D.C.," Martin said. "I didn't believe them, but it's actually happening."

Another incident involved Dr. Robert Johnson, a heart surgeon in up-state New York and a retired lieutenant colonel in the U.S. Army Reserve who had served during the time of the first Gulf War. In early 2006 when he arrived at a Syracuse airport for a flight, he was barred and told he was on the federal no-fly list as a possible terror suspect.

"Why would a former lieutenant colonel who swore an oath to defend and protect our country pose a threat of terrorism?" Johnson asked a local newspaper. Johnson speculated that he was placed on the list because in 2004, as a Democrat, he had challenged Republican representative John McHugh for his Twenty-third District congressional seat. The colonel also had been an outspoken critic of the invasion and occupation of Iraq.

Like many other citizens who find themselves on the no-fly list, Johnson is demanding answers as to who decides which name goes on the list and how someone like him can get off. By 2010, the TSA's Terrorist Identities Datamart Environment (TIDE), the intelligence community's central repository of information on known and suspected international terrorists, had grown to more than a half-million persons. But that was not the worst of it.

On Christmas Day 2009, a twenty-three-year-old Nigerian, Umar Farouk Abdulmutallab, allegedly tried to set off explosive powder hidden in his undergarments aboard Northwest Flight 253, becoming known as "the underwear bomber." It was big news at the time, but after more facts about the event surfaced, the whole episode began to accrue a nasty smell.

Other passengers told how Abdulmutallab was escorted by a well-dressed man who talked his way past airline employees despite the fact that Abdulmutallab had only carry-on bags and had paid cash for the transatlantic flight to Detroit. Additionally, Abdulmutallab's father, Al-haji Umaru Mutallab, former chairman of First Bank Nigeria and a former Nigerian minister, had reported his son's militant activities to the U.S. embassy and Nigerian security agencies six months before the incident. Then, during a January 2010 session of the House Committee on Homeland Security, it was found that U.S. intelligence agencies had prevented the State Department from revoking Abdulmutallab's U.S. visa, a move that would have prevented him from boarding the plane. Patrick

F. Kennedy, an under secretary for management at the State Department, said intelligence officials asked his agency not to deny a visa to the suspected terrorist because they felt it might have hampered an investigation into al Qaeda. Others saw the action as evidence that elements within intelligence agencies had paved the way for Abdulmutallab. Another problem for this story was that reports stated Abdulmutallab had attempted to ignite pentaerythritol tetranitrate (PETN), an explosive used by the U.S. military. Persons familiar with PETN said a blasting cap, not simple fire from a match or lighter, is required to detonate PETN. No blasting caps or primers were found on Abdulmutallab, and the chemicals he carried were inadequate to generate an explosion, leading conspiracy researchers to suspect that the whole episode was another false-flag attack, one engineered to cast blame on others rather than the real culprits. But what could have been the purpose?

It is possible that Congress needed the new terrorist scare to coincide with a debate over rescinding some PATRIOT Act measures. After all, the measures were continued. Also, following the terrorist scare, existing plans to equip major airports with full-body scanning devices went into high gear alongside a public relations blitz by former Homeland Security chief Michael Chertoff, cofounder of the Chertoff Group, a security and risk-management firm whose clients include Rapiscan Systems, a manufacturer of the body-imaging screening machines, 150 of which were purchased by the TSA for $25 million in early 2010.

"Mr. Chertoff should not be allowed to abuse the trust the public has placed in him as a former public servant to privately gain from the sale of full-body scanners under the pretense that the scanners would have detected this particular type of explosive [PETN]," said Kate Hanni, a founder of FlyersRights.org, an airport passengers' rights group opposed to the use of the full-body scanners. In January 2010, about forty body scanners were in use at nineteen U.S. airports. This number was expected to climb to more than three hundred machines by the end of that year, mostly due to the publicity over the Christmas Day incident. Despite Chertoff's claim that scanners could have detected PETN, Ben Wallace, a member of Parliament who formerly had worked on developing such scanners for airport use, told newsmen that trials had shown such low-density materials as PETN would not show up on scanners anyway. Wallace said scanners picked up shrapnel, heavy wax, and metal, but the scanners missed plastic, chemicals, and liquids.

PEEPING TSA

OTHER CONCERNS OVER THE full-body scanners involved health and privacy. The terahertz radiation waves used in body scanners penetrate nonconducting material like clothing, but also deposit energy in the human body. Boian Alexandrov, heading a team of researchers at Los Alamos National Laboratory in New Mexico, announced they found that the terahertz radiation used in the full-body scanners damages human DNA. "Based on our results, we argue that a specific terahertz radiation exposure may significantly affect the natural dynamics of DNA, and thereby influence intricate molecular processes involved in gene expression and DNA replication," they reported. The team said that while terahertz produces only tiny resonant effects, it nevertheless allows terahertz waves to unzip double-stranded DNA, creating bubbles in the double strand that may significantly interfere with normal processes such as gene expression and DNA replication. Such subtle changes may explain why evidence of damage has been so hard to find. According to Alexandrov's team, ordinary resonant effects are not powerful enough to do this kind of DNA damage but nonlinear resonances can.

Not only is health a concern, but so is privacy. Many privacy advocates claim the TSA was not being truthful when its officials said that scanning machines cannot clearly show an individual's genitalia and, certainly in all media stories on the machine, one can only see blurry outlines of the body. However, alternative media such as PrisonPlanet.com exposed the TSA statement as untrue when it quoted Melbourne Airport's Office of Transport Security manager Cheryl Johnson, who admitted, "It is possible to see genitals and breasts while they're going through the machine. . . . It will show the private parts of people, but what we've decided is that we're not going to blur those out, because it severely limits the detection capabilities."

In England, where full-body scanning has been declared mandatory (although it violates U.K. child pornography laws against the depiction of the genitals of underage children), opponents have declared the images so graphic that they amount to "virtual strip-searching." They called for more safeguards to protect passengers' privacy.

Paul Joseph Watson, a reporter for PrisonPlanet.com, said examples sent in by the website's readers confirmed that by simply inverting some of the pictures produced by body scanners, one can create a near-perfect replica of a naked body in full color. The inverting process, available in most image editing software, simply changes an indistinct negative image to a clear positive one. "Airport screeners will have access to huge HIGH DEFINITION [original emphasis] images that, once inverted, will allow them to see every minute detail of your body," noted Watson.

While TSA officials tried to assure the public that flyers' naked images will not be saved, printed, or transmitted, government documents obtained by the Washington-based Electronic Privacy Information Center (EPIC) told a different story. The documents showed that the TSA specifies that body scanners must have the ability to store and send images when in "test mode." EPIC executive director Marc Rotenberg said such a requirement makes it possible for the machines to be abused by TSA insiders and even hacked by outsiders. "I don't think the TSA has been forthcoming with the American public about the true capability of these devices," said Rotenberg. "They've done a bunch of very slick promotions where they show people—including journalists—going through the devices. And then they reassure people, based on the images that have been produced, that there's not any privacy concerns. But if you look at the actual technical specifications and you read the vendor contracts, you come to understand that these machines are capable of doing far more than the TSA has let on."

Official assurances that full-body scans would be seen only by the necessary airport authorities and quickly destroyed were shattered in

February 2010 when the BBC revealed that Indian actor Shahrukh Khan had passed through a body scan and later had the image of his naked body printed out and circulated by female security staffers at Heathrow. "You walk into the machine and everything—the whole outline of your body—comes out," said Khan. "I was a little scared . . . and I came out. Then I saw these girls—they had these printouts. I looked at them. I thought they were some form you had to fill. I said 'give them to me'—and you could see everything inside. So I autographed them for them."

It is not just body-scanning machines that conjure up images of a *1984* Orwellian techno-society.

SECURITY ABUSES

ALTHOUGH POPULAR BELIEF HOLDS that those with Middle Eastern names are most susceptible to being detained by the government, the Secret Service considered Robert Lee "Bob on the Job" Lewis enough of a threat to arrest him on the basis of an offhand remark. Lewis is a fervent Christian who has spent decades researching government scandals and worked with airline lawyers during the investigation of the bombing of Pan Am Flight 103 over Lockerbie, Scotland. In April 1998, Lewis was in a restaurant in Houston, Texas, regaling waiters with his knowledge of government skulduggery and little-reported information of former president George H. W. Bush. Lewis admitted he made a remark about Bush along the lines of "I'll have his ass."

Secret Service agent Tim Reilly was sitting in the restaurant near Lewis and promptly placed Lewis under arrest for threatening the former president. The next day, in a short hearing, federal magistrate Marcia Crone avoided any First Amendment issue and instead accepted the hearsay testimony of Agent Reilly. Because Lewis did not have enough money to post bail, he was held for nearly a year in federal custody. During this time, he was sent to the Fort Worth Federal Correctional Institution and was placed in the cell where Whitewater scandal figure James McDougal reportedly committed suicide. Lewis knew who McDougal had been and felt his placement there was a form of intimidation. Some months later, Lewis was transferred to a federal hospital in Springfield, Missouri, where he was involuntarily drugged until letters from journalists and academic contacts protesting his drugging gained him a release. There was never a court trial or even an adversarial hearing in the case.

For anyone who thinks that the DHS's abuse of power might stop over time, consider this 2006 story from Bethesda, Maryland: Two uniformed men wearing baseball caps with the words "Homeland Security" on them walked into the Little Falls Library and loudly announced that the viewing of pornography was forbidden. They then asked one library Internet user to step outside.

After complaints were lodged against the two "security" officers, Montgomery County chief administrative officer Bruce Romer stated that the two officers were members of the security division of Montgomery County's Homeland Security Department, an unarmed unit charged with patrolling about three hundred county buildings. He added that this group was not tasked with seeking out pornography and that the incident was "unfortunate" and "regrettable." Romer said the two officers had "overstepped their authority" and had been reassigned.

To illustrate the ease with which a TSA employee can impact a person's life, just consider the experience of Rebecca Solomon, a twenty-two-year-old University of Michigan student. On January 5, 2010, she was stopped in an airport by a TSA agent who pretended to find a bag of white powder in her carry-on computer bag. The man then demanded to know where she had gotten the powder and, as the student stood in shock, proceeded to wave the bag in front of her and said he was just kidding. "You should have seen the look on your face," the TSA man told her, laughing. When this incident was made public, the TSA said the agent was no longer with the agency. But if the TSA man had not admitted it was a joke, Solomon would still be behind bars somewhere. This small incident should be a sobering example to any thinking American of unwarranted and unsupervised power.

Such incidents, of course, illustrate the ease with which persons of authority can abuse that authority. It also begs the question of how many other HSD employees "overstep" their authority and how many other such stories never make it to the public.

PHOTOGRAPHERS UNDER FIRE

APPARENTLY EVEN TRADITIONAL AMERICAN activities such as taking pictures around town are not exempt from the scrutiny of Homeland Security enforcers. Amateur photographer Mike Maginnis was intrigued by all the activity around Denver's Adams Mark Hotel in early December 2002, which was surrounded by Denver police, army rangers, and rooftop snipers. Maginnis, who works in information technology and frequently shoots photos of corporate buildings and communications equipment, took a few snapshots. He was then confronted by a Denver policeman who demanded his camera. When he refused to hand over his expensive Nikon F2, he was pushed to the ground and arrested.

After being held in a Denver police station, Maginnis was interrogated by a Secret Service agent. Maginnis learned that Vice President Cheney was staying in the area and that he was being charged as a terrorist under the PATRIOT Act. According to Maginnis, the agent tried to make him confess to being a terrorist and called him a "raghead collaborator" and "dirty pinko faggot."

After being held for several hours, Maginnis was released without explanation. When Maginnis's attorney contacted the Denver police for an explanation, they denied ever arresting Maginnis.

The website PhotographerNotaTerrorist.org proclaimed, "Photography is under attack. Across the country it seems that anyone with a camera is being targeted as a potential terrorist, whether amateur or professional, whether landscape, architectural or street photographer. Not only is it corrosive of press freedom but creation of the collective visual history of our country is extinguished by anti-terrorist legislation designed to protect the heritage it prevents us recording. This campaign is for everyone who values visual imagery, not just photographers. We must work together now to stop this before photography becomes a part of history rather than a way of recording it."

In early 2009, David Proeber, photo editor for the central Illinois newspaper the *Pantagraph,* was stripped of his camera's memory card and threatened with arrest after he took a photo of the police in a shoot-out with a gunman. Proeber recovered the memory card more than three hours later after complaining to a sheriff's department supervisor who was an acquaintance. After the supervisor contacted the state police, Proeber's memory card was returned with apologies. His photo of the gunman was published on the Internet, garnering more than 1.2 million page views during the first thirty-two hours. However, Proeber later learned that the police had made a DVD of his photos, which he claimed they had no legal right to do.

In 2007, Carlos Miller, a Miami freelance photographer, was arrested, tried, and sentenced for photographing Miami police officers on a public street. Miller was found not guilty of disobeying a police officer and disorderly conduct but was convicted of resisting arrest. The prosecution recommended three months of probation, fifty hours of community service, anger management classes, and court costs. But the presiding judge, Jose Fernandez of Miami's county court, was apparently angered that Miller had documented his trial on an Internet blog and sentenced Miller to one year of probation, a hundred hours of community service, anger management classes, and more than $500 in court costs.

The severity of Miller's sentence upset the Society of Professional Journalists (SPJ), which had initially donated some funds for Miller's defense. In a news release, SPJ president Clint Brewer said, "The fact that Mr. Miller was arrested for taking pictures in a public place was the first violation of his First Amendment rights. Those rights were violated again when Mr. Miller's statements in his blog became factors in Fernandez's sentence. The Society fully defends Mr. Miller's right to speak freely in his blog."

Even after leaving the scene of a photo opportunity, apparently photographers today are still susceptible to raids by the authorities. In September 2009, Laura Sennett, a photojournalist specializing in protests and demonstrations, filed a federal complaint stating that both the federal government and local law enforcement violated her rights under the First and Fourth Amendments after coming into her home and seizing computer hardware and data, digital cameras, memory cards, a still camera, digital storage devices, and a digital voice recorder along with other work materials and personal belongings. Only her son was home at the time of the raid.

According to Sennett, this happened because she photographed protesters at a meeting of the International Monetary Fund on April 12, 2008. In her complaint, Sennett claimed to have suffered extreme emotional and mental distress and humiliation. She sought an injunction ordering the DOJ to return her belongings plus pay $250,000 in compensatory damages and $1 million in punitive damages. No criminal charges were filed against her.

Sennett named Attorney General Eric Holder, the FBI Joint Terrorism Task Force, Prince William, the Police Department of Arlington Counties, and the Department of Justice in her complaint. Sennett says she was not a target of any criminal investigation and her work has been published by several media outlets, including CNN and the History Channel.

One of the long-standing tenets of journalism is that in reporting the news, both film, video, and still photographers have the right to shoot pictures, especially on public property. This right appears in grave danger as news reporting today already is limited to handouts from the authorities, reporters being held behind yellow tape blocks from the scene, and duplicate TV coverage on most channels coming from pool cameramen.

YOU'RE ON CAMERA

IT'S NOT JUST THE photographers who are having concerns over cameras. Recently there has been an explosion in the number of surveillance cameras being used in cities small and large that perturbs libertarians.

Instead of a conventional welcome sign outside the small city of Medina, Washington, visitors today are greeted by one reading YOU ARE ENTERING A 24-HOUR VIDEO SURVEILLANCE AREA. Police chief Jeffrey Chen declined to say how many cameras had been installed at intersections using "automatic license plate recognition" technology to record license numbers. Should a database search turn up an outstanding warrant, police immediately dispatch units to track the car. Chen said information gathered by the cameras is stored for sixty days, which allows police to keep searching if a crime occurs.

"These cameras provide us with intelligence," explained Chen. "It gets us in front of criminals. I don't like to be on a level playing field with criminals."

Chen told newsmen that in 2008 there were eleven burglaries in this town of thirty-one hundred, which boasts an average household income of more than $220,000. "Some people think [eleven burglaries] is tolerable. But even one crime is intolerable," Chen said.

Doug Honig, a spokesman for the American Civil Liberties Union in Washington, was troubled by the new surveillance system, saying it smacks of privacy violations. "Government shouldn't be keeping records of people's comings and goings when they haven't done anything wrong," he said. "By actions like this, we're moving closer and closer to a surveillance society."

Despite Honig's statements, many believe that intrusive measures are necessary today. Former Washington, D.C., mayor Anthony A. Williams is among the believers, and he warned his constituents when he was mayor that "We are in a new . . . really dangerous world now, and we have to maintain a higher level of security."

Williams planned to increase Washington, D.C., security by emulating cities like London and Sydney that have thousands of video cameras throughout the city linked to a central command office. England currently has more than two million cameras in airports, train stations, streets, and neighborhoods.

Asked if such a scheme would seriously impact individual civil rights, Williams admitted, "There will be trade-offs."

The United Kingdom is a great example of a modern surveillance society, where companies can thrive on a citizen's penchant for voyeurism. A new company called Internet Eyes offers up to 1,000 pounds to citizen volunteers who stay at home watching several video monitors connected to some of Britain's ubiquitous surveillance cameras. As part of this "instant event notification system," the viewers are to report any "alert"—a suspicious activity—which according to company literature most commonly includes shoplifting, burglary, vandalism, and "anti-social behavior." The alerts are passed to the camera owners, subscribers to the Internet Eyes service, who evaluate the alerts and decide who gets the reward money. Monitors cannot designate or control the video camera feeds nor are they allowed to know the location of the cameras.

How long will it be before some new terrorist threat, whether real, imagined, or fabricated, enlists well-meaning American television viewers to report anything they feel is suspicious behavior on the part of their neighbors? Don't think it cannot happen. It's already happening in what once was called "the Mother Country."

PART IV

HOW TO FREE ZOMBIES: THE THREE BOXES OF FREEDOM

It does not take a majority to prevail... but rather an irate, tireless minority, keen on setting brushfires of freedom in the minds of men.

—SAMUEL ADAMS

WITH THE CORPORATE MASS media centering their news programs on stories of fires, wrecks, murder, mayhem, and scandals, one must ask: Is there any good news?

Yes, there is.

A few thoughtful people believe the United States is undergoing an exciting, if uncomfortable, maturation. Although they admit that growth and change may be unsettling, some Americans feel current advances in technology and environmentalism will eventually lead to a brighter and more harmonious future replete with alternative fuels, engines that run on water, and natural energy sources such as solar, tidal, and geothermal.

But the public must be cautious. They have been bamboozled for too long by the plutocrats who dominate finance, corporate life, and the mass media. For many years, authors, filmmakers, radio and TV commentators, and even some street corner speakers have warned of a coming New World Order, that socialist globalization desired by a small group of plutocrats and their hirelings centered within secretive societies. In the past, these same types of harbingers have warned that there was no "light at the end of the tunnel" in the Vietnam War, that Nixon was a crook and shouldn't serve out his term, that George H. W. Bush was lying when he said "Read my lips, no new taxes," and that the events of Ruby Ridge and Waco were not just attacks on cult members, but upon the rights of all Americans.

In hindsight, the harbingers were right.

Today, the American public hears of untested vaccines, corporate drug companies influencing government policy, totalitarian martial law, and restrictions to liberties promised by the Constitution.

Perhaps it is time for the public to listen to the "conspiracy theorists" and the youthful activists. Andrew Gavin Marshall, a research associate with Canada's Centre for Research on Globalization, asked, "In light of the ever-present and unyieldingly persistent exclamations of 'an end' to the recession, a 'solution' to the crisis, and a 'recovery' of the economy, we must remember that we are being told this by the very same people and institutions which told us, in years past, that there was 'nothing to worry about,' that 'the fundamentals are fine,' and that there was 'no danger' of an economic crisis. Why do we continue to believe the same people that have, in both statements and choices, been nothing but wrong?" Marshall's question could be applied to many of the problems that exist in America. This, in turn, begs a larger question—why do we listen to anything these institutions say?

If America is to again experience the individual freedom and capitalist initiative that once brought this nation to new heights of technological and social success, it is obvious some things must change. Simply bouncing back and forth between conservative and liberal presidential administrations, both controlled from the shadows by the same globalists, will not do the job. Americans must unite, as during World War II.

But rather than uniting against a foreign enemy like Nazi Germany, today our enemy is domestic. This formidable enemy is one that tries to control the nation's federal government, the financial system, the education system, and even the lifestyles of America's citizens.

But the enemy force is few in number while Americans number nearly 304 million.

It is time for individual Americans to become proactive. It is time to remember the three boxes of freedom—the Soap Box, the Ballot Box, and the Ammo Box.

THE SOAP BOX

PRESIDENT JOHN F. KENNEDY once said, "We are not afraid to entrust the American people with unpleasant facts, foreign ideas, alien philosophies, and competitive values. For a nation that is afraid to let its people judge the truth and falsehood in an open market is a nation that is afraid of its people."

Today, freedom of speech is under attack by political correctness and even so-called hate speech legislation. To protect against these attacks, we must ensure a free and investigative news media, one that truly serves as the public "watchdog."

AN UNFETTERED NEWS MEDIA

CORRUPTION AND TYRANNY HAVE pervaded human history.

In the United States, governmental and corporate avarice has been combated historically by a free-ranging and unfettered investigative news media—media that once were privately owned. Yet today, the mass media are controlled by only a handful of multinational corporations. Furthermore, the federal government has continuously impeded incisive journalism by operating under secrecy and disregarding the Freedom of Information Act signed into law by President Lyndon Johnson on September 6, 1966. The purpose of the law is to declassify governmental documents. It is also the subject of ongoing conflicts between government officials and both news organizations and private citizens. Various administrations have differed in their interpretations of the law, which also contains several specific exemptions.

It is especially troubling that media ownership is so concentrated when one considers that more than 98 percent of Americans have a television. Of that 98 percent, 82 percent watch prime-time TV and 71 percent watch cable programming in an average week. Additionally, 84 percent of Americans listen to radio regularly, while 79 percent are newspaper readers. Nearly half of the American population has access to the Internet, and certain demographic groups reach close to 70 percent. These totals suggest that most of America spends an inordinate amount of time staring at a screen, which might be bad enough. But when one realizes that everything these citizens see and hear emanates from a mere five major media corporations, the threat of potential propagandizing and mind control becomes clear.

In addition to broadcasting watered-down content because of corporate ownership, staffers at the White House and Pentagon manipulate the media through "perception management." Although government propagandists cannot tell the audience how to think, they can tell them what to think about as they set the agenda and frame the arguments. They cleverly craft the perception. Too many news reporters simply regurgitate government press handouts. Washington-based investigative journalist Wayne Madsen pointed out that "It is not the job of a journalist to participate in propagandizing the news. Journalists report the basic facts of a story. The reality that the U.S. occupation of Iraq has been an unmitigated disaster is not the fault of the news media. The fact that the Iraq war has gone badly for the United States is news."

Neither the Bush nor the Obama administrations have permitted news reports from Iraq to be aired before being filtered by officials. Additionally, the government has forced the media to "embed" reporters within U.S. and Iraqi military units, limiting their view of the hostilities and forging personal relationships within their assigned units that cannot fail but tinge their reporting. The government has also employed contractors like the Lincoln Group and the San Diego–based Science Applications International Corp. (SAIC) to place pro-U.S. propaganda in Iraqi newspapers. Additionally, the Pentagon gave SYColeman Inc. contracts to develop slogans, advertisements, newspaper articles, radio spots, and television programs to promote support for U.S. policies overseas. Naturally, the head of SYColeman is a retired general who at one point was a top official in the Defense Department agency that gave SAIC its Iraqi media contract.

"The mainstream media should bolster their independent reporting of the Iraq war," wrote Wayne Madsen in the *San Diego Union-Tribune*. "They should reject the lies consistently fed to journalists in Baghdad and at the Pentagon, State Department, and White House. And editors must encourage journalists to publish 'off-the-record' interviews with U.S. military members," advised Madsen. "This is contrary to the Pentagon's media policy, but the military is not the final arbiter of First Amendment freedom of the press rights. The military is responsible for defending those rights."

In July 2009, the PEN American Center, an eighty-seven-year-old organization dedicated to defending the freedom of writers around the world, joined the American Civil Liberties Union in court to challenge the FISA Amendments Act (FAA). Both PEN and the ACLU said the FAA greatly expanded the ability of the U.S. government to spy on Americans without a warrant and granted retroactive immunity to telecommunications companies who aided in government spying on citizens.

"We are plaintiffs in this lawsuit first and foremost because we believe our own communications, which include sensitive phone calls and emails with writers facing persecution in countries from Afghanistan to Zimbabwe, are vulnerable under the program," wrote Larry Siems, director of the PEN American Center's Freedom to Write program, in the Huffington Post. "We know from the experiences of our colleagues in countries where governments had unchecked surveillance powers (including the United States as recently as the 1970s) that programs that allow governments to spy on their own citizens are often directed against writers and intellectuals, and that surveillance in general poses a serious threat to the intellectual and creative freedoms of all citizens."

But there are organizations other than PEN and the ACLU fighting back. The World Press Freedom Committee is an international group composed of members from forty-five news organizations that have fought for more than thirty years against the licensing of journalists, mandatory codes of conduct, mandatory tasks for journalists, and other news controls. The World Press Freedom Committee created a Charter for a Free Press that lists ten principles to guarantee the "unfettered flow of news and information both within and across national borders." The committee said such a charter deserves the support of "all those pledged to advance and protect democratic institutions." The principles are as follows:

1. Censorship, direct or indirect, is unacceptable; thus laws and practices restricting the right of the news media freely to gather and distribute information must be abolished, and government authorities, national or local, must not interfere with the content of print or broadcast news, or restrict access to any news source.

2. Independent news media, both print and broadcast, must be allowed to emerge and operate freely in all countries.

3. There must be no discrimination by governments in their treatment, economic or otherwise, of the news media within a country. In those countries where government media also exist, the independent media must have the same free access as the official media have to all material and facilities necessary to their publishing or broadcasting operations.

4. States must not restrict access to newsprint, printing facilities and distribution systems, operation of news agencies, and availability of broadcast frequencies and facilities.

5. Legal, technical and tariff practices by communications authorities which inhibit the distribution of news and restrict the flow of information are condemned.

6. Government media must enjoy editorial independence and be open to a diversity of viewpoints. This should be affirmed in both law and practice.

7. There should be unrestricted access by the print and broadcast media within a country to outside news and information services, and the public should enjoy similar freedom to receive foreign publications and foreign broadcasts without interference.

8. National frontiers must be open to foreign journalists. Quotas must not apply, and applications for visas, press credentials and other documentation requisite for their work should be approved promptly. Foreign journalists should be allowed to travel freely within a country and have access to both official and unofficial news sources, and be allowed to import and export freely all necessary professional materials and equipment.

9. Restrictions on the free entry to the field of journalism or over its practice, through licensing or other certification procedures, must be eliminated.

10. Journalists, like all citizens, must be secure in their persons and be given full protection of law. Journalists working in war zones are recognized as civilians enjoying all rights and immunities accorded to other civilians.

Organizations other than the World Press Freedom Committee also have contributed to the establishment of a free and unfettered media. Phil Donahue, the talk-show host who lost his job shortly after questioning the official story of 9/11, urged the public to support "the *Los Angeles Times,* the Society of Professional Journalists, the National Press Club, and other organizations (not to mention the Framers of our Constitution) and help keep journalists free to be pushy, unpopular and inelegant—sticking a nose under the tent to learn what the righteous have decided is good for us." Donahue rightly proclaims that "There is no substitute for free and unfettered news gathering. Journalists are not cops nor are they public relations people. They are reporters and there is no substitute for them."

One of the largest problems affecting freedom of the press is the corporatization of the media, which dilutes news content in order to make it more appealing to larger audiences. Yet one organization, named the StopBigMedia.com Coalition, is attempting to halt this corporate takeover of an American tradition. The organization is composed of a number of politically diverse groups that have banded together without government or corporate funding to "stop the FCC from allowing a handful of giant corporations to dominate America's media system." According to information on its website, "Corporate media giants are silencing diverse voices, abandoning quality journalism and eliminating local content (we've got the evidence). Our democracy needs better media. Bad policies made in Washington could have a big impact on the news in your community." Elsewhere, they state, "We believe that a free and vibrant media, full of diverse and competing voices, is the lifeblood of America's democracy. We're working together to see that our media system remains, in the words of the Supreme Court, 'an uninhibited marketplace of ideas in which truth will prevail.'"

And as with any profession, laziness largely contributes to the recent flux of poor media content. As the British novelist and critic Kingsley Amis once put it, "Laziness has become the chief characteristic of journalism, displacing incompetence."

This laziness is partially due to the prevalence of public relations press releases in the last half century. It has always been easier to rewrite a government or corporate news release than conduct the legwork necessary for documenting a good story. Good stories require a reporter to go out into his community, not sit around at a desk. Yet when a reporter is not at his desk all day, his editor usually becomes upset, as businesses (after all, media companies are businesses) like employees on the premises, at their desks. But sitting behind a desk all day does not promote good daily news or investigations.

University of Illinois communications professor and media reformer Robert McChesney and many others believe journalism may be one of the greatest issues facing the American public, because if the public is not informed about current events, then it is almost impossible for the public to make electoral decisions. Thus, democracy becomes impossible in a journalism-free society.

In the future, journalists may be private citizens who, making use of cell-phone cameras and the Internet, take it upon themselves to find and report the news. It's happened before (albeit without cell phones and the Internet). According to McChesney, perhaps the greatest of such journalists was I. F. Stone, the iconoclastic and once-blacklisted editor of *I. F. Stone's Weekly,* a self-published newsletter in the 1950s and 1960s with far-reaching influence. In 1999, ten years after his death, Stone's newsletter was named among "The Top 100 Works of Journalism in the United States in the 20th Century." "Stone is currently celebrated by professional American journalism schools as a great hero. But for most of his life, Stone was an anathema to those that relied on official sources," explained McChesney. "Stone refused to have any relationship with people in power because he knew that relationship would corrupt his ability to be a real journalist. He knew that this would limit his capacity to get at the truth of what the government does and whose interests it serves."

"I want a thousand I. F. Stones, combing Washington and Wall Street, investigating power," said McChesney. "To do this well, [the journalists] would need a decent salary, professional training, and a newsroom to protect them from the powerful. They would need much more time. If I work at an office or a factory all day, go home, feed my kids and make their lunch for the next day, clean the house and do the laundry, and then sit down to blog at 11 p.m., it is going to suck.

"What people can do, though, let's say if they've studied some economics and become really interested in economic issues, is this. They can actively search for, collect and read numerous pieces by journalists on the economy. They can compare different points of view, fact-check, and scrutinize sources. Then they can blog on all of this. They can actively participate in the media debate. But this does not mean trained journalists are no longer important. I view the blogosphere (the part-time or volunteer citizen-journalist) as a number of musicians improvising on a melody written by journalists. Bloggers may contribute to the melody in interesting ways. But without journalism, there is just a lot of noise. Journalism should be there to make sure that blogging is not just a lot of noise, but a beautiful song."

BACK-TO-BASICS EDUCATION

IT MAY PERHAPS BE the case that the media's blandness is only a mere reflection of the blandness and conformity that public school systems instill in the nation's citizens. Mark Taylor, currently a teacher at Olathe South High School, in Olathe, Kansas, believes that "In order to win the struggle in the classrooms of America, teachers must first realize that today's public education system was designed by powerful economic elites, whose true intention was for students to think of themselves as employees in a system designed to dumb them down to be good little consumers of the goals that the purveyors of that system have arbitrarily chosen. This system has been sold to educators all across America as the values of a Democratic Republic. These teachers, whether they are university professors, or secondary and elementary school teachers, should begin by addressing that lie and begin to use the elite cover story of a Democratic Republic as a weapon to defeat the imbedded design of subservience."

Taylor has spent his teaching career trying to instill critical thinking skills in his students. "The essence of education is not what we learn, but questioning what we learn. The first step to questioning what we learn is to realize that in order to think outside the box, we must first know we are in it. In order to know we are in it, we must learn how to think, and in order to learn how to think, we must learn primarily from the four horsemen of intellectual enlightenment: philosophy, economics, political science, and history. From philosophy we must learn about what is real, what is true and what is good. From economics we must learn that wealth creation is an illusion created by powerful economic elites. From political science, we must learn that the pursuit of politics is the pursuit of power and it is tied to the secrets of economic wealth creation. Finally, we must learn that today's news is tomorrow's history and since most of today's news is based on lies and deceit, most of history is a lie," he added.

There are teachers across America who teach their students to question the system through the pursuit of philosophy, economics, political science, and history. "Every teacher who uses this formula will contribute to a moment in time when the critical mass of history will implode the designs of those who seek to enslave us all," Taylor said.

Unfortunately, such teachers illuminate their charges in spite of the system, not because of it. To produce literate and functioning members of society, education must first offer all students a basic grounding in the three R's—reading, 'riting, and 'rithmetic. Past this, they also should be grounded in the history and philosophy of American freedom. Not in simply memorizing names and dates but, more important, understanding why revolutions and wars were fought and what results came of them. And, over all, students must be shown how to think and reason critically, how to research and examine issues on their own, and, last, how to speak out for what is right and just. Only with a truly educated and responsible citizenry can American regain its place as a leader among nations. With an increasing number of parents beginning to understand the worth of real education, not simply passing tests or fulfilling state-required curricula, they are turning to an alternative.

HOMESCHOOLING

To ENSURE A BETTER education, some parents remove their children entirely from the public education system. Once considered highly controversial, homeschooling seems to gradually be gaining favor across America. In the past, homeschooling was thought to be only for xenophobes and religious fanatics. Now, by some estimates, two million American kids are getting their education at home. And there is a growing belief, backed by studies and statistics, that the education of homeschooled children often outstrips those in the public system.

Even well-respected collegiate institutions are recognizing homeschooling as a legitimate educational practice. According to a study by the Virginia-based advocacy group National Center for Home Education (NCHE), 68 percent of colleges were accepting parent-prepared transcripts or portfolios in place of an accredited diploma. Those universities accepting homeschoolers included Stanford, Yale, and Harvard. The NCHE said such colleges "generally require SAT I (one) and/or ACT scores, a high school transcript, letters of recommendation, and writing samples."

"Homeschoolers bring certain skills—motivation, curiosity, the capacity to be responsible for their education—that high schools don't induce very well," said Jon Reider, the director of college counseling at San Francisco University High School who at one time was Stanford's senior associate director of admissions.

Isabel Shaw, a writer and homeschooling researcher, wrote: "On average, homeschooled kids score one year ahead of their schooled peers on standardized tests. The longer the student homeschools, the wider this gap becomes. By the time homeschooled children are in the eighth grade, they test four years ahead of their schooled peers." Isabel and her husband, Ray, homeschooled their two daughters for fifteen years. "Of course, these results translate into better American College Test (ACT) scores. Research shows that high achievement on the ACT strongly indicates a greater likelihood of success in college. According to official

ACT reports, homeschooled students repeatedly outperform publicly and privately educated students in the ACT assessment test." Kelley Hayden, a spokesman for ACT, said, "What you can say about the homeschoolers is that homeschooled kids are well-prepared for college."

Those leery of homeschooling say that the very absence of "real world" experiences may put homeschoolers at a disadvantage in later life. "Public school students learn how to deal with a system, no matter how capricious it may be," said one Texas public school teacher. "They learn how to put up with the incompetents (including administrators) they will have to deal with in the real world."

According to Hal Young, a past education vice president for North Carolinians for Home Education, "One of the most common objections levied against home education is that homeschool students lack exposure to different social settings." But Young said that "graduates integrate well into the campus environment. Homeschooling is individual, but it's not isolated. Most homeschoolers that we hear from are pretty well networked in support groups, church activities, Scouting programs, and sports programs . . . so when they get to the college campuses where there are other groups around, that's just another day in life." Young noted that, as a result, "A lot of colleges are saying that [homeschoolers] are a good population to pursue. They've had positive results dealing with home-educated students, and so they actively go out and look for them. . . ."

The late Chris Klicka, as senior counsel for the Home School Legal Defense Association, also addressed the idea that homeschooled kids are poorly socialized:

> *[P]ublic school children are confined to a classroom for at least 180 days each year with little opportunity to be exposed to the workplace or to go on field trips. The children are trapped with a group of children their own age with little chance to relate to children of other ages or adults. They learn in a vacuum where there are no absolute standards. They are given little to no responsibility, and everything is provided for them. The opportunity to pursue their interests and to apply their unique talents is stifled. Actions by public students rarely have con-*

sequences, as discipline is lax and passing from grade to grade is automatic. The students are not really prepared to operate in the home (family) or the workplace, which comprise a major part of the "real world" after graduation.

Homeschoolers, on the other hand, do not have the above problems. They are completely prepared for the "real world" of the workplace and the home. They relate regularly with adults and follow their examples rather than the examples of foolish peers. They learn based on "hands on" experiences and early apprenticeship training. In fact, the only "socialization" or aspect of the "real world" which they miss out on by not attending the public school is unhealthy peer pressure, crime, and immorality. Of course, the average homeschooler wisely learns about these things from afar instead of being personally involved in crime or immorality or perhaps from being a victim.

With the advances in information technology, online education has also grown popular. According to the *Wall Street Journal,* "Roughly 100,000 of the 12 million high-school-age students in the U.S. attend 438 online schools full-time [in 2009], up from 30,000 five years ago, according to the International Association for K–12 Learning Online, a Washington nonprofit representing online schools. Many more students take some classes online, while attending traditional schools. The National Center for Education Statistics, part of the U.S. Department of Education, says 1.5 million K–12 students were home-schooled in 2007, a figure that includes some who attended online schools. That is a 36% increase from the 1.1 million in 2003."

As with regular homeschooling, one major concern about online schooling has been that strictly sitting at a computer will stunt a student's social skills. Raymond Ravaglia, deputy director of Stanford's Educational Program for Gifted Youth, pointed out, "For online high schools, the biggest obstacle is addressing the social interaction for the students. At that age, people really crave social interaction."

Others believe that online students, with their access to multimedia Internet content, will gain an advantage in our increasingly digital world.

"What they learn while in the online high school will make them more adaptable thinkers," said Rand Spiro, a professor in education psychology at Michigan State University.

On the Oklahoma Council of Public Affairs website, Matthew Ladner, vice president of research at the Goldwater Institute, offered a possible alternative to traditional public education that could be applied nationwide: "John Stuart Mill once observed that if government would simply require an education, they might save themselves the trouble of providing it (or in this case, unsuccessfully trying to provide it). . . . State lawmakers could make the passing of a civic knowledge exam a precondition for receiving a driver's license, and simply make the necessary study materials available online and at public libraries."

The costs of such a system would be a fraction of what taxpayers are currently spending, and it would likely prove much more effective. Ultimately, the American public must see to it that children learn civics, for as Thomas Jefferson said, "If a nation expects to be ignorant and free, in a state of civilization, it expects what never was and never will be."

CARING FOR HEALTH

ASIDE FROM DISHING OUT the usual bland media content full of stories about celebrities, political scandals, and petty crimes, the mass media gives the public the false impression that only experts and medical doctors can determine what constitutes good health and how to achieve it. Every talk show and newscast turns to an expert from the globalist-controlled government agencies or corporations to present their version of health news. Their advice is constantly validated by ubiquitous drug advertising, emanating from the same corporations.

Yet change is in the air. Many people are taking charge of their own health and seeking alternative means of ensuring a satisfying and productive life. Even some medical professionals are turning away from profit-driven corporate medicine and finding new ways to improve public health.

Dr. Len Saputo, a practicing physician for more than forty years, encourages a paradigm shift in how medicine should be practiced. Over the years, Saputo saw the quality of health care in the United States sink to new lows as the medical community shifted from concern for the patient to a concern for profit. In 1994, Saputo founded the Health Medicine Forum, which changed the outlook and practices of many health-care practitioners in the San Francisco Bay Area.

"I entered the profession aspiring to be a healer, as did most of my colleagues," wrote Saputo in his 2009 book *A Return to Healing.* "We wanted to attend to the health and medical needs of *whole persons;* we were inspired to serve our patients through our aspiration to provide genuine healing and to promote healthy living based on science and common sense. Sadly, this ideal has been replaced by the corporate bottom line, resulting in a dysfunctional system focused almost entirely on what I prefer to call *disease care* [original emphasis].

"The physician's natural focus on the health needs of a unique, living person embedded in his family and society has today been largely replaced by a model that reduces each person to his body, his body to a machine, and his health needs to a set of symptoms to be treated mainly with drugs—too often ignoring the patient's mind, emotions, spirit, environment, and lifestyle."

Today, Dr. Saputo and many other physicians are turning to natural biochemical solutions to treat health problems. The base premise of this type of treatment lies in a simple recognition—if all of a body's cells are functioning properly, there is no cause for sickness. "The restoration of good health and vitality is accomplished by supporting the body and allowing the natural healing process to take charge," explained Saputo. Those who undertake this type of progressive medicine "boldly acknowledge the importance of treating body, mind, and spirit—the imperative of caring for the whole person, not just the disease. . . . They are choosing prevention, wellness, natural solutions, and the integrative model—and they are blazing the path to the integral-health medicine of the future."

Many medical professionals have followed Dr. Saputo's standard, asking not how to fund the current American health-care system but how to find better ways of securing and maintaining better health. Instead of asking which drugs should be used to cure an illness, they are asking whether or not drugs should be used as primary treatment. In 2007, the National Health Interview Survey reported that approximately four out of every ten Americans had used some form of complementary or alternative medicine during that year. Reportedly, complementary and alternative therapies now account for 11.2 percent of total out-of-pocket health-care expenses—approximately $33.9 billion a year.

Before the 2008 presidential election, more than five thousand U.S. physicians signed an open letter to then candidate Barack Obama urging him and Congress "to stand up for the health of the American people and implement a nonprofit, single-payer national health insurance system." The physicians noted: "A single-payer health system could realize administrative savings of more than $300 billion annually—enough to cover the uninsured and to eliminate co-payments and deductibles for all Americans." Single-payer national health insurance is a system in which a single public or quasi-public agency organizes health financing, yet delivery of the care remains largely private. Such systems are currently in use in Canada, Great Britain, and other nations.

The doctors said incremental changes to health-care policy by the Democrats would not solve health-care problems, and that Republican plans to pursue market-based strategies would only exacerbate the situation. "What needs to be changed is the system itself," they wrote. One of the letter's signers, Dr. Oliver Fein, a professor of clinical medicine and public health at Weill Cornell Medical College in New York, stated, "With the sudden economic downturn, more people than ever before are worried about how to pay for health care. A single-payer system—an improved Medicare for all—would lift those worries, provide care to all who need it and require no new money. It's the only morally and fiscally responsible approach to take."

Despite spending more than twice as much as other industrialized nations on health care (more than $7,000 per person), America's health-care system is not only expensive but inadequate. The United States ranks below fifty other nations in life expectancy, including Canada, Bermuda,

Norway, Jordan, South Korea, Bosnia, Herzegovina, and Puerto Rico. In child mortality (the death of a child under one year old per 1,000 live births), the United States shamefully dropped behind 180 other nations, including Serbia, Chile, Russia, Fiji, Botswana, Jamaica, Thailand, China, Mexico, and Libya.

"The reason we spend more and get less than the rest of the world is because we have a patchwork system of for-profit payers," explained the website for the Physicians for a National Health Program, an organization of more than seventeen thousand physicians. "Private insurers necessarily waste health dollars on things that have nothing to do with care: overhead, underwriting, billing, sales and marketing departments as well as huge profits and exorbitant executive pay. Doctors and hospitals must maintain costly administrative staffs to deal with the bureaucracy. Combined, this needless administration consumes one-third (31 percent) of Americans' health dollars. Single-payer financing is the only way to recapture this wasted money."

John C. Goodman, president of the Dallas-based National Center for Policy Analysis, said, "The only sensible alternative to relying on a welfare state to solve our health care needs is a renewed reliance on private sector institutions that utilize individual choice and free markets to insure against unforeseen contingencies. In the case of Medicare, our single largest health care problem, such a solution would need to do three things: liberate the patients, liberate the doctors, and pre-fund the system as we move through time.

"By liberating the patients I mean giving them more control over their money—at a minimum, one-third of their Medicare dollars. Designate what the patient is able to pay for with this money, and then give him control over it. Based on our experience with health savings accounts, people who are managing their own money make radically different choices. They find ways to be far more prudent and economical in their consumption."

Dr. John Geyman, professor emeritus of family medicine at the University of Washington and author of *Do Not Resuscitate: Why the Health Insurance Industry Is Dying and How We Must Replace It,* said the private health system is obsolete and the insurance industry is to blame. "While there is widespread consensus that the nation's health care system is broken and in urgent need for reform, too little attention has been paid to the role of the private insurance industry in perpetuating our problems. Over the past 40 years, private insurance has evolved from a not-for-profit activity into a $300-billion-a-year, for-profit, investor-owned industry. The six biggest insurers made over $10 billion in profits in 2006. They did so by enrolling healthy people, denying claims, and screening out the sick, who are increasingly being shunted into our beleaguered public safety net programs. . . . These for-profit companies have burdened our system with enormously wasteful administrative costs and skyrocketing CEO salaries, while leaving tens of millions uninsured and underinsured. The risk pool has been badly fragmented among more than 1,300 private insurers, defeating the goal of insurance, which is to provide coverage by sharing risk across a broad population. Premium prices continue to climb at a double-digit rate alongside other health costs.

"Thus, the average family premium for employer-based coverage was $11,500 in 2006, an increase of 87 percent from 2000. At the rate we are going, health insurance premiums will consume almost one-third of average household income by 2010 and all of household income by 2025. This clearly is not sustainable."

With all the incredible advances in medical science in recent years, the problems with health care obviously are not in the technology. It all boils down to who will make the decision ultimately affecting any national health-care plan. Unfortunately, this turns out to be the Congress, that collection of do-nothings, adulterers, tax cheats, liars, mercenaries, and arrogant windbags. By 2010, the global corporatists and their lobbyists had won out. Single-payer health insurance was off the table. What's to be done?

THE BALLOT BOX

WHILE ADDRESSING THE NATIONAL Democratic Convention in 1896, thrice presidential nominee William Jennings Bryan declared, "If they ask us why we do not embody in our platforms all the things that we believe in, we reply that when we have restored the money of the Constitution, all other necessary reform will be possible; but that until this is done, there is no other reform that can be accomplished." Mr. Bryan surely would be spinning in his grave if he saw the abuses being practiced today under the name of government finance.

"Constitutional money" is clearly spelled out in the U.S. Constitution under Article I. In Section 8, the Constitution reads, "The Congress shall have the power . . . to coin money, [and] regulate the value thereof. . . ." It is important to note that fiat money—the Federal Reserve paper dollars that are considered the nation's legitimate currency—is not mentioned. But, in Section 10, the Constitution makes it clear that "No state shall . . . make anything but gold and silver coin a tender in payment of debts. . . ." As explained earlier in this work, fiat money—the U.S. paper dollar—is rapidly becoming worthless due to the amounts in circulation and the ballooning debt behind it.

America will not experience any genuine reform until there is a meaningful overhaul of the financial system. Because our economy is part of a global economy, an overhaul may have to include the entire world.

Even accounting for the Obama administration's plans and budget, Michel Chossudovsky, a professor of economics at the University of Ottawa and director of the Centre for Research on Globalization, predicted, "There are no solutions under the prevailing global financial architecture. Meaningful policies cannot be achieved without radically reforming the workings of the international banking system."

Chossudovsky suggested a complete "overhaul of the monetary system including the functions and ownership of the central bank, the arrest and prosecution of those involved in financial fraud both in the financial system and in governmental agencies, the freeze of all accounts where fraudulent transfers have been deposited, the cancellation of debts resulting from fraudulent trade and/or market manipulation.

"People across the land, nationally and internationally, must mobilize. This struggle to democratize the financial and fiscal apparatus must be broad-based and democratic encompassing all sectors of society at all levels, in all countries. What is ultimately required is to disarm the financial establishment: confiscate those assets which were obtained through fraud and financial manipulation; restore the savings of households through reverse transfers; return the bailout money to the Treasury, freeze the activities of the hedge funds; freeze the gamut of speculative transactions including short-selling and derivative trade."

Economist William K. Black has supported the idea that bailouts are pernicious to overall economic health. In an interview with PBS commentator Bill Moyers, Black said, "Now, going forward, get rid of the people that have caused the problems. That's a pretty straightforward thing, as well. Why would we keep CEOs and CFOs and other senior officers that caused the problems? That's . . . nuts. . . . So stop that current system. We're hiding the losses, instead of trying to find out the real losses. Stop that, because you need good information to make good decisions. . . . Follow what works instead of what's failed. Start appointing people who have records of success, instead of records of failure. . . . There are lots of things we can do. Even today, as late as it is. Even though they've had a terrible start to the administration. They could change, and they could change within weeks."

According to Dr. Charles K. Rowley, Duncan Black Professor of Economics at George Mason University and general director of the Locke Institute, "The prognosis is catastrophic if projected government policies are not cut back. According to the White House's own estimates, the federal budget deficit in 2009 will be $1.6 trillion, approximately 11.2 percent of the overall economy, the highest on record since the end of the Second World War. In 2019, the national debt will represent 76.5 percent of the US national economy, the highest proportion since just after the Second World War. In such circumstances, the international reserve status of the US dollar will not survive. As it fades, so interest rates on government securities will rise and the real burden of servicing the debt will increase. In such circumstances, the US economy will teeter on the edge of a black hole."

To prevent an American economic collapse, Rowley argued against socialism, stating, "Prosperity and full employment in the US will only be restored by a return to laissez-faire capitalism . . . on the micro-economic side, tariffs and other trade barriers should be repealed unilaterally; a 'Right-to-Work' Act should reduce the minimum wage and curtail the powers of unions; and business regulation should be reduced. Individual banks and their counterparties should not be bailed out, although the system should be protected by ensuring that failing banks are wound up in an orderly fashion—this is the only way to restore market discipline." Laissez-faire is a French term loosely meaning "let it be." Laissez-faire capitalism generally is defined as a system that allows the marketplace to regulate and police itself, that the law of supply and demand will smooth all production and distribution problems. This system works fine unless, as has happened in modern American, the marketplace devolves down to a handful of multinational corporations under the control of the globalist fascists.

Rowley's argument in favor of laissez-faire capitalism is acceptable only if one assumes the economic playing field is level and that everyone has an equal chance at commercial prosperity. But, historically, prosperity has never been within equal reach to everyone. The history of the United States is the story of groups prospering at the expense of others, whether in the name of the trust, syndicate, cartel, or corporation.

In fact, the big just get bigger in a laissez-faire system. Case in point: the consolidation of media, banking, and automobile corporations. To use media again as an analogy, today freedom of the press belongs only to those who own the presses. But if this is the case, how are unique points of view expressed to the public? There are plenty of dedicated and well-intentioned journalists still working in the United States, but hardly any can afford to purchase and run a major news outlet. Consequently, news and information is left in the hands of the large corporations, where a pecking order demands acceptance of the boss's demands.

If the United States is to have a truly free-enterprise marketplace, legislators must find a way to balance the laws and regulations necessary to prevent monopolies and to curtail freewheeling capitalist systems so that anyone with the intelligence and ambition can succeed. They must find a way to break apart the giant multinational corporations so that true free enterprise can once again assert itself.

AUDIT THE FED

MUCH OF THE NATION'S monetary problems come from the Federal Reserve System.

In a study entitled "Is the Federal Reserve System a Governmental or a Privately Controlled Organization?" the American Monetary Institute (AMI) explained the confusion and ambiguity over the ownership and purpose of the Federal Reserve System, which is neither wholly a federal agency nor a completely private company.

The AMI asserted that "ambiguity of control has resulted in the monetary power being misused. It has allowed great power to be wielded without responsibility. No amount of false PR will change that. The money power vested in Congress by the Constitution has been improperly delegated to private interests without sufficient public interest benefit, if any. Congress must resume the power vested in it. Had such delegation of power been shown to work in the public interest, one could consider maintaining or adjusting the present system. But look what it has done. This calls for a major shifting of how our money system operates and is controlled. Anything less, with minor benefits that merely alleviate the problems temporarily, will allow the destructive process to eventually resume. . . . The ambiguity must cease."

AMI director Stephen Zarlenga wrote that the institute has been working on comprehensive legislation called "The American Monetary Act," designed to resolve ambiguity over who controls the Fed. Rather than resorting to simple abolishment, the AMI's plan posits that the federal government should incorporate the Fed.

"Monetary reform is achieved in three parts, which must be enacted together for it to work. Any one or any two of them alone won't do it, but could actually further harm the monetary situation," Zarlenga explained.

"First, incorporate the Federal Reserve System into the US Treasury where all new money is created by government as money, not interest-bearing debt, and spent into circulation to promote the general welfare; monitored to be neither inflationary nor deflationary.

"Second, halt the banks' privilege to create money by ending the fractional reserve system in a gentle and elegant way. All the past monetized private credit is converted into US government money. Banks then act as intermediaries accepting savings deposits and loaning them out to borrowers; what people think they do now.

"Third, spend new money into circulation on infrastructure, including education and healthcare needed for a growing society, starting with the $1.6 trillion that the American Society of Civil Engineers estimate is needed for infrastructure repair; creating good jobs across our nation, reinvigorating local economies and re-funding government at all levels."

Zarlenga noted that AMI's plan would not be supported except under emergency conditions. "The idea is to have it ready and to inform enough citizens and lawmakers around the country about it," he wrote. "At the same time, it is necessary to begin action now and there is a 'small step' called the Monetary Transparency Act. . . . It starts the process of making the Fed more accountable to the Congress, by requiring the compilation of certain statistics which are otherwise difficult to get. These are numbers which almost automatically point the way toward better public policy decisions."

The statistics that Zarlenga noted are accessible through the little-known Comprehensive Annual Financial Reports (CAFRs). More than eighty-four thousand CAFRs are completed each year by local governments in the United States, but only rarely does the public get a view of them. According to veteran Wall Street commodity trader Walter Burien, every state, county, and major metropolitan city is keeping two sets of books. One—the budget—is commonly available and tracks each governmental entity's costs and tax revenue. "The Budget is the financial record that's seen by the public and used by politicians to justify new governmental services and higher taxes," he said.

However, the second set of books—the CAFR—is virtually unknown to the public but contains the real record of total governmental income. The budget gives an accurate account of government costs, according to Burien, but only the CAFR gives an accurate account of the government's income. "The CAFR is the accounting Bible for all local government. It shows the total gross income, investment structure, and also shows the general purpose operating budget as is 'selectively' created by your local government. I note that the selectively created operating budget usually amounts to one-third of the gross income and is where 100 percent of tax income is shown. The other two-thirds of the gross income is shown only in the CAFR report and . . . is derived from return on investments and enterprise operations of which said enterprise operations will have their own CAFR or Annual Financial Report listing their own investments and gross income separate from the local government they are under (many games are played here)."

Burien explained tax cuts and CAFR hidden assets with this metaphor: "The foxes have been writing the laws on how many chickens they can eat from the hen house. At first, out of our 3,000 chickens . . . we gave [the foxes] 100 per year. They ate them and said they need 200. So we gave them 200. They ate them and then said they needed 400. So we gave them 400, but we started complaining saying enough is enough. So the foxes said they needed 440, justifying 440 with any logic available to them but realizing we were complaining about giving them 100, then 200, then 400, so they, in their wisdom, started to put 150 aside each year in their own hen house held by them and undisclosed to us. Well, after many a year, in the foxes own hen house they have collected 6,500 chickens (total available revenue not tied in directly with the publicly known operating budget) as they continue to collect the now 510 (the disclosed operating Budget) as the foxes cry to us saying they are barely getting by on the 510, but since we are complaining about the 510 they will cut back the annual take to 490 at great sacrifice to themselves, the foxes. . . ." Under this method of cooking the books, government entities, from the federal to the state and even county level, can store much more money than is reflected in the publicly available budgets.

"[I]t is obvious that the inside players' crucial element for success was to make sure the people did not review, understand, or comprehend their financial game plan as it grew," explained Burien. "To be able to pull this off government required the full cooperation of the syndicated media, organized education, and the political parties. . . . It is obvious they got it and got it due to the money involved. If you cooperated, you were on easy street. If you did not, you were marginalized or worse."

To correct this hidden government theft, citizens must first learn how much money their local government is hiding. "Make sure all know to carefully look at, review, and examine their local government CAFRs. Avoidance or refusal by your local government to do so in plain language is treason and financial fraud by intentional non-disclosure of the worst sort. When all the people know to look, I am confident there are a few sharp cookies out there that can take the corrective measures necessary to reverse the game back into the benefit and control of the people."

Bruce Wiseman, the U.S. national president of the Citizens Commission on Human Rights, compared President Barack Obama's 2009 agreement to join the Financial Stability Board (FSB), which he signed at the G-20 meeting in London, to the Bretton Woods Agreements approved by representatives of forty-four Allied nations in July 1944. The result of Bretton Woods was an established monetary management system and the rules for international commercial and financial relations. Additionally, the agreements served as the foundation for creating the World Bank and the International Monetary Fund.

Wiseman stated that President Obama's approval of the FSB, the global monetary authority connected to the Bank for International Settlements, must be scrutinized carefully. "Let your Representatives and Senators know the Financial Stability Board must be approved by Congress and must be subject to oversight by elected officials of the countries involved. Personal visits, followed by calls and faxes to both Washington and local offices, are the most effective. Don't be surprised if they don't know what you're talking about. Politely insist they find out and take action. And understand this when dealing with legislators or their staffs: they are focused almost exclusively on legislation that has already been introduced—a bill with a number on it. That is not the case here. You want them to take action on this matter by introducing legislation that brings the approval and structure of the Financial Stability Board under congressional control."

Wiseman noted that there is nothing inherently evil about an international financial organization: "It is a global world today, and a body that oversees the smooth flow and interchange of currencies and other financial instruments [such as the FSB] is needed in today's world. . . . But the organization cannot be controlled by international bankers who are not answerable to the citizens of the countries in which they operate. It should be overseen by a senior level group which itself is organized as a liberal republic, following the original model of the United States.

"The point is not to get Congress to approve what has been done. It is to first get them to recognize that agreements have been made that affect our entire financial system and that it is their responsibility to shape these agreements in a way that is beneficial to our Republic AND [original emphasis] to provide a mechanism for real oversight of this international body. Central bankers should not be making decisions about international finance without oversight and a system of checks and balances that are reflective of those provided by a republican form of government."

For those states whose governments have mismanaged finances, history offers suitable lessons for revitalizing local economies. Most people probably haven't heard of a small island off the coast of England called Guernsey. After the blitzkrieg, the Germans occupied the island and deported nonnative islanders to German concentration camps. According to Toby Birch, managing director of Birch Assets Limited in Guernsey, the little-known history of Guernsey includes a great deal of monetary ingenuity. "As weary troops returned from a protracted foreign war [the Napoleonic Wars], they encountered a land racked with debt, high prices and a crumbling infrastructure, whose flood defenses were about to be overwhelmed. While 1815 brought an end to the conflict on the battlefront, however, severe austerity ensued on the home front. The application of the Gold Standard meant that loans issued over many years were then recalled to balance the ratio of money to precious metals. This led to economic gridlock as labor and materials were abundant, but much-needed projects could not be funded for want of cash. . . . This led to a period of so-called 'poverty amongst plenty.'" A committee was formed to find a way out of the situation.

"Like all great ideas, the principles were straightforward," Birch noted. "The committee realized that if the Guernsey States issued their own notes to fund the project, rather than borrowing from an English bank, there would be no interest to pay. This would lead to substantial savings. Because as anyone with a mortgage should understand, the debtor ends up paying at least double the amount borrowed over the long-term. . . . The irresponsible creation of credit is a dangerous game that temporarily benefits the current generation but steals from the next; a lesson that has been forgotten yet again in modernity. To bring balance to the equation, therefore, the people of Guernsey had to find a way to neutralize such deficits while neither contracting nor expanding the money supply.

"On a purely practical level, this was achieved by adding a sell-by date to the notes in issue, rather like a maturity date on a bond. For example, on a note issued 21 November 1827, it 'Promises to pay the bearer One Pound on the first of October 1830'. This begs the question as to how the future obligation was to be honored, but again, a simple mechanism was implemented whereby rent from the resulting infrastructure and tax revenues on liquor was set aside into a sinking fund to pay off the interest-free borrowing.

"The end result of the Guernsey Experiment was spectacular—new roads, sea defenses and public buildings were established, fostering widespread trade and prosperity. Full employment was achieved, no deficits resulted and prices were stable, all without a penny paid in interest. What started as a trial led to a string of construction projects, which still stand and function to this day. Money was used in its purest form: as a convenient mechanism for oiling the wheels of commerce and development."

But Birch also noted that there was a fly in the ointment. "One would have thought that everyone would be happy with such a success story but this was not the case. When you open a closed shop to competition, those with vested interests become highly protective. In those days it was the private banks who were threatened, because they were cut out of the equation. No loans meant no interest and no profit margin. So they may well have been the source of a mysterious complaint made to England's Privy Counsel which put a ceiling on the issuance of Guernsey notes for the next century."

Why should we pay attention to a situation on a small British isle almost two hundred years ago? "Whenever stimulus packages, tax rebates or bank bail-outs are paraded as solutions to the credit crisis they are actually part and parcel of its very cause," explained Birch. "It all stems from the quick-fix approach of producing money out of thin air and leaving it for the next generation to pay-off. This has been on-going in the United States since [at the very least] the Vietnam War, when the last vestige of monetary restraint was cast aside; in abandoning gold as a check on the money supply, the US freed the world from financial discipline. The dissolution of the Dollar has been evident ever since."

Birch said banks still have a role to play in providing liquidity by matching investors with borrowers, but they can no longer be trusted with the unrestrained creation of credit. "The Guernsey Experiment . . . shows that simple ideas can work wonders," he said. "They simply require an unselfish philosophy and a desire to do the right thing for future generations, much like America's Founding Fathers."

To disengage from the inflated national economy and to bolster local businesses, some Americans are experimenting with their own money. One instance of this is the BerkShare system, a local currency that has circulated in the Berkshires area of Massachusetts since 2006. According to the BerkShare website, nearly four hundred businesses in the Berkshire area accept BerkShares, which are printed on special paper including security features.

Labeled a "great economic experiment" by the *New York Times,* Berk-Shares are "a tool for community empowerment, enabling merchants and consumers to plant the seeds for an alternative economic future for their communities." The BerkShare website proclaims, "Five different banks have partnered with BerkShares, with a total of thirteen branch offices now serving as exchange stations. For BerkShares, this is only the beginning. Future plans could involve BerkShare checking accounts, electronic transfer of funds, ATM machines, and even a loan program to facilitate the creation of new, local businesses manufacturing more of the goods that are used locally."

FIRE CONGRESS

Our government, partially modeled after that of the Greeks, once flourished as a republican democracy. Yet, under recent authority, our government has devolved into a dichotomy of socialism and capitalism—melding public ownership and private ownership. Capitalism brings wealth to individuals who work hard while socialism brings wealth to those in control, who lie to get elected; for example, "No new taxes"—G.H.W. Bush, 1988; "Change you can believe in," Obama, 2008. In virtually every case, capitalism yields more wealth for an individual than needed. To resolve the discrepancy between those who have too much and those who have nothing, the capitalists invented charity, a product of religious morality.

The globalists devalued individual charity decades ago by suppressing the free exercise of religion and by replacing private charity with government charity. Those who promised the most charity to the people got elected to public office. This leads to a nation where a majority of nonproducers are in charge of the producers. As a growing number of people realize that life is easier as a nonproducer, more and more people strive for jobs as nonproducers. In the end, there are fewer people left to produce. This is the ultimate failure of socialism, and appropriately accounts for the collapse of Soviet communism.

America's elected Congress has allowed more and more nonproducers to live off the largesse of fewer and fewer producers. Today, adding government retirees, the disabled, Medicare, and Social Security to the welfare recipients, there are more Americans living off the government than paying into it.

Although no compassionate person is advocating cutting programs to those truly in need, the national budget must be trimmed and Congress appears unwilling or unable to do so.

In 2009 and 2010, a plan was offered to send an indelible message to Congress that the taxpayers want serious change—fire Congress. Many believe that such a plan may be the only way to effect real change in government. Several websites and organizations sprang up advocating voting out every incumbent in Congress. To borrow famous words, generally attributed to Mark Twain: "Politicians are like diapers; they need to be changed often and for the same reason."

Anticipation for drastic change has been coming for decades. Wright Patman, who was chairman of the House Committee on Banking and Currency for more than sixteen years, predicted in 1941 that the public would demand a drastic change in Congress due to its monetary policies. Patman said, "I have never yet had anyone who could, through the use of logic and reason, justify the Federal Government borrowing the use of its own money. . . . I believe the time will come when people will demand that this be changed. I believe the time will come in this country when they will actually blame you and me and everyone else connected with the Congress for sitting idly by and permitting such an idiotic system to continue."

In early 2009, Rasmussen Reports, a firm that distributes public opinion polling information, reported that corporate CEOs were the least favorably regarded professionals among a list of professional groups that included bankers, lawyers, and small business owners. But in September, Congress took the honor of being the least favored. "Seventy-two percent (72%) view them unfavorably," stated a Rasmussen news release. "There's some intensity in that perception, too. Only four percent (4%) have a very favorable view of congressmen, while 37 percent view them very unfavorably. Even 56 percent of Democrats have an unfavorable view of Congress although their party controls both the House and the Senate. Of course, their opposition pales next to the 86 percent of Republicans and 81 percent of adults not affiliated with either party who have an unfavorable opinion of Congress. But then voters are evenly divided over whether a group of people randomly selected from the phone book would do a better job than the current Congress."

As Obama's promised "Change we can believe in" failed to materialize in 2009, a movement to throw out Congress began to gain strength. The website for an organization called Kick Them All Out reads, "Presidents have no Constitutional authority to do most of the things they claim they can do. They can only ask the Congress to do what they want. The Congress could have stopped everything that's happening; the wars, the Wall Street takeover, the trillion-dollar defense budget they just passed. Our so-called representatives have sold us out so many times it makes my head spin and what do we all do? We not only let them keep their jobs, but you watch, they will most likely give themselves a raise, like they always do.

"The Congress critters work for *us,* not the central bankers and transnational corporations. What would you do if you owned a company and none of your employees listened to you, they lied to you, didn't do the jobs you gave them to do, and in fact, were actually working for your competition and selling your company down the river as fast as they could? I don't think you'd keep them on and give them a raise! Well, that's exactly what we've been doing, only in this case, your company is our Federal Government, and your employees are the 435 members in the House of Representatives and the 100 members of the Senate, virtually all of them working for the transnational corporations (the competition) and they have already achieved a hostile takeover of our government on every level and are using the powers of our own government against us in order to take over our entire nation. What the heck happened to that thing called 'the wisdom of the American people'? You don't reward employees that betray you. YOU FIRE THEM [original emphasis]!"

The Kick Them All Out website offers free posters of the famous Uncle Sam painting by James Montgomery Flagg. But in this rendition, an artist has changed the slogan to read "I want you! To kick them all out! Do your patriotic duty and show Congress who the boss is!"

A similar group from Texas is calling to empty Congress of its incumbents. Formed by Houston native Tim Cox, the group is called GOOOH (Get Out of Our House) and, as of 2009, had two thousand members in Houston and a hundred thousand outside the city. According to one Houston TV station, ABC affiliate KTRK, the group might succeed in its goal as a poll conducted by the station showed 47 percent of respondents reported they were no longer aligned with a political party. Some observers feel this number may be reflected in other parts of the country and signifies a movement away from the two-party system.

WorldNetDaily.com, a conservative online website headquartered in Washington, D.C., offered its readers the opportunity to send actual "pink slips" to specific members of Congress, warning that "if they vote for more spending, socialized medicine, cap-and-trade legislation and a hate-crimes measure" they would not be reelected in 2010. World Net Daily claimed to have distributed as many as three million slips in a two-week period. "I believe this campaign, already tremendously successful beyond my wildest expectations, can have a real impact on politicians whose first priority is getting re-elected," said WND's editor and CEO Joseph Farah.

Slightly less radical groups are trying to shake up the status quo through legislation. The Fire Congress Meetup Groups effort looks for members to join "more than two hundred thousand Americans and impose 'de facto' term limits on all U.S. Congressmen and Senators, regardless of party affiliation or whatever they promise. . . . Kick them all out, so the new ones finally hear us," stated the Fire Congress's website. One Internet wag recommended, "Limit all U.S. politicians to two terms—One in office, one in prison."

Term limits is another idea that has been brought up in the past as a means to curtail congressional power. Though credible legislators have made such proposals in the past, the very people affected by the change—the members of Congress—have always voted the bills down.

U.S. Term Limits (USTL), headquartered in Fairfax, Virginia, claims to be the leader of what the organization's creators describe as the largest grassroots movement in American history. According to the USTL website, the organization has placed term limits proposals into fifteen state legislatures. "[E]ight of the ten largest cities in America adopted term limits for their city councils and/or mayor, and 37 states place term limits on their constitutional officers," stated USTL literature. "American politicians, special interests and lobbyists continue to combat term limits, as they know term limits force out career politicians who are more concerned with their own gain than the interests of the American people. . . . Remember, every town councilman wants to be a congressman; every congressman wants to be a senator; and every senator wants to be president."

At WeShouldFire Congress.com, the message is the same. "It's time we send the message straight to Congress—do your job or you're fired!" states literature on the website. This site raises money to place billboards across the nation urging voters to fire Congress.

All of these organizations and websites implore Americans to vote for America, not a political party. Though the imperatives from these organizations resonate with many voters, when election time rolls around, many voters will continue to vote for the same old faces and political parties.

The question naturally arises, why don't more progressives and independents run for public office? It would seem as if the progressives or independents would receive votes from those dissatisfied with Congress.

Yet this does not occur because progressives and independents are tied down trying to survive in a society in which the love of money has superseded the love of their fellow human being. To be specific, to win public office in any large city or state, a candidate must have television and radio broadcast time. Purchasing this time is expensive and often media outlets want cash in advance for political ads. Additionally, there are the costs of producing a professional-looking and effective ad. This is an expense that goes far beyond hand-painted posters and yard signs. It can run into the thousands, if not tens of thousands, of dollars.

If candidates still want to succeed, they must ally themselves with one of the two major political parties and look for corporate or political action committee (PAC) money. This need for a huge stockpile of cash prevents most honorable and honest people from competing in the political campaign process. Most candidates, especially at the local and state level, simply do not have the kind of money it takes to produce and air an influential advertisement.

George Green once raised campaign funds for Jimmy Carter and was asked to be Carter's campaign finance chairman. "I remember being flown to Aspen in a private jet and then being asked to be the Democratic Finance Chairman for the Carter election," Green said in a 2006 interview. "I remember then saying I was a Republican and then Paul Volcker [former chairman of the board of governors of the U.S. Federal Reserve System and North American chairman of the Trilateral Commission] leaned over and said, 'That's okay, kid. It doesn't matter, we control them both.'"

Based on Green's words as well as the evidence detailed earlier in this work, it should be clear now that the same secret society globalists control both the Democratic and Republican parties. This may be why people often call our members of Congress "the best representatives that money can buy."

"The biggest problem in our government is corporate power, and with that, the huge amount of resources and political power taken by the military. Until we deal with those issues, we will go nowhere in this country on health care, the environment, social justice or anything else of importance," said Harvey Wasserman, an author, a journalist, and an energy activist. "People should now understand that while it's been monumentally important to finally have an African-American as president (a woman will come next), it's now more important to have someone who is not a Republican or a Democrat, and who is committed to the welfare of the public rather than that of the corporations."

One possible way to curtail the abuses on the election process would be to outlaw TV ads for prospective candidates, which would in many ways take money out of the equation. This would allow interested voters to learn about candidates through debates, newspaper articles, or printed flyers outlining candidate positions and policies. Political candidates would get radio and TV airtime through talk-show or journalistic interviews open to all candidates. Such interviews could open political debates to alternative ideas and less mudslinging.

Another good way of culling out greedy or financially sponsored politicians is to vote for the candidate with the least money. This person may not be any less susceptible to corruption, but it is a sure sign the individual has not sold out for campaign funds. As Bernard Baruch, the financier and political consultant to Presidents Woodrow Wilson and Franklin Roosevelt, once advised, "Vote for the man who promises least; he'll be the least disappointing."

Once the incumbent politicians have been turned out of office and a new crop arrives in Washington, the public must scrutinize their every move. The public must force them to consider term limits and to do away with their private retirement funds. Place Congress on Social Security and watch how fast it is cleaned up and well funded. Only when Congress members act suitably for the public should they be voted back into Congress.

This is not a revolutionary idea—it's the way the system is supposed to work. Unfortunately, this system is predicated on the idea that there is an alert and educated electorate and that the voting mechanism is honest and fair.

POLL WATCHERS AND PAPER BALLOTS

Zombies don't vote. Only about half of the eligible electorate cast ballots in recent presidential elections. In 2008, this dismal record was turned around when 62 percent of eligible voters cast ballots, the highest turnout since the 1960 campaign between John F. Kennedy and Richard M. Nixon. Though this percentage looks impressive, one must consider the average voter turnout in comparable European nations: Italy, 93 percent; Germany, 81 percent; Spain, 77 percent; and the United Kingdom and Ireland, 75 percent.

The comedian W. C. Fields once said, "Hell, I never vote for anyone. I always vote against." Nowadays, people still don't vote for anyone—they simply pull the lever or touch the screen for their political party, holding little regard for the issues or the quality of their party's candidate. This method of voting may be due to the fact that far too many voters feel that neither of the two candidates in an election stand for their ideals. Instead of voting for a person, many voters feel they must vote against the lesser of two evils, which still means they are voting for an evil. "Once you don't vote your ideals . . . that has serious undermining effects. It erodes the moral basis of our democracy," opined unsuccessful presidential candidate Ralph Nader.

Consider the presidential election of 2004. Voters had the option of the Republican candidate George W. Bush, the scion of a rich family and a member of the secret society Skull and Bones, or Bush's cousin, Democratic candidate John Kerry, the scion of a rich family and a member of the secret society Skull and Bones. Most informed and thoughtful people did not consider this much of a choice.

What could be worse than having two bad presidential choices? Not even being able to choose between the two. President Franklin Roosevelt said, "Nobody will ever deprive the American people of the right to vote except the American people themselves—and the only way they could do this is by not voting." But then, Roosevelt had no way of knowing that voters could be disenfranchised by computers and voting machine fraud.

As Boris Bazhanov notes in *Memoirs of Stalin's Former Secretary,* Joseph Stalin once proclaimed, "I consider it completely unimportant who . . . will vote, or how; but what is extraordinarily important is this— who will count the votes, and how."

Until very recently, votes were cast with paper ballots under the watch of poll watchers—someone appointed by a candidate, a political party, or supporters/opponents of a particular measure to observe the election procedures in a given precinct, watching for any voting irregularities. Watchers and voters may not converse within the polling place, nor are watchers permitted to interfere with the orderly conduct of the election or influence any voter.

Poll watchers have largely been outmoded by electronic voting machines, which are fundamentally just computers. The Help America Vote Act was signed into law by President Bush in 2002. It was intended to streamline and improve voting methods, such as eliminating the punch-card ballots that had caused so much trouble in the 2000 Florida election, setting standards for the training of poll workers and upgrading to electronic voting machines. But effecting these changes was left up to the individual states, which resulted in varying interpretations and effectiveness.

There has been a great deal of controversy over the use of electronic voting machines that display ballots and record and tabulate votes. Advocates of using machines claim such machines are fast, accurate, and easy to set up for disabled and non-English-speaking voters. Yet there are problems with the machines. Critics claim voting machines have many technical problems that could lead to inaccuracy and hacking. The touch-screen models are a special concern since some models do not provide a paper record of the votes, which might be necessary in the case of a manual recount.

Researcher Bev Harris, founder of the national nonpartisan, nonprofit elections watchdog group Black Box Voting Inc., wrote, "Our voting system, which is part of the public commons, has recently been privatized. When this happened, the counting of the votes, which must be a public process, subjected to the scrutiny of many eyes of plain old citizens, became a secret."

In 2003, Bev Harris obtained internal memos from Diebold, which used to be one of the major manufacturers of electronic voting systems. Some of the internal memos documented that uncertified software was being used in its voting machines and that Diebold programmers intentionally bypassed the certification system. She posted the memos on the Internet. Though Diebold claimed Harris's action constituted copyright infringement, a California U.S. district judge forced Diebold to relent in October 2004, when the judge ruled that Diebold had abused its copyright privileges while trying to suppress the embarrassing memos.

In 2007, Diebold changed the name of its election division to Premier Election Solutions, Inc. (PES), following a spate of bad publicity. On September 3, 2009, Election Systems & Software (ES&S) announced that it would purchase PES, which means that America is now provided voting machines by only three companies—ES&S, Sequoia Voting Systems, and Hart InterCivic. Many viewed ES&S's acquisition as creating a near monopoly over the voting machines widely used throughout the country. "Election Systems & Software's $5 million acquisition of Diebold Inc.'s voting-machine company amounts to a near monopoly," cried an editorial in the *Miami Herald*. "The state [of Florida, during the 2000 presidential election,] learned the hard way that touch-screen voting did not reassure voters that their ballots were being counted because the machines left no independently verifiable paper trail."

Of the ES&S purchase, Harvey Wasserman said, "The ES&S purchase of Diebold [PES] is indicative of a larger problem . . . between the two of them, they control 80% of the touchscreen machines in the U.S. Both are corrupt GOP-dominated corporations. So, the idea that just one of them will be in control doesn't matter that much, although it has been a positive to see so much attention paid to the situation." Party politics aside, it should be clear that the consolidation of the nation's voting process into only a few hands offers the appearance of opportunity for, if not actual, vote manipulation.

According to Wasserman, what is more troublesome than a voting machine monopoly is "the use of the machines in the first place." Wasserman believes that "All electronic voting machines, tabulators, etc. should be banned. We need universal automatic voter registration, and universal paper ballots that are hand-counted. Simple as that. Until we get there, there is no reason to believe any election in this country will be a reliable reflector of the popular will." Wasserman also advocated universal automatic registration and a national holiday for voting and for vote counting, "to give working people an equal opportunity to vote."

Diebold's voting machines have long been controversial. Following investigations over Diebold's voting machines, California banned one Diebold model from the state in 2004. California decertified some voting machines again in 2007. After it was learned through an open source ballot-counting program that 197 ballots had been silently dropped from voting machines in Humboldt County, investigators conducted a "top-to-bottom review" of voting machines. At the conclusion of the investigation in 2009, Secretary of State Debra Bowen decertified Diebold's Global Election Management System (GEMS) version 1.18.10 software program and three other electronic voting systems, meaning they cannot be used in California.

In March 2009, Diebold/PES's problems became much larger when the firm admitted in a Sacramento hearing that audit logs produced by its tabulation software could miss significant events such as the deletion of votes. The company acknowledged that the problem existed with every version of its tabulation software, even those used in other states. Vote-counting GEMS software is used to tabulate votes cast on every Premier/ Diebold touch-screen or optical-scan machine in more than fourteen hundred election districts in thirty-one states.

"Today's hearing confirmed one of my worst fears," said Kim Alexander, founder and president of the nonprofit California Voter Foundation. Alexander noted, "The audit logs [a program that monitors additions and deletions to the operating program] have been the top selling point for vendors hawking paperless voting systems. They and the jurisdictions that have used paperless voting machines have repeatedly pointed to the audit logs as the primary security mechanism and 'fail-safe' for any glitch that might occur on machines. To discover that the fail-safe itself is unreliable eliminates one of the key selling points for electronic voting security."

In 2007, the Maryland General Assembly voted for paper ballots counted by optical scanners to replace paperless touch-screen voting machines. But the plan fell apart in 2008 when a vote in the U.S. House of Representatives didn't approve an Election Assistance Commission program to provide the necessary states funds for the purchase of paper ballots as a backup to voting machines. In other words, efforts to return to paper ballots have been blocked at the federal level. Could this be because the New World Order socialists (sometimes National Socialists, sometimes Marxist Socialists) have gained control over the federal apparatus? Wits have said that if God intended for us to vote, he would have given us candidates. It can likewise be said that if we were intended to have fair voting, we would have hard-copy paper ballots that could remain for years in case of the need for a recount.

ENFORCE THE TENTH AMENDMENT

"THE FEDERAL GOVERNMENT TODAY can wage wars without the consent of our congressional representatives, overthrow foreign governments, tax nearly half of national income, abolish civil liberty in the name of 'homeland security' and 'the war on drugs,' legalize and endorse infanticide ('partial-birth abortion'), regulate nearly every aspect of our existence, and there's little or nothing we can do about it. 'Write your congressman' is the refrain of the slave to the state who doesn't even realize he's a slave (thanks to decades of government school brainwashing)."

These were the observations of Thomas J. DiLorenzo, a professor of economics at Loyola College in Maryland and the author of *How Capitalism Saved America* and *Hamilton's Curse: How Jefferson's Archenemy Betrayed the American Revolution—and What It Means for America Today.* DiLorenzo noted that "until 1865, the Supreme Court's opinion was just the Supreme Court's opinion. The citizens of the states reserved the right to offer their own opinions on constitutionality, which they often considered to be every bit as valid as the Court's." President Woodrow Wilson, who one might recall was placed into power by Wall Street financiers, the forerunners of today's globalists, argued against states having the power to determine constitutionality in his 1908 book *Constitutional Government in the United States,* writing, "the War between the States [which ended in 1865] established . . . this principle, that the federal government is, through its courts, the final judge of its own powers."

Beginning with the 2008 election of Barack Obama, state legislators began acting less subservient to the federal government as many citizens joined the Tenth Amendment Movement to rally against too much federal control. Members of the movement argue that the Constitution's Tenth Amendment clearly states: "The powers not delegated to the United States by the Constitution, nor prohibited by it to the States, are reserved to the States respectively, or to the people."

In late September 2009, the Ohio State Senate passed Senate Concurrent Resolution 13 (SCR 13), which was meant to "claim sovereignty over certain powers pursuant to the Tenth Amendment to the Constitution of the United States of America, to notify Congress to limit and end certain mandates, and to insist that federal legislation contravening the Tenth Amendment be prohibited or repealed." The Ohio State Senate was the eighth state senate behind Alaska, Idaho, North Dakota, South Dakota, Oklahoma, Louisiana, and Tennessee to pass a resolution reaffirming state sovereignty. By October 2009, Tenth Amendment resolutions had been introduced in thirty-seven state senates.

Oklahoma state representative Charles Key compared the resolution he authored for Oklahoma to a cease-and-desist order given by a landlord to a nonpaying tenant. "If you've got a tenant that's not paying rent, you don't just show up one day with an empty truck," said Key. "First, you serve notice. That's how we see these resolutions, as a notice to the federal government. And there definitely will be follow-up." Supporters of the resolutions say that they are a long-overdue first step in moving the country toward a constitutional government.

The Tenth Amendment is similar to a portion of the Articles of Confederation, which were written before the Constitution. A provision of the articles state, "Each state retains its sovereignty, freedom, and independence, and every power, jurisdiction, and right, which is not by this Confederation expressly delegated to the United States, in Congress assembled."

Although states have long grumbled about the enforcement of federal laws, the U.S. Supreme Court has ruled only twice on Tenth Amendment cases in modern times. In 1992, the court found that the Low-Level Radioactive Waste Policy Amendments Act of 1985 was unconstitutional in forcing the states to retain and assume liability for radioactive waste. In 1997, the Supreme Court ruled that the Brady Handgun Violence Prevention Act unconstitutionally required state and local law enforcement officials to conduct background checks on persons attempting to purchase handguns. Both cases involved only narrow and defined interpretations of the Tenth Amendment, indicating the high court will not hasten to clarify the overall intention of this basic constitutional revision. This is similar to the Court's refusal to hear arguments that the penalties imposed by the IRS on those who fail to file a 1040 tax form, which can be used by the prosecution in tax cases, are a direct violation of the Fifth Amendment, which states that persons cannot be compelled to give evidence against themselves. As noted by both Professor DiLorenza and President Wilson, the War Between the States temporarily settled the argument over whether local representatives elected by the citizens or some federal bureaucrat in Washington would rule over the public. Today, there is virtually no law or ordinance passed anywhere in the United States that cannot be overturned or superseded by federal authorities. If one questions this, just ask the medical marijuana shops in California that were raided by the feds even after California voters approved such sales for medicinal purposes in 1996.

Rather than passing resolutions to simply reaffirm their sovereignty, some states pushed for specific freedoms. In 2009, Montana and Tennessee passed Firearms Freedom Act legislation to "declare that any firearms made and retained in-state are beyond the authority of Congress under its constitutional power to regulate commerce among the states." Ten other states considered similar legislation. After the legislation passed, officials from the Bureau of Alcohol, Tobacco, Firearms and Explosives (BATFE) sent letters to gun dealers and federal firearm permit holders in both states. The letters stated that the dealers and permit holders should ignore the state law. Clearly, the contest for state regulation of firearms will continue in higher courts.

Disputes between the government and the states aren't just limited to firearms. Voters in Alaska, California, Colorado, Hawaii, Maine, Maryland, Michigan, Montana, Nevada, Oregon, Rhode Island, Vermont, and Washington voted on and passed legislation permitting marijuana use for medical purposes, but federal authorities have looked down on these laws, even though they were voted on by the majority of those states' citizens.

In 1996, medical marijuana was legalized in California after Proposition 215 passed by a 56 percent citizen vote. Regardless, marijuana remained illegal at the federal level by the Controlled Substances Act, which has led to a number of disputes. In 2005, a California woman sued the Drug Enforcement Administration after her medical marijuana crop was seized and destroyed by federal agents. Citing a constitutional clause that grants the federal government the power to regulate interstate commerce, the U.S. Supreme Court ruled that even though the woman grew pot strictly for her own consumption and had never sold any, growing one's own marijuana affects the interstate market of marijuana. The Court warned that homegrown marijuana for medical purposes could nevertheless, even inadvertently, enter the stream of interstate commerce. On the basis of this argument, the judges deemed that the federal actions of the Drug Enforcement Agency were warranted.

Regardless of the Supreme Court's decision, an editorial in the September 21, 2009, edition of the *San Francisco Examiner* dealing with concerns over the REAL ID Act, firearms, and marijuana laws and even the health-care debate proclaimed, "State sovereignty supporters stand on solid historical ground. . . . James Madison's 'Virginia Plan,' which would have given Congress veto power with state laws and allowed the federal judiciary to hear all disputes, was soundly defeated by the signers of the Constitution. A needed check on an overreaching federal government that grows bigger by the day, the reassertion of state sovereignty should be a welcome development to Americans concerned about losing their liberties—just like the Founders were."

On February 1, 2010, five Democrats in the Virginia State Senate broke ranks with their party to endorse bills prohibiting compulsory government health care. Three bills protecting Virginians from being forced to buy federally mandated health care were approved on 23–17 votes in the Virginia Senate, where Democrats have a 22–18 majority. If approved, Virginia would join Arizona as the second state to pass measures in defying such federal legislations. Senator Frederick M. Quayle, sponsor of one of the three proposals, argued that the federal government does not have the constitutional authority to require individuals to buy anything. "This is not a bill that deals with health care. It is a bill that attempts to reinforce the Constitution of the United States," he explained.

It remains to be seen how successful states will be in regaining their sovereignty. Regardless, there are some encouraging signs. Beginning with Maine in 2007, nearly twenty-five states have passed legislation opposing the REAL ID Act, which mandated federally approved identification. The act was passed in 2005 and was to go into effect in 2008 but was not enforced by 2010. Many governors scorned the responsibility and cost of ensuring that those who hold driver's licenses are citizens or legal residents of the United States.

It is clear to many that more state sovereignty is achievable in the near future. Supporters of the Tenth Amendment Movement point to successful actions against the REAL ID Act, as well as the legalization of medical marijuana in thirteen states, as proof that with enough state-level resistance, the federal government may have no option but to back off—with or without judicial approval.

There is also a good chance that when states are able to freely practice sovereignty, we will find practical, profitable, and safe alternatives to our current dependence on petrochemicals. As new energy sources become available, the globalists who profit from monopolies on gas and oil will have to diversify their products and begin to market alternatives.

Twenty years ago we were told that solar energy was a viable alternative but that the necessary harnessing technology wouldn't be available for twenty years. Now that twenty years have passed, one must ask, where's the solar energy? It has been estimated that the sun provides between 10,000 and 20,000 times more energy than we use on a given day. In order to use this energy, we need to learn how to collect it and put it to work.

The fact that we haven't learned how to collect this energy should no longer be blamed on technology—rather, what is at fault is stubbornness, the lack of will on the part of corporate business and its hired politicians in Congress. There is even a fundamental disconnect in the thinking of schooled energy experts. For example, one solar expert explained that it would take solar-collector panels covering the state of Arizona to produce enough electricity to power the city of Los Angeles. Although this may be true, the expert based his conclusion on the faulty assumption that central generation was necessary to produce electricity for the city. Few power experts can visualize that by simply placing solar collectors on every rooftop in Los Angeles, the city could become largely energy independent. This independence could mean that electric bills would be cut in half or more. The only real problem would be for the monopoly utility companies. They could not place a cloud over a home owner who failed to pay the monthly electric bill.

But advances in alternative energy slowly continue to move forward. In October 2009, Suniva announced plans to transform some Michigan farmland near Saginaw into a 200,000-square-foot solar manufacturing facility. The announcement was made by Michigan governor Jennifer Granholm, who said that the $250 million project could create five hundred jobs during the coming years. Local business leaders call the project a much needed economic boost for the whole region. "We have generations of skilled manufacturers here and we have people that understand the manufacturing industry," said Saginaw Future Inc. president JoAnn Crary. Though Suniva was hoping to break ground in 2010, it was having trouble securing financing for the project.

Even some of the corporate giants seem to be jumping on the alternative energy bandwagon. In October 2009, Dow Chemical announced its innovative Powerhouse solar shingle, which company officials hoped would boost solar energy use by home owners in the coming years. The Powerhouse solar shingle incorporates photovoltaic solar collecting/generating technology into a roof shingle. This allows people to use their entire rooftop to generate electricity at a reasonable cost. Dow officials said the new solar shingles will be on the market in limited quantities in 2010 and more widely available in 2011.

With the advent and implementation of these new technologies, Americans must commit to new ways of thinking about energy. With apologies to Edmund Burke—the only thing necessary for the triumph of evil is for good zombies to do nothing.

NONVIOLENT NONCOMPLIANCE

STRENGTHENING THE POWER OF the central government will not solve many of the country's central public issues, especially that of public health care. The government has failed over and over with so many federal programs. How can the public remain confident in a health-care program built by the government amid a financial crisis? One unsigned message circulating on the Internet bluntly presented the truth in this manner:

"The U.S. Postal Service was established in 1775. They've had 234 years to get it right. It is broke, and even though heavily subsidized, it can't compete with private sector FedEx and UPS services. The U.S. Postal Service will lose over $7 BILLION this year and will require yet another bailout.

"Social Security was established in 1935. They've had 74 years to get it right. It is broke. There is nothing in the Social Security Trust Fund except IOUs from the government.

"Fannie Mae was established in 1938. They've had 71 years to get it right. It is broke. Freddie Mac was established in 1970. They've had 39 years to get it right. It is broke. Together Fannie and Freddie have now led the entire world into the worst economic collapse in 80 years.

"The War on Poverty was started in 1964. They've had 45 years to get it right. One trillion dollars of our hard earned money is confiscated each year and transferred to 'the poor'. It hasn't worked.

"AMTRAK was established in 1970. They've had 39 years to get it right. [In 2008, the government] bailed it out as it continues to run at a loss!

"Medicare and Medicaid were established in 1965. They've had 44 years to get it right. They are both broke. And now our government [the Obama administration] dares to mention them as models for all US health care. . . . This is the government at work and they now want to run the most complex economic program they have ever tackled—our health care system."

The secret here is that the government is not the problem. Nor is the Constitution on which it's founded. The problem lies in the people who control the government—the New World Order global fascists who now have a chokehold on the government. Their corporate money controls all three branches of government while their associates are appointed to cabinet-level positions of authority. Recall that the Obama administration is top-heavy with members of the Trilateral Commission, Council on Foreign Relations, and Bilderberg group.

The states do not have to follow the schemes and dictates of these plutocratic globalists. Nor does the general public. The path to independence from government control does not need violence. The people of India did not gain independence by battling the British army in the field. African Americans did not gain freedom by waging violent war against the government. In both instances, they simply practiced nonviolent noncompliance.

Though it is not often mentioned in the corporate mass media, the more literate citizens are opting out of the New World Order control box. There is a widespread but underreported antitax movement, with millions of American simply refusing to "voluntarily" pay taxes for purposes they can't support. Naturally, the number of these resisters are rarely, if ever, given in the mass media, which nevertheless routinely reports on IRS crackdowns, usually about tax time. Convicted tax cheats garner major headlines while victories over the IRS get scant coverage if any at all.

Yet small victories over the New World Order are taking place all the time. Citizens in both big cities and small towns are growing neighborhood gardens, supplementing fast food with organic and healthy vegetables and fruits. Most religious institutions today stock large pantries where food can be distributed to the poor, relieving strain on the welfare system. Those concerned about the environment are serving as examples to others in ways to lessen human impact—recycling trash, riding bicycles, supporting mass transit, and driving the new hybrid or totally electric cars. Their demand for nonpolluting, energy-efficient vehicles is now being met by customer-seeking corporations. Everyone can make a difference. If one person stops to pick up some trash on the street and places it in a receptacle, others notice and some will be prompted to action.

THIRTY-SIX REMEDIES FOR A BROKEN SOCIETY

FOLLOWING IS A LIST of recommendations and suggestions, compiled from various sources, for bringing a zombie nation back to being a free and functioning democratic republic. Some of the recommendations are self-evident, others perhaps wistful, but all should be given consideration:

1. The Federal Reserve System, a collection of privately owned banks, should be audited immediately. Privatization of U.S. money is unconstitutional, because the Constitution states that only Con-

gress shall coin and regulate money. Now, privatization has led to economic disaster. The printing of the dollar should be approved through Congress and issued through the U.S. Treasury as U.S. Treasury notes. Notes should be distributed gradually so as not to significantly inflate the worth of the currency in circulation. U.S. debt through fractional reserve lending has been created by sleight of hand; it can be abolished by sleight of hand.

2. Only those who pay into Social Security should be able to benefit from the system. Placing Congress under the Social Security plan that the remainder of the nation must live under would swiftly bring needed repairs. The members of Congress have exempted themselves from Social Security as well as from any future mandatory health-care plan. Their current and generous private congressional retirement program should be ended.

3. The National Security Act of 1947 should be reviewed and perhaps rescinded. Currently, the law allows the president and his National Security Council handlers to bypass the elected representatives in Congress, the media and the public in serious policy-making decisions involving war, technology, and even issues of outer space.

4. Executive Order #13233, which allows the incumbent president to classify and keep from the public the libraries and documents of his predecessors, should be rescinded.

5. No U.S. intelligence employee, whether civilian or military, who has attained the status of "officer" should be allowed to run for or serve as president of the United States. Years of intelligence work expose a person to the seamy world of lies, deceit, and misdirection. For some instances of national defense, this may be necessary, but such work leaves a person in public office open to blackmail and control from former superiors and their loyalty oaths.

6. Unlike the current system where sometimes a dozen or more lobbyists can seek communication with legislators, all corporations should be allowed to have only one lobbyist per congressman. They should also have to visit that congressman with a public advo-

cate who can argue on the side of the people. Fact-finding junkets and entertainment for Congress members should come solely from closely monitored public expenses.

7. Limit senators to three terms and representatives to no more than six. Legislators who remain in office too long become political professionals, more concerned with getting reelected and maintaining their power than with the problems of the public. Most start their career with a genuine desire to serve the people. They should be turned out as this desire is turned to cynicism by the temptations of money and power.

8. A term limit of twelve years should be set for Supreme Court justices, to prevent old-age infirmities and experience based on life thirty years ago from occluding their judgment. Furthermore, all federal district judges should be elected by the public and limited to two terms of five years, to prevent the loading down of federal benches with political hacks who primarily vote party politics or the wishes of those who put them in power.

9. The Pledge of Allegiance should be said every day at school and every day in Congress to remind both young and old of the basic tenets of U.S. sovereign freedom and democracy.

10. Legislation should prohibit any person who has membership in any secretive organization—the Council on Foreign Relations, the Trilateral Commission, Bilderberger group, and so on—from holding public office. One cannot have dual allegiance. It is clear that individuals cannot support state sovereignty while supporting the globalist agenda of their fellow society members.

11. The classification process has gotten out of hand. Today, routine documents are classified, sometimes due to holdover policies of the cold war, sometimes just to cover up bungling or neglect. The current practice of classifying any nonclassified document if it can be connected to one that is classified must be stopped. Unless information clearly jeopardizes national security, it should remain open to public scrutiny. A citizen review board, composed of academics, journalists, and others—not just government insiders—should

oversee this process to protect both security and the public's right to know.

12. The PATRIOT Act should be rescinded. It was passed by a panicked Congress that was not given time to even read it and has led to infringements on the public's civil liberties.

13. To prevent a repetition of the deficient Warren and 9/11 commissions, both of which in 2010 continued to draw criticism from a wide swath of the American public, any future investigation of a national tragedy should be formed from citizens representing a wide cross-section of regional, political, philosophical, and professional expertise. A 1991 Gallup poll showed almost 75 percent of the public disbelieved the Warren Commission's lone-assassin theory of the JFK assassination. All major pieces of evidence against the accused Lee Harvey Oswald—his fingerprints on the rifle, neutron activation analysis of the bullet metal, and testimony taken at the time—have proven deficient or untrue. The entire JFK assassination case has been riddled with fabrication of evidence, suppression of evidence, alteration of evidence, and intimidation of witnesses (read Jim Marrs's *Crossfire: The Plot That Killed Kennedy* for full details). By 2010, even top officials of the 9/11 commission, tasked with finding out what happened to America on September 11, 2001—including commission cochairman Lee Hamilton and senior counsel John Farmer—had publicly questioned the conclusions of their own commission. Farmer, a former New Jersey attorney general, in his 2009 book *The Ground Truth: The Story Behind America's Defense on 9/11* even wrote, "In the course of our investigation into the national response to the attacks, the 9/11 Commission staff discovered that the official version of what had occurred [the morning of September 11, 2001]—that is, what government and military officials had told Congress, the Commission, the media, and the public about who knew what when—was almost entirely, and inexplicably, untrue . . . at some level of the government, at some point in time . . . there was an agreement not to tell the truth about what happened."

14. A committee composed equally of professionals and ordinary citizens from separate states should be formed to oversee government health agencies such as the FDA, the NIH, and the CDC, to ensure that decisions affecting the public, particularly those dealing with research and conflicts of interest concerning employees and contract personnel, are impartial.

15. No state law passed by popular vote should be superseded by any federal statute except for those found in the U.S. Constitution. No federal official should tell the people of a state to ignore their own laws, as happened in the case of new Tennessee firearms legislation. If a state law is bad, federal officials should simply work to see that law revised or rescinded.

16. Citizens should regularly request Comprehensive Annual Financial Reports (CAFRs) from all school, local, county, and state offices, including federal agencies. This may be done by submitting Public Information Requests (PIRs). In this manner, citizens could see precisely how much money is being held and how it is being spent.

17. All regional and global trade agreements, pacts, and treaties, such as the World Trade Organization (WTO), the North American Free Trade Agreement (NAFTA), and the Central America Free Trade Agreement (CAFTA), should be reviewed to determine if they violate the U.S. Constitution or the rights of Congress to regulate commerce and trade. Today, some trade agreements have been used to supersede U.S. laws. As a signatory nation, the United States has committed itself to conforming its laws and policies to WTO dictates and, as the WTO has exhibited strong enforcement of its policies, the mere threat of a WTO challenge usually results in changes of the national laws or policies. For example, in 2002, a WTO appellate panel ruled that U.S. tax rules exempting some corporate income earned overseas from taxation constituted an "illegal subsidy." The tax rules were changed. According to Representative Ron Paul, a 2008 presidential candidate, "Incredible as it seems to liberty-minded Americans, the

WTO and the Europeans are now telling us our laws are illegal and must be changed. It's hard to imagine a more blatant example of a loss of U.S. sovereignty. Yet there is no outcry or indignation in Congress at this naked demand that we change our laws to satisfy the rest of the world. I've yet to see one national politician or media outlet even suggest the obvious, namely that our domestic laws are simply none of the world's business. . . . Congress may not object to being pushed around by the WTO, but the majority of Americans do."

18. The Posse Comitatus Act, which prevents the military from policing the U.S. public, should be upheld by the executive branch of government.

19. The current practice of outsourcing the production of military hardware to foreign countries must be stopped. Any arms and equipment, particularly computers vulnerable to hacking, being used by the U.S. military should be produced in the United States by American companies using American workers. Under present outsourcing policies, an enemy of the United States could gain intelligence, if not outright control, over our defense systems, particularly through third parties. One friendly country makes our weapons, then sells or trades the technology to an enemy nation. The benefits of such action should be self-evident, especially in view of the number of former friends who later turned enemy—for example, Saddam Hussein. The current system, of course, is compatible with the one-world plans of the globalists.

20. Nonimmigrant visas should be discouraged. Temporary foreigners, working for lower wages, take jobs from the U.S. labor pool, today plagued by rising unemployment.

21. The government should rescind all so-called hate crime legislation. Such laws cannot truly stop individuals from holding hateful beliefs. Furthermore, these laws contradict the Bill of Rights and can be abused to silence political dissidents and enemies. The mass media has been quite successful in changing prejudicial attitudes in the past. No laws, susceptible to misuse, need be made.

22. America's prison systems should be overhauled so that nonviolent offenders are able to move through a series of increasingly lenient punishments (fines, community service, etc.) without going to jail. Career criminals and gang members should be separated and placed in supervised work projects outside the public.

23. The government must end the failed war on drugs and legalize marijuana, which has been proven less harmful than legal drugs, cigarettes, and alcohol. According to DrugWarFacts.org, "[T]here are simply no credible medical reports to suggest that consuming marijuana has caused a single death." Furthermore, the National Commission on Marihuana [sic] in 1972, after making a study of pot smokers, concluded, "No significant physical, biochemical, or mental abnormalities could be attributed solely to their marihuana smoking. . . . Neither the marihuana user nor the drug itself can be said to constitute a danger to public safety. . . . [Marijuana's] actual impact on society does not justify a social policy designed to seek out and firmly punish those who use it." President Richard Nixon, who appointed this commission, disavowed its findings and launched the first war on drugs. Like Prohibition before it, the current prohibition of many drugs has only promoted organized criminals and injustices by authorities. Surveys have shown that a large portion of the nation's inmates are in prison because of drug-related offenses. By ending the prohibition of certain drugs, the overcrowding in U.S. prisons could immediately be relieved. This change, along with a tax on marijuana, should increase government income without the need for taxes to support more police and prisons.

24. At the very least, marijuana should be federally decriminalized. There should only be misdemeanor fines for abuse of the drug. Drug abuse should be seen for what it is—a health problem. The criminalization of drugs has only created a legacy of corruption and violence, just as Prohibition did before it. Industrial hemp, which has no psychoactive properties, should be legalized so that American farmers can once again make use of this profitable and

exceptional rotational crop useful for making clothing, rope, biodegradable plastics, and paper. Hemp, which was a major crop in the United States until after World War II, must now be imported from other countries. The U.S. government cannot seem to distinguish between nonpsychoactive industrial hemp and marijuana.

25. The export of arms from the United States should be significantly curtailed. As the largest arms-exporting nation in the world today, the United States must take some responsibility for the armed violence wracking the planet.

26. Farmers who now collect payment for not planting crops, an attempt to keep crop surpluses down and prices up to protect the growers, instead should be allowed to plant whatever they desire. Any surplus should be purchased by the government and exported for profit under the reasoning that few nations will bite the hand that feeds them.

27. The government should not be allowed to confiscate private assets unless they are taken from someone who has been convicted and sentenced to have assets forfeited in a court of law. Under current asset forfeiture policies, discussed previously and which vary widely between jurisdictions, government agencies, including local police, can confiscate private property without charging anyone with a crime. Today, the asset forfeiture policies are increasingly unfair and being misused. Should a home be raided by police and any amount of drugs found, the house can be confiscated despite the objections of the owner who may have been absent or even renting the property. Yet if drugs are found in a corporate-owned facility such as a large hotel, the hotel is not forfeited. A nonprofit organization called Forfeiture Endangers American Rights (FEAR) claimed $7 billion has been forfeited to the federal government since 1985 and that 80 percent of the forfeited property during the past ten years was seized from owners who were never charged with a crime. Although asset forfeiture was initially tolerated by the public because it was attached to drug laws, today more than

two hundred federal forfeiture laws are now applied to non-drug-related crimes.

28. Public school systems should allow students the freedom to gain experience outside the classroom. Experience is the greatest teacher in life. Students, as with most humans, tend to act more responsibly if treated like mature persons rather than as children or inmates. Public money now spent on massive football stadiums and Astroturf could be better spent on field trips to libraries and museums.

29. For most students, less attention should be given to becoming prepared for college and more emphasis should be placed on vocational training, which will prepare students to make a living in the real world. Such preparation would place a large number of graduates into a meaningful and profitable workforce of those with needed skills, such as auto mechanics, plumbers, carpenters, masons, welders, and others.

30. Each student should be taught that English is the official language of America but with all due consideration given to other languages and ethnic cultures. Americans can speak the same language and the country will still remain the "melting pot" of the world. Just try going to any other country in the world and trying to get them to speak English as the official language. In a nation filled with traffic signs, commercial signage, and media material in English, non-English speakers are at a distinct disadvantage. But it is up to them to correct this, not the nation.

31. Students should be encouraged to think critically rather than simply to regurgitate names and dates that will never be relevant in their lives. Additionally, they should learn to question authority rather than blindly obey it, as this could prevent another devastating experience such as Hitler's Germany or Stalin's Russia.

32. After a thorough grounding in reading, writing, and arithmetic, students should be encouraged to follow their own interests without being straitjacketed by government curricula. No one can be taught if they are not willing to learn. People learn when they are motivated to learn. Schools should make available the materials

while the students, aside from the basics, should be allowed to pursue their own interests. By the way, whatever happened to studies in philosophy?

33. Large pharmaceutical corporations should not be able to hold proprietary information and patents on discoveries made in publicly supported academic institutions. Today, should a new drug be discovered by a university department, it frequently is licensed to a drug manufacturer who, through mass marketing, gains great profits while the school merely makes the licensing fee.

34. Direct-to-consumer drug ads should again be banned from visual and electronic media. Only the patient and his or her doctor should be able to influence that patient's decision to take a certain medicine. Drug companies should provide doctors with full factual information on any given drug. Furthermore, there should be a ban on lobbying activities that border on being bribes, such as paid seminars to luxury resorts or any form of expensive gift from drug companies.

35. The Codex Alimentarius, which sets standards to regulate or prohibit vitamins, minerals, and other forms of homeopathic therapies, should be done away with on the grounds that it unjustly limits personal liberties. Rather than simply dismissing homeopathic treatments (alternative medicine) because they are not sanctioned by pharmaceutical companies, government agencies like the FDA should order experiments to determine if any homeopathic therapies produce positive results in health. That said, in order to reduce the power of the giant pharmaceutical corporations, physicians should return to more natural and homeopathic remedies.

36. Any new health-care plan must eliminate the waste and cost of nonproductive intermediaries. Doctors who actually treat patients, and their support systems such as testing laboratories, should be the only ones who get paid. There should be direct responsibility and obligation between the treating physician and the patient.

DEFEAT FASCISM

THE WEBSITE FREEPEOPLEONTHELAND.WORDPRESS.COM HAS pledged to "Defeat Fascism" and stop the abuse of power by the government and top corporations. In Nazi Germany, the state gained control over the corporations. In modern America, the corporations have gained control over the state. The end result is the same.

For all those zombies capable of awakening from their media-induced daze, the "Defeat Fascism" pledge may prove a rousing rallying call:

I WILL TURN OFF ALL Mainstream Media NOW and question EVERYTHING I see and hear! Ask yourself what information they are editing out and why.

I WILL Study the history of our Founding Fathers, our Declaration of Independence and our Constitution.

I WILL let the Fed,—an un-Constitutional banking system—pay their own bills! I vow to get off their money system NOW and establish local monetary systems for exchange of goods and services.

I WILL donate $10 in cash to [local monetary systems] and in 6 months we will have a new interest-free Constitutional economic system for a free people.

I WILL Take the Oath Keeper Oath to defend the Constitution.

I WILL Join the Constitutional local militia. I understand the local militia is our greatest Constitutional deterrent against a tyrannical government and I will no longer hold them in disdain, but will serve in any capacity even if I don't have a gun. Let "Don't Forget Katrina" be your battle cry!

I WILL teach my children their Constitutional rights by standing up for these rights at every turn.

I WILL leave a legacy of Liberty and Freedom for our Children and future generations.

I WILL Start a Victory Garden and GET OFF THE GRID!!

I PLEDGE, along with ALL Constitution-loving, Free People on

the Land my life, my fortunes and my sacred honor to one another and
will attack any "brownshirts," or thugs, who come to my doors in the
middle of the night. [All emphases in the original.]

Even without a pledge, some zombies are awaking from their hypnotic
state and changing their lifestyles. Former business executive Chris Martenson explained his personal awakening from living a life of material
dependence and media saturation:

"Before: I am a 40-year-old professional who has worked his way up to
Vice President of a large, international Fortune 300 company and is living
in a waterfront, five-bathroom house in Mystic, CT, which is mostly paid
off. My three young children are either in or about to enter public school
and my portfolio of investments is being managed by a broker at a large
institution. I do not really know any of my neighbors, and many of my
local connections are superficial at best.

"After: I am a 45-year-old who has willingly terminated his former
high-paying, high-status position because it seemed like an unnecessary
diversion from the real tasks at hand. My children are now homeschooled
and the big house in Mystic was sold in July of 2003 in preference for a
1.5-bathroom rental in rural western Massachusetts. In 2002, I discovered that my broker was unable to navigate a bear market and I've been
managing our investments ever since. Since that time, my portfolio has
gained 166 percent. . . . I grow a garden every year; preserve food, know
how to brew beer & wine, and raise chickens. I've carefully examined
each support system (food, energy, security, etc), and for each of them
I've figured out either a means of being more self-sufficient or a way to
do without. But, most importantly, I now know that the most important
descriptor of wealth is not my dollar holdings, but the depth and richness
of my community."

Moving from the city and living off the land is not possible for everyone. But simple changes in lifestyle can be accomplished with a minimum
of disruption.

One of Newton's laws of physics states that for every action, there is
an equal and opposite reaction. Though the analogy is not exact, imagine

how it may work in a just world: Whatever pernicious plans tyrants create to hold down the citizens of their zombie nation, an equally powerful and beneficial force will surface to thwart such efforts.

What must the people of the zombie nation do to achieve personal freedom and contentment? Resist the impulse to give way to anxiety, fear, or depression when you read and view the depravations of those who would enslave humanity. Citizens must know they are not alone in their dissatisfaction. Millions of thoughtful and good-hearted people are working diligently each day to bring enlightenment and peace to the planet. But these citizens must also recognize that when they hear of only suffering and hardship, they are not getting the whole story from the mass media. In fact, most times, news of a beneficial and positive nature will rarely be found in the newspapers or on TV. It simply has to become a local reality in our lives.

These citizens must remember that what one person can make, another person can break, and whatever is broken by one can be fixed by another.

America does not need a violent revolution. The goodness of its people and the Constitution are still in place. If enough citizens simply wake up to the treachery of the New World Order, the situation will change. After all, almost no one truly wants to live in servitude or under a tyrannical police state. And to avoid future tyranny, it will take a united citizenry dedicated to truth, justice, tolerance, and equality of opportunity. To work together, we do not need to resort to a socialist government, which could easily be transformed into a tyranny.

What is even more terrifying is what could happen if the current administration continues with its policies unabated: Taxpaying Americans will become so disenchanted and disgusted with the government's attempts to turn America into a socialist government that they will accept an inevitable right-wing backlash. Again, America will oscillate back to a National Socialist administration as an answer to the country's problems. As the economy deteriorates and the police state tightens its grip, the corporate mass media will present to the public a new leader as the nation's savior. He, or she, will mimic the words of Hitler, and essentially say, "Give me the power and I will save and protect you." Americans must be on guard against the effort to swing the electorate back to a more right-wing version of socialism. If the out-of-control government spending, a cessation of civil rights abuses, and a restructuring of the financial system cannot be resolved in a few years, the alternative is unthinkable.

Well into 2010, a remarkable number of Americans even continue to question President Obama's constitutional qualification to serve. Obama's legal, long-form birth certificate still had not been made public and the controversy over his birth was continuing, despite a lack of coverage in the mass media. Even some members of the military were questioning the legality of their commander in chief.

First Lieutenant Scott Easterling, an active-duty soldier stationed in Iraq, in an open letter, wrote, "To Whom It May Concern: As an active-duty officer in the United States Army, I have grave concerns about the constitutional eligibility of Barack Hussein Obama to hold the office of President of the United States." Easterling added, "Until Mr. Obama releases a 'vault copy' of his original birth certificate for public review, I will consider him neither my Commander in Chief nor my President, but rather a usurper to the Office—an imposter." Easterling also noted that his officer's oath contains the phrase "I will support and defend the Constitution of the United States against all enemies, foreign or domestic." Others who have echoed Easterling's challenge include Major General Carroll Childers; Lieutenant Colonel Dr. David Earl-Graif; police officer Clinton Grimes, formerly of the U.S. Navy; and two state legislators, New Hampshire state representative Timothy Comerford and Tennessee state representative Frank Nicely.

Should it be found that Obama is indeed not a natural-born citizen, a constitutional crisis would follow because every command and law issued by the Obama administration would be called into question as illegal. Lawyers would have a field day.

The birth certificate issue is not really about President Obama nor is it a political or race issue. It touches on the most basic foundation of the United States by posing the question: Are we a nation of law or a lawless nation? Must we all abide by the Constitution or are our laws only applicable when we choose to obey them? If any chief executive of the nation can disregard the law of the land, what's to hold in check the criminal who chooses to disobey the law? Should everyone choose which law to obey, the result would be chaos. Perhaps such confusion is part of the globalist plan to deconstruct the United States.

THE AMMO BOX

AMERICAN CITIZENS HAVE A rich heritage of individual freedom and liberty that is legally reinforced by the Constitution. But perhaps more important, they have guns—the means to ensure their individual liberty. The early American colonists' petitions for meaningful change fell on deaf ears in England. It was only after armed clashes that they were able to gain their independence. Thomas Jefferson clearly understood the importance of the right to bear arms when he said, "Those who hammer their guns into plows will plow for those who do not."

During World War II, the Japanese generals scrapped plans to invade America once they realized that a great many of the American citizenry possessed guns. "You cannot invade the mainland United States. There would be a rifle behind every blade of grass," warned Admiral Isoroku Yamamoto. According to *Injury Prevention Journal,* there are 308 million guns in the hands of citizens, which is an average of one per adult in the United States. Today, with nuclear weapons as a deterrent, and millions of guns in the hands of the American public, it is highly improbable that any outside enemy will successfully invade the continental United States. If the danger cannot come from the outside, then it can only come from within.

One of our nation's looming threats is the specter of martial law. Today, the federal government is adding firepower to its existing armed military. Even the IRS is arming. In February 2010, the IRS solicited bids on the purchase of sixty Remington Model 870 Police 12-gauge pump-action shotguns for its Criminal Investigation Division agents.

But ordinary citizens might consider putting a stop to armed government intervention before it starts. In *The Gulag Archipelago 1918–1956,*

Nobel Prize winner Aleksandr Solzhenitsyn explained how Russians held in detention camps bemoaned the fact that nothing was done to prevent government terrorism until it was too late: "And how we burned in the camps later, thinking: What would things have been like if every Security operative, when he went out at night to make an arrest, had been uncertain whether he would return alive and had to say good-bye to his family? Or if, during periods of mass arrests, as for example in Leningrad, when they arrested a quarter of the entire city, people had not simply sat there in their lairs, paling with terror at every bang of the downstairs door and at every step on the staircase, but had understood they had nothing left to lose and had boldly set up in the downstairs hall an ambush of half a dozen people with axes, hammers, pokers, or whatever else was at hand? . . . The Organs would quickly have suffered a shortage of officers and transport and, notwithstanding all of Stalin's thirst, the cursed machine would have ground to a halt!" It is hoped that the rising American police state does not force its citizenry to respond in the way that Solzhenitsyn suggests. But as President John Kennedy observed, "Those who make peaceful revolution impossible will make violent revolution inevitable."

Any thinking person fervently wishes that any serious change in America should come about through the peaceful exercise of the basic rights contained in the U.S. Constitution and Bill of Rights, the supreme law of the land. But after recent Democratic and Republican regimes shredded individual rights and stripped the U.S. economy, firearm and ammunition sales went through the roof. According to Federal National Instant Criminal Background Check system statistics, between January and March 2009, Americans bought 3,818,056 firearms. This is enough weaponry to arm both the Chinese and Indian armies. In reality, this number is quite low since it does not include the significant number of denials issued or private gun sales that bypass paperwork.

Is it possible that the nation is arming for something other than self-protection? Is a violent revolution inevitable?

GUN AND AMMO SALES BOOMING

IN AN ATTEMPT TO mollify gun owners during a 2008 campaign rally in Lebanon, Virginia, presidential hopeful Barack Obama said, "I don't want any misunderstanding when you all go home, and you're talking to your buddies, and they say, 'Aw, he wants to take my gun away.' You've heard it here; I'm on television, so everybody knows it. I believe in the Second Amendment. I believe in people's lawful right to bear arms. I will not take your shotgun away. I will not take your rifle away. I won't take your handgun away. . . . There are some common-sense gun safety laws that I believe in. But I am not going to take your guns away. So if you want to find an excuse not to vote for me, don't use that one. . . . It just ain't true." Despite President Obama's assurances, presidential promises are often weak. Many Americans still recall that President George H. W. Bush pledged "no new taxes" during his campaign, then raised them after taking office. Many are not taking chances to see whether Obama goes back on his promise—guns and ammunition sales boomed during the economic crisis.

"President Barack Obama is the best thing to happen to American gun and ammunition manufacturers since they invented the Defense Department," wrote Eric Sharp in the April 9, 2009, edition of the *Detroit Free Press*. No one could image a business increasing its sales by 60 percent in these times. But this appears to be the case with firearm stores, online ammo sites, and regional gun shows, where sales exploded beginning in late 2008.

Joe DeSaye opened a family sporting goods store in Montana in 1946. In 1977, he moved the business to Prescott, Arizona, and began to sell guns and ammo under the name J&G Sales. His son, Brad DeSaye, said that since 2008 their business has tripled normal sales. "It's unprecedented," remarked DeSaye.

Roy Eicher of Hunter's Den in Cincinnati told newsmen, "The issue with ammo is a pretty simple one, supply and demand, and it's not so much that people are shooting it . . . it's that they're buying it. You can talk to gun ranges around town, around the country, nobody is shooting the ammo, people are just buying the ammo, in fear of the fact they won't be able to get the ammo." John Woniewski, operations manager at Cabela's (the nation's largest sporting goods store) in Dundee, Michigan, said ammunition there was "selling like wildfire." "Anything that's a center fire round is selling," he said.

Although many argued that the rise in gun sales and ammunition is mostly due to an expectation that Obama will go back on his promise and will restrict firearms in some way, there was nevertheless a darker undercurrent to sales trends: Americans were arming themselves at an alarming rate—and with guns that aren't necessary for hunting.

Many stores' ammo shelves, especially surplus outlets, were depleted of military-type munitions, such as AK-47s, large .380-caliber rounds, 7.62 x 54 Russian rounds, and .223 rifle rounds (the caliber for the AR-15, the civilian model of the military's M-16 rifle in common use around the world). Stocks of 9mm and 8mm Mauser ammo were dwindling.

The public ammo consumption has caused problems for law enforcement officers. Arizona sheriff Darren White expressed concern that ammunition shortages caused by the public's buying could curtail police training. He complained that ammunition for his own sidearm had been on back order for nearly three months. "I've never seen it like this in my more than two decades of law enforcement," said White.

Possibly in an effort to keep ammunition out of the public's hands, the Pentagon, under orders from the Obama administration, in 2009 sent letters to the nation's ammunition retailers stating that it would no longer sell spent shell casing brass and would instead reduce expended ammunition to scrap metal, virtually useless to ammunition reloaders. Normally, these spent shells are recast by manufacturers and resold to law enforcement agencies, gun shops, and other retail outlets. Curtis Shipley, owner of the ammunition manufacturer Georgia Arms, said, "The distressing part of it was that the government was going to lose money. They were going to lose $2 a pound and accomplish nothing. We felt like it was just an option to bring in ammunition control rather than gun control."

The Defense Department directive, however, was rescinded about ten days later following a deluge of letters, calls, and e-mails from irate gun owners and manufacturers. "Upon review, the Defense Logistics Agency has determined the cartridge cases could be appropriately placed in a category of government property allowing for their release for sale," stated the Pentagon in a statement in March 2009.

"It just restores my faith that the system works," said a relieved Shipley. "If enough people are motivated and say 'Hey, that is wrong,' the system does still work."

The system can work if the zombies of modern America refuse to remain in a dazed and drugged state. They can make the system work by seeking alternative sources of news and information and then acting on such information. Most important, they can regain their freedom and sovereignty by seeking a government that goes beyond mere lip service and provides a true democratic republic. Otherwise the nation could slip into collapse and chaos, perhaps even revolution.

To avert such a future, the American public must gain control over their country. National politicians no longer refer to the "Republic," because modern America has ceased to be one. Today, it is the American empire and like Rome and Hitler's Third Reich, it has spread its corporate and military tentacles throughout the world. Political and corporate leadership continually swap roles, creating a merger of the state and industry— the very definition of fascism. This change to socialist fascism—whether from the right or left—has been engineered by the globalist elite who hold monopolies over basic resources, energy, pharmaceuticals, transportation, and telecommunications, including the news media.

It appears that the "New World Order" is really just the "Old World Order," a continuing game of the wealthy minority against the working majority confounded by debt, a controlled mass media, and political confabulation. Today, thanks to amazing media technology, the game is packaged with modern advertising slickness—new names, logos, and slogans. But it still remains a matter of the haves lording over the have-nots.

These self-styled globalists are now attempting to subdue the American population through a maze of government policies, drugs, a dumbed-down education system, and a controlled corporate mass media. Mergers and leveraged takeovers have concentrated corporate power into fewer and fewer hands. The weakening of the national economy and corporate downsizing have placed undue stress on workers, resulting in the gradual destruction of the nuclear family. Even the fields of religion, education, and entertainment are being used to transform whole generations of formerly free Americans into cowed and subservient zombies in a system increasingly under the control of the globalist elite.

The current socialist fascism in America is the way it is simply because somewhere, someone wants it that way. If no one truly wanted the problems that beset the nation, they wouldn't be there. These problems have been created, or excoriated, by globalists—many of them not even Americans—and their secret societies in the hope of molding the entire world into a few competing socialist blocs. They view the United States as the biggest stumbling block to their plans. This is due to America's tradition of individual freedom, its Constitution that guarantees such freedom, and the fact that so many Americans possess firearms to protect their freedom. But true freedom is a transient quality. It must be continually nurtured by a people unified in their dedication to liberty. Americans must seek common ground if the nation is to progress and prosper. America still has millions of competent workers and an abundance of natural resources. If these assets were put to proper use, a unified America could once again become a shining beacon of liberty, justice, and production. And the formula for unity is quite simple—men and women of good intention and faith all should just agree to disagree but do so without being disagreeable. They must approach disagreements in a thoughtful and considerate manner. A return to civility is long overdue.

To prevent such thoughtful unity, the globalist fascists have attempted to break the United States into divisions of race, sex, age, generation, and culture. They pit bureaucrats, politicians, academics, corporate leaders, and the public against one another in an agenda of divide and conquer. They maintain control in a society fragmented by combative ideologies and philosophies as well as competing corporate interests by using their corporate mass media assets to degrade the popular culture, downgrade the education process, permit acceptance of a steady flow of illegal immigrants, and divide the population over peripheral issues such as party politics, abortion, sexual relationships, stem-cell research, so-called hate crimes, and the like.

These globalist fascists scoff at the concepts of true individual freedom and multicultural egalitarianism, for they have no faith in the innate goodness of humankind or its ability for self-government. They have no real faith in a god and use religious ideals and concepts merely as another tool for social control. These globalists see their agenda for worldwide socialism as the only means of maintaining their power and control, the only way in their view to maintain the purity of their race and class. They are in it for the long haul. The owners of the multinational corporations with their membership in secretive societies and their well-paid administrators know their goals will not be achieved overnight, although since the attacks of 9/11 they seemed to have redoubled their efforts.

The struggle against such steadfast will to power and its attendant control will not be easy. All areas of society will require sacrifice and change. Lifestyles will have to be altered. But it can be done—hopefully before the United States falls into depression, anarchy, and then a police state. New energy sources and technologies are on the horizon. Technological breakthroughs await only the change of attitude on the part of conventional politics, commerce, and finance. An aroused public could push this attitude change along.

Though seldom reported in the corporate-controlled mass media, there is a rising consciousness well under way in the public mind. Informed consumers are beginning to realize they can improve their health by changing their diet and seeking alternative health remedies. Individuals are taking the initiative by listening to voices outside the mainstream media; writing their representatives and local news media; taking part in peaceful demonstrations; and conducting study groups and book review discussions in their homes. They can also vote with their spending habits. If enough people refuse to buy a certain product—whether it's a brand of car, gasoline, or some federal policy proposal—it can force a change of direction in the corporate controllers, who, after all, must respond to the bottom line.

Many Americans are hopeful. They sincerely believe the system can be changed nonviolently and will begin to work for the benefit of all citizens. But just in case, they retain the right to hold on to their guns. It's not as if America has never experienced a revolution before.

If ye love wealth greater than liberty, the tranquility of servitude greater than the animating contest for freedom, go home from us in peace. We seek not your counsel, nor your arms. Crouch down and lick the hand that feeds you; and may posterity forget that ye were our countrymen.

—SAMUEL ADAMS, Founding Father and revolutionary

SOURCES

Introduction

"Zombie" demeanor from psychiatric drugs: http://www.adhdtreatment.org:10500/adhd/adderall_stimulant_treatment_for_adhd.html

Zombie ants: http://www.kxan.com/dpp/news/local/Fire_ants_turn_into_zombies

Zombie banks: http://www.lewrockwell.com/sardi/sardi116.html

Largest single increase in debt: http://www.cbsnews.com/blogs/2009/10/01/politics/politicalhotsheet/entry5355738.shtml

A trillion dollar bills to the sun and back: http://www.thewisdomjournal.com/Blog/how-much-is-a-trillion/

Obama administration raises deficit projection: http://www.reuters.com/article/wtMostRead/idUSTRE57K4XE20090821?feedType=RSS&feedName=wtMostRead

U.S. assets less debt: http://answers.yahoo.com/question/index?qid=20080818114023AAQ5bHU

PART I—A ZOMBIE NATION

Economic Decline

$5 trillion of wealth evaporated: http://www.ustreas.gov/press/releases/tg296.htm

One in eight mortgages in default: http://www.npr.org/blogs/money/2009/11/mortgage_defaults_hitting_reco.html

Foreclosures by 2012: http://www.realtytrac.com/contentmanagement/realtytraclibrary.aspx?channelid=8&ItemID=6675

Adults and children on food stamps: http://www.nytimes.com/2009/11/29/us/29foodstamps.html?_r=2

Unemployment figure second highest since WWII: http://finance.yahoo.com/news/What-recovery-Unemployment-apf-563122944.html?x=0

Economic collapse like Argentina: http://www.telegraph.co.uk/finance/comment/6146873/Adam-Smith-would-not-be-optimistic-in-todays-economic-world.html

Loss of income tax revenues: http://money.cnn.com/2009/10/07/news/economy/tax_revenue_falling/?postversion=2009100711

Massive federal loans for jobless: http://www.usatoday.com/money/economy/employment/2008-09-08-unemployment_N.htm

No pauper funerals in Indiana: http://www.tribstar.com/news/local_story_138231725.html

Public housing demolition in Atlanta: http://www.atlantaprogressivenews.com/news/0141.html

Reverting paved road to gravel in Michigan: http://www.wwmt.com/articles/roads-1363526-mich-counties.html

Family-owned businesses closed their doors: http://online.wsj.com/article/SB125478399429765967.html

Peter Schiff on increase in household debt despite savings: http://www.lewrockwell.com/schiff/schiff49.1.html

Socialism and Loss of Individuality

Lenin on subservience to the State: http://www.fff.org/freedom/0197d.asp; http://www.time.com/time/magazine/article/0,9171,729546,00.html

Lenin's globalist quotes: http://quotes.liberty-tree.ca/quotes_by/vladimir+ilyich+lenin

Paul Craig Roberts on how a private individual had no need for forms: http://www.firstprinciplesjournal.com/articles.aspx?article=465&theme=home&loc=b

Norman Thomas on "liberalism": http://quotes.liberty-tree.ca/quote/norman_thomas_quote_ffb1

"We Are All Socialists Now": Cover, *Newsweek,* February 16, 2009

New World Order

Adolf Hitler on New World Order: http://hitlersdiaries.com/HitlersDiaries
Philosophy2.html

Globalists as "flexians": http://www.huffingtonpost.com/arianna-huffington/the-first-huffpost-book-c_b_412999.html

U.S. capitalists funded Bolsheviks: Jim Marrs, *Rule by Secrecy* (New York: Harper-Collins Publishers, 2000), pp. 192–193

U.S. capitalists funded Nazis: Jim Marrs, *The Rise of the Fourth Reich* (New York: HarperCollins Publishers, 2008), pp. 25–32

Nick Rockefeller quote: http://www.jonesreport.com/articles/210207_rockefeller_friendship.html

Catherine Austin Fitts on global financial coup d'état: http://solari.com/blog/?p=2058

Walter Cronkite on America's ruling class: http://www.newswatch.org/ (August 28, 2009)

Dissension in the Ranks

Dissension at the University of Chicago: http://www.nytimes.com/2009/09/06/magazine/06Economic-t.html?pagewanted=1&_r=2&th&emc=th

Paul Krugman on failure of economists and Ben Bernanke quote: http://www.nytimes.com/2009/09/06/magazine/06Economic-t.html?pagewanted=1&_r=2&th&emc=th

PART II—HOW TO CREATE ZOMBIES

POLITICAL HACKING

Foreign Trade and Bonds

Easy credit "forces" permanently altered: http://www.nytimes.com/2009/05/10/business/economy/10saving.html?_r=1

Grand Net bonds inflow drop: http://www.financialsense.com/fsu/editorials/willie/2009/0820.html

Weimar territory lies ahead: Ibid.

Dollar losing status as world currency: http://www.bloomberg.com/apps/news?pid =20601087&sid=aeD0JMxdEA_c

China may default on derivatives: http://www.economist.com/businessfinance/ displaystory.cfm?story_id=14365060

Peter Schiff on Chinese not buying Treasury debt: http://www.lewrockwell.com/ schiff/schiff49.1.html

Liars' Loans

William K. Black on "liars' loans": http://www.pbs.org/moyers/journal/04032009/ watch.html

Black on pig in a poke: Ibid.

Bush administration refused to replace five hundred agents: Black, op. cit.

Gramm and Deregulation

Brooksley Born on opaque market: http://www.dcbar.org/for_lawyers/resources/ legends_in_the_law/born.cfm

Phil Gramm as meltdown culprit: http://www.guardian.co.uk/business/2009/ jan/26/road-ruin-recession-individuals-economy; http://ac360.blogs.cnn.com/2008/ 10/14/culprits-of-the-collapse-7-phil-gramm/; http://www.time.com/time/ specials/packages/article/0,28804,1877351_1877350,00.html

CFTC prevented from asking questions: Born, op. cit.

Black on Swiss bank UBS: Black, op. cit.

AIG execs failed to return bonuses from taxpayer bailout: http://washingtontimes. com/news/2009/oct/15/less-than-half-of-aig-bonuses-returned/

All people who have failed: Black, op. cit.

Overwhelmed by regulation quotes by Bill Moyers and William K. Black: http:// www.pbs.org/moyers/journal/04032009/watch.html

Colonial BancGroup fails: http://www.bloomberg.com/apps/news?pid=newsarchi ve&sid=aOTAckySeznw

Robert Auerbach and Fed banks: http://www.reuters.com/article/company NewsAndPR/idUSN0756271320090407

They think Americans are a bunch of cowards: http://www.pbs.org/moyers/ journal/04032009/watch.html

Downsizing America

CFR explains the president's agenda for Paulson: http://www.cfr.org/publication/11165/what_the_boss_wants_from_hank_paulson.html

Henry Paulson on safe banking system: http://cbs5.com/national/henry.paulson.economy.2.775329.html

Paulson's face: http://www.time.com/time/specials/2008/personoftheyear/article/0,31682,1861543_1865103,00.html

Scandal of such proportions due to relatively few: Black, op. cit.

Martin D. Weiss on three government reports and insanity: http://www.marketoracle.co.uk/Article13977.html

Conspiracy theorists are right: http://www.youtube.com/watch?v=70G7hQov6dI

Representative Kay Granger's comments: http://granger.houseenews.net/mail/util.cfm?gpiv=2100045681.12317.38&gen=1

Debt Slaves

Obama on a multiplier effect: http://abcnews.go.com/Politics/Business/WireStory?id=7330274&page=3

Stimulus Package

A spiraling public debt crisis: http://www.globalresearch.ca/index.php?context=va&aid=12517

Charles Millard and PBGC: http://money.cnn.com/2009/05/20/news/economy/pbgc_refuses_testify.reut/index.htm

Before the Crash

Franklin Raines on home ownership expansion: http://www.nytimes.com/1999/09/30/business/fannie-mae-eases-credit-to-aid-mortgage-lending.html?sec=&spon=&partner=permalink&exprod=permalink

Steven Holmes on Fannie Mae risk: http://www.nytimes.com/1999/09/30/business/fannie-mae-eases-credit-to-aid-mortgage-lending.html?scp=1&sq=fannie+mae+eases+credit+to+aid+mortgage+lending&st=cse

Larry Summers convinces Bill Clinton: http://www.financialsense.com/editorials/engdahl/2009/0330.html

F. William Engdahl on Geithner's dirty little secret: Ibid.

Bank Stress Tests

Ben Bernanke on comprehensive effort: http://www.cnbc.com/id/30619915

Douglas Elliott on swapping a loan for ownership: http://www.npr.org/templates/story/story.php?storyId=103842153&ft=1&f=1001&sc=YahooNews

Engdahl on bankers' coup d'état: http://www.financialsense.com/editorials/engdahl/2009/0330.html

Chuck Collins on plutocracy: http://blog.buzzflash.com/interviews/154

Improprieties and Death

David Kellermann's death: http://www.housingwire.com/2009/04/23/kellerman-the-scapegoat-of-a-self-fulfilling-prophecy/

Lyndon LaRouche on Kellermann's right to justice: http://www.larouchepac.com/node/10172

The Rich Get Richer

G. William Domhoff on power can lead to wealth: http://sociology.ucsc.edu/whorulesamerica/power/wealth.html

Tax receipts decreasing: http://www.whitehouse.gov/omb/budget/fy2008/pdf/apers/receipts.pdf

Public Debt, Private Profit

Bank failures and low reserve fund: http://news.yahoo.com/s/ap/20090829/ap_on_bi_ge/us_meltdown101_bank_failures

William M. Isaac's conversation with Don Regan: http://www.kitco.com/ind/schoon/aug172009.html

Darryl Robert Schoon: Ibid. http://www.kitco.com/ind/schoon/aug172009.html

Fed monetizing U.S. debt: http://seekingalpha.com/article/158330-how-the-federal-reserve-is-monetizing-debt

How It All Began

Joan Veon and the transfer of America's sovereignty: http://www.newswithviews.com/Veon/joan156.htm

$23.7 trillion bailout costs: http://www.scoop.co.nz/stories/HL0907/S00250.htm

Dennis Lockhart on job losses: http://www.breitbart.com/article.php?id=CNG.44 52bed82adf3124e5884678e236d7fb.361&show_article=1

Greenspan on gold: http://www.restore-government-accountability.com/greenspan-on-gold.html

Greenspan stands by assessment: Ibid.

GATA on hidden gold swaps: http://finance.yahoo.com/news/Federal-Reserve-Admits-Hiding-bw-2550373789.html?x=0&.v=1

Any financial instrument subject to seizure: http://www.gata.org/node/5606

Usury

The mechanism converting debt into money: G. Edward Griffin, *The Creature from Jekyll Island* (Westlake Village, CA: American Media, 1994), p. 207

Greider and Henry Ford: William Greider, *Secrets of the Temple: How the Federal Reserve Runs the Country* (New York: Simon & Schuster, 1987), pp. 12, 55

Barry Goldwater on Americans have no understanding: Barry M. Goldwater, *With No Apologies* (New York: William Morrow, 1979), p. 281

Money for Faith and Debt

Money as function of faith: Greider, p. 53

Massive debt at every level: William Bramley, *The Gods of Eden* (San Jose, CA: Dahlin Family Press, 1990), p. 432

The Federal Reserve Anomaly

The ultimate control by the Fed: Greider, p. 12

Marco Polo and fiat money: Griffin, p. 156

Khazar Empire: http://www.khazaria.com/

Khazars as progenitors of eastern European Jews: James Mitchell, editor in chief, *The Random House Encyclopedia* (New York: Random House, 1977), p. 2318

Benjamin Franklin on prime reason: http://www.quoty.org/tag/currency

Hamilton on debt as a national blessing: Griffin, p. 329

Thomas Jefferson on the evils and unconstitutionality of banks: Martin A. Larson, *The Essence of Jefferson* (New York: Joseph J. Binns, Publisher, 1977), pp. 185–186, 192, 196

Andrew Jackson and the central banks as a curse to the republic: Epperson, p. 134

Joan Veon on English origins of a central bank: http://www.newswithviews.com/Veon/joan2.htm

Joan Veon on deceitful legislators: http://www.newswithviews.com/Veon/joan156.htm

Frank Vanderlip as furtive conspirator: Mullins (1983), p. 8

Woodrow Wilson on committee of men like J. P. Morgan: http://www.atimes.com/atimes/Global_Economy/JA30Dj02.html

Aldrich plan as Wall Street plan: Mullins (1983), p. 11

Globalists had faith in Colonel House: W. Cleon Skousen, *The Naked Capitalist* (Salt Lake City, UT: Self-published, 1970), p. 21

Congress outflanked and outfoxed: Griffin, p. 469

Federal Reserve Bank of New York directors: http://www.newyorkfed.org/aboutthefed/org_nydirectors.html

Gary Allen on inflation and deflation as an exact science: Gary Allen, *None Dare Call It Conspiracy* (Seal Beach, CA: Concord Press, 1971), p. 53

Lindbergh on Fed as gigantic money trust: Mullins (1983), p. 28

Public allowed erroneous assumptions: http://www.monetary.org/federalreserveprivate.htm

Bruce Wiseman on how a central bank works: http://canadafreepress.com/index.php/article/10954

Financial Stability Board

Dick Morris on European power over U.S. finances: http://www.dickmorris.com/blog/2009/04/06/the-declaration-of-independence-has-been-repealed/

Communiqué turns over financial control: http://canadafreepress.com/index.php/article/10954

Undue foreign influence such as BIS: http://www.monetary.org/federalreserveprivate.htm

Into a BIS

Joan Veon on controlling the world's monetary system: http://www.newswithviews.com/Veon/joan2.htm

BIS report on unstable system: http://www.telegraph.co.uk/finance/newsbysector/ banksandfinance/6184496/Derivatives-still-pose-huge-risk-says-BIS.html

William White on surprise if there is a rapid and sustainable recovery: http://www. ft.com/cms/s/0/e6dd31f0-a133-11de-a88d-00144feabdc0.html?nclick_check=1

BIS as a Nazi bank: Jim Marrs, *The Rise of the Fourth Reich*, pp. 24–28

World system of financial control: Carroll Quigley, *Tragedy and Hope: A History of the World in Our Time* (New York: MacMillan, 1966), p. 50

A money funnel: Charles Higham, *Trading with the Enemy: An Expose of the Nazi-American Money Plot, 1933–1949* (New York: Delacorte Press, 1983)

BIS's power and committees: http://www.newswithviews.com/Veon/joan2.htm

Paul Warburg on Federal Reserve notes: Griffin, pp. 466–467

Attempts to Audit the Fed

Nader on unperturbed control: http://www.nader.org/index.php?/archives/1038-Fed-Needs-Auditing.html

The fallacy of need for a central bank: www.forbes.com/2009/05/15/audit-the-fed-opinions-contributors-ron-paul.html

Rep. Ron Paul introduces audit the Fed bill: http://www.house.gov/list/speech/ tx14_paul/AudittheFedBill.shtml

Gallup poll on Fed: http://current.com/items/90551867_gallup-poll-americans-turning-against-federal-reserve.htm

Rasmussen poll: http://www.washingtonexaminer.com/opinion/blogs/beltway-confidential/Poll-Public-wants-to-rein-in-the-Fed-52084492.html

Barney Frank letter: http://www.campaignforliberty.com/profile.php?member= GroverWasGreat

Appropriations bill passes without amendment: http://tekgnosis.typepad. com/tekgnosis/2009/07/breaking-news-on-audit-the-fed-hr-2918-as-senate-amendment-1367.html

Thomas F. Cooley: http://www.forbes.com/2009/05/12/federal-reserve-bernie-sanders-ron-paul-opinions-columnists-talf.html

Fed Arrogance

Grayson was shocked: http://www.youtube.com/watch?v=cJqM2tFOxLQ

Bernanke doesn't know whereabouts of dollars: http://www.prisonplanet.com/bernanke-i-dont-know-which-foreign-banks-were-given-half-a-trillion.html

Black on regulators with no power: http://www.reuters.com/article/companyNewsAndPR/idUSN0756271320090407

Conrad DeQuadros on lapses: http://www.reuters.com/article/companyNews AndPR/idUSN0756271320090407

Fed's opaqueness: http://www.reuters.com/article/companyNewsAndPR/idUSN 0756271320090407

Bloomberg suit and Judge Loretta Preska's decision: http://www.bloomberg.com/apps/news?pid=20601087&sid=a7CC61ZsieV4

DEBILITATING FOOD AND WATER

Bad Food and Smart Choices

Smart Choices food industry program: http://www.smartchoicesprogram.com/

Jim Hightower on Smart Choices: http://www.jimhightower.com/node/6932

Amish have better health: http://www.healthnewsdigest.com/news/Cancer_Issues_660/Amish_Have_Lower_Rates_of_Cancer.shtml

False Claims and Recalls

CSPI class action suit against Coca-Cola: http://www.cspinet.org/new/200901151.html

Sparks with more alcohol than beer: http://www.cspinet.org/litigation/

Processing steps bring hazards: http://www.answers.com/topic/food-safety

Foodborne contamination unprecedented: http://www.answers.com/topic/food-safety

Growing Hormones

Estrogenic effects on the body: http://well.blogs.nytimes.com/2009/05/04/earlier-puberty-in-european-girls/

No adequate studies on growth hormones: http://envirocancer.cornell.edu/Factsheet/Diet/fs37.hormones.cfm

Test animals all our lives: http://historymatters.gmu.edu/d/5090

The Rise of the FDA

Citizen petition to the USDA in 2001: http://www.dfwnetmall.com/veg/pcrmpetition.htm

Michael Taylor as example of revolving door: http://www.washingtonpost.com/wp-dyn/content/article/2010/01/13/AR2010011304402.html?wprss=rss_politics

Genetically Modified Foods

Rice growers sue over Bayer seeds: http://www.bloomberg.com/apps/news?pid=newsarchive&sid=aT1kD1GOt0N0

Monsanto seeks patents on pigs and Christoph Then's comments: http://www.greenpeace.org/international/news/monsanto-pig-patent-111

Monsanto's GMO banned: http://www.psrast.org/bghcodex.htm

Jessica Long on agrarian martyrs: http://www.globalresearch.ca/index.php?context= va&aid=6522

Monsanto offers denial: http://www.monsanto.com/monsanto_today/for_the_record/india_farmer_suicides.asp

Brian Thomas Fitzgerald warns of unprecedented control of food supply: http://www.greenpeace.org/international/news/monsanto-pig-patent-111

Monsanto argues over rat study: http://www.huffingtonpost.com/2010/01/12/monsantos-gmo-corn-linked_n_420365.html

Dr. Stanley Ewen and GM-caused cancer: http://www.biotech-info.net/cancer_risk.html

50 harmful effects and Harvard's Dr. George Wald: http://groups.yahoo.com/group/bdresearchers/message/3835

Steve Wilson and Jane Akre fired over GMO report: http://www.goldmanprize.org/node/65

Codex Alimentarius

170 nations are members of the Codex Alimentarius Commission: http://www.fsis.usda.gov/Codex_Alimentarius/index.asp

Codex Alimentarius Commission: http://en.wikipedia.org/wiki/Codex_Alimentarius

John Hammell on remedies that will be gone: http://www.genesisradio.co.uk/index.php?option=com_content&task=view&id=157&Itemid=102

A shady, secretive organization: http://www.natural-health-information-centre. com/codex-alimentarius.html

Dr. Joel Wallach on unfairness of mineral-deficient infant formula: http://www. kingmaker.net/DEADDOCTORStxt.html#28.%20Diabetes

WTO will apply trade sanctions against noncompliance: http://www.genesisradio. co.uk/index.php?option=com_content&task=view&id=157&Itemid=102

Llewellyn H. Rockwell on WTO forcing changes in U.S. laws: http://vaclib.org/ news/wto.htm

Mike Adams on rigged FDA game: http://www.naturalnews.com/027303_the_ FTC_America_vaccines.html

Fluoridated Water

Charles Perkins on sodium fluoride: Eustace Mullins, *Murder by Injection: The Story of the Medical Conspiracy Against America* (Staunton, VA: The National Council for Medical Research, 1988), pp. 353–354

Christian Science Monitor survey: http://www.battery-rechargeable-charger.com/ water-filter-fluoride-poisoning-info.html

Dr. Ted Spencer on it cannot be both ways: http://articles.mercola.com/sites/ articles/archive/2008/01/02/fluoride-controversy.aspx

German scientists not silenced: Mullins, p. 158

Dr. Perry Cohn and Dr. Dean Burk on correlation of fluoride to cancer: http:// homepage.eircom.net/~fluoridefree/campaign_update/bonecancer.htm

Dr. Ted Spencer on fluoride studies: http://articles.mercola.com/sites/articles/ archive/2008/01/02/fluoride-controversy.aspx

Oscar Ewing: http://www.trumanlibrary.org/oralhist/ewing3.htm

Congressman A. L. Miller's quote: Mullins, pp. 153–154

NYC leaflet: http://www.trumanlibrary.org/oralhist/ewing3.htm

Health-Care Blues

Texas Dr. Michael Truman on insurance parasites: e-mail to author, October 14, 2009

Medical turnstiles will be the same: http://online.wsj.com/article/ SB123993462778328019.html

Figures too huge to be acknowledged: http://www.frontpagemag.com/readArticle.
aspx?ARTID=35215

Scale back or prepare for tax tsunami: http://www.hillsdale.edu/news/imprimis/
archive/issue.asp?year=2009&month=03

THE MYCOPLASMA ATTACK

Nazi and Japanese Biological Warfare

Dulles brothers create partnership: Dr. Leonard G. Horowitz and Dr. Joseph S.
Puleo, *Healing Codes for the Biological Apocalypse* (Sandpoint, ID: Tetrahedron
Publishing Group, 2000), p. 209

CFR as manpower pool and chairman of the Establishment: http://www.nytimes.
com/1989/03/12/obituaries/john-j-mccloy-lawyer-and-diplomat-is-dead-at-93.htm
l?scp=1&sq=john+mccloy&st=cse

Walter Emil Schreiber: Dr. Leonard Horowitz, *Emerging Viruses: AIDS and Ebola*
(Rockport, MA: Tetrahedron, Inc., 1998), p. 331

Kurt Blome: http://www.conspiracyarchive.com/NWO/project_paperclip.htm

General Ishii Shiro: http://en.wikipedia.org/wiki/Shiro_Ishii

Mycoplasmas and Prions

Stanley B. Prusiner: http://www.pnas.org/content/95/23/13363

Horrible weapons of mass destruction: Garth L. Nicolson and Nancy L. Nicolson,
Project Day Lily: An American Biological Warfare Tragedy (Bloomington, IN:
Xlibris Corp., 2005), p. 25

Delightful weapon: Nicolsons, p. 29

Dr. Maurice Hilleman: http://hubpages.com/hub/degenerativedisease

HIV and the chimeric: e-mail correspondence with Dr. Nancy Nicolson, August
18, 2009

Vaccine support by Merck: Randy Shilts, *And the Band Played On: Politics, People
and the AIDS Epidemic* (New York: Penguin Books, 1987), pp. 201–202

Man-altered brucellosis: Donald W. Scott and William L. C. Scott, *The Brucellosis
Triangle* (Sudbury, Ontario: The Chelmsford Publishers, 1998), p. iii

Essential raw material: Scott and Scott, pp. 11, 99.

Gary Tunsky on treatment: http://www.rense.com/general62/molecularterrorism. htm

Gulf War Syndrome

Veterans with brucellosis symptoms: Ibid.

No other compelling explanation: http://www.gulfwarvets.com/ijom.htm

Military medical records missing: Ibid.

"Iraqibacter": http://www.google.com/hostednews/canadianpress/article/ALeqM5 g2Jia9Lu7ynGJqfhyCobPUavfdYQ

CDC acknowledges chronic fatigue syndrome: http://www.cdc.gov/cfs/ cfsbasicfacts.htm

Dr. Martin Lerner: http://www.cfsviraltreatment.com/

Treatment Center for Chronic Fatigue Syndrome: http://www.investinme.org/ Article-334%20Martin%20Lerner%20October%202009.htm

A very real physical disease: Scott and Scott, p. 115

Depopulation Efforts

NSSM 200: http://pdf.usaid.gov/pdf_docs/PCAAB500.pdf

Maxwell Taylor quote: Editors, "Maxwell Taylor: 'Write Off a Billion,'" *Executive Intelligence Review* (September 22, 1981), p. 56

Prince Philip on long-term threat to survival: http://www.people.com/people/ archive/article/0,,20080998,00.html

Prince Philip as a virus: http://www.prisonplanet.com/Pages/100604_prince_ philip.html

Justice Ruth Bader Ginsburg on population concern and *Roe vs. Wade:* http://www. cnsnews.com/public/content/article.aspx?RsrcID=50819

William Norman Grigg: http://www.lewrockwell.com/grigg/grigg-w102.html

G. Edward Griffin on Holdren's plans for population reduction: http://www. heartcom.org/choice4health.htm

Catherine Austin Fitts on swine flu as depopulation method: http://www.scoop. co.nz/stories/HL0907/S00250.htm

Manufactured AIDS

Don't discount conspiracy theories on AIDS: http://www.nation.co.ke/oped/Opinion/-/440808/815898/-/5ohsob/-/

Dr. D. M. MacArthur's testimony: Department of Defense Appropriations for 1970, Hearings Before the Subcommittee of the Committee on Appropriations, House of Representatives, 91st Congress, June 9, 1969, p. 129

Boyd Graves's biography and flowchart: http://www.boydgraves.com/flowchart/download.html; http://www.boydgraves.com/

A population control weapon: www.agoracosmopolitan.com/home/Frontpage/2007/10/19/01898.html

Overpopulation ad: Advertisement, the Council on Foreign Relations *Foreign Affairs* 75, issue 2 (March/April 1996)

Overpopulation a priority concern: http://www.timesonline.co.uk/tol/news/world/us_and_americas/article6350303.ece

History is repeating: Horowitz and Puleo, p. 226

Orders from Kissinger: http://www.cfr.org/publication/18515/remarks_by_national_security_adviser_jones_at_45th_munich_conference_on_security_policy.html

Rockefeller interests: Scott and Scott, p. 12

Dr. Rife's Discovery

Nothing can convince a closed mind: http://www.rense.com/health/rife.htm

Harry Hoxsey driven insane: Ibid.

James Folsom's trial and conviction: http://www.rifewiki.org/wiki/Jim_Folsom_Trial; http://www3.signonsandiego.com/stories/2009/feb/18/bn18convict-medical-scam/?zIndex=55119

James Folsom conviction: http://www.usdoj.gov/usao/cas/press/cas90219-Folsom.pdf

FDA to evaluate medical devices: http://www.fda.gov/NewsEvents/Newsroom/PressAnnouncements/ucm149560.htm

Supporter's comment: http://www.rifewiki.org/wiki/Jim_Folsom_Trial

DRUGGING THE POPULATION

Big Pharm

Drug dealers supplying 195 million doses: http://articles.mercola.com/sites/articles/archive/2009/10/13/Dr-Oz-Helps-Shill-the-Flu-Vaccine.aspx

Dr. Marcia Angell and rise of Big Pharm: http://www.nybooks.com/articles/17244

Angell on watershed year of 1980: Ibid.

Disease-mongering: Dr. Michael Wilkes, "Inside Medicine: Some 'diseases' invented for profit," *Sacramento Bee* (May 26, 2007)

Dr. Sharon Levine on molecule change: Peter Jennings Special, "Bitter Medicine: Pills, Profit, and the Public Health," *ABC Television,* May 29, 2002

Angell on taxpayer research for private companies: http://www.nybooks.com/articles/17244

DTC Ads

Biggest sales per ad dollar: www.consumerreports.org/health/prescription-drugs/adwatch/overview/adwatch-hub.htm

2008 direct-to-consumer advertising: http://www.biojobblog.com/tags/ddmac/

Vioxx as example of DTC marketing damage: http://64.143.177.241/HealthNews/MedicineOnTheHorizon/MedicineOnTheHorizon.htm

Big Pharm R&D versus promotion: http://www.piribo.com/publications/general_industry/world_pharmaceutical_market_2007.html; http://www.piribo.com/publications/general_industry/pharmaceutical_market_trends_ 2008_2012.html;

Direct-to-consumer drug advertising (lack of fair balance): http://www.biojobblog.com/2009/07/articles/biobusiness/several-us-legislators-begin-to-seriously-scrutinize-directtoconsumer-advertising/

Brand-name drug prices compared with cost of ingredients: http://liberty.hypermart.net/voices/2003/Actual_Cost_Of_Making_These_Popular_Prescription_Drugs.htm

Pfizer faces largest criminal fine in U.S. history: http://www.newsday.com/business/pfizer-to-pay-record-2-3b-penalty-over-promotions-1.1416017?localLinksEnabled=false

Lord Oliver Franks and Wellcome Trust: Mullins, p. 345

Round Tablers fanned out over the world: Dr. John Coleman, *Conspirators'*

Hierarchy: The Story of the Committee of 300 (Carson City, NV: America West Publishers, 1992), p. 153

Carroll Quigley on an international Anglophile network: Carroll Quigley, *The World Since 1939: A History* (New York: Collier Books, 1968), p. 290

Rockefeller medical monopoly: Mullins, p. 342

Committee on the Costs of Medical Care and health-care crisis: http://www.innominatesociety.com/Articles/The%20Committee%20%20On%20The%20Costs%20Of%20Medical%20Care.htm

William Rockefeller as carnival medicine show barker: Mullins, p. 321

Standard Oil and I. G. Farben: Charles Higham, *Trading with the Enemy: An Expose of the Nazi-American Money Plot 1933–1949* (New York: Delacorte Press, 1983), pp. 46–48

Aspartame

Dr. Louis J. Elsas on phenylalanine: http://www.dorway.com/dr-elsas.txt

Dr. Madelon Price on rodents ingesting aspartame: http://www.myaspartameexperiment.com/index.php?page=7

Dr. Adrian Gross on who will protect the public: http://www.newswithviews.com/NWVexclusive/exclusive15.htm

Dr. H. J. Roberts and aspartame interaction: http://www.wnho.net/aspartame_interacts.htm

Dr. Betty Martini on aspartame release: http://www.newswithviews.com/NWVexclusive/exclusive15.htm

FDA commissioner Arthur Hull Hayes: http://americanfraud.com/arthurhayes.aspx

Searle salesperson Patty Wood Allott: http://www.soundandfury.tv/pages/rumsfeld.html

Donald Rumsfeld and Gilead Sciences: http://money.cnn.com/2005/10/31/news/newsmakers/fortune_rumsfeld/?cnn=yes

Big Pharm co-opts every institution: http://www.nybooks.com/articles/17244

Big Pharm political contributions: http://www.opensecrets.org/pres08/select.php?Ind=H04

An aroused and determined public: Ibid.

Drugging the Kids

Connecticut House vote: http://www.namiscc.org/newsletters/Sept01/Alternative. htm#connecticut

University of Wisconsin study: http://omnihealthcaregroup.com/ADD.htm

Alan Larson on teachers who cannot tolerate active children: Bruce Wiseman, *Psychiatry: The Ultimate Betrayal* (Los Angeles: Freedom Publishing, 1995), p. 287

Failure to find real disease: P. R. Breggin, *Toxic Psychiatry* (New York: St. Martin's Press, 1991), chapters 12 and 13

Growth of clinical, consulting, and school psychologists: http://stats.bls.gov/oco/ocos056.htm#projections_data

Dr. Loren R. Mosher on way to get paid: http://usa.mediamonitors.net/content/view/full/24876

Psychiatry may still be suspect among the public: Wiseman, p. 31

Dr. Helmut Remschmidt: Dr. Thomas Röder, Volker Kubillus, and Anthony Burwell, *Psychiatrists—The Men Behind Hitler* (Los Angeles: Freedom Publishing, 1995), pp. 136–137, 142–143

105 adverse reactions to Ritalin: Wiseman, p. 285

WHO compares Ritalin to cocaine: Kelly Patricia O'Meara, "New Research Indicts Ritalin," *Insight on the News* (October 1, 2001)

Dr. Breggin and Luvox at Columbine: http://www.wnd.com/news/article. asp?ARTICLE_ID=55310

Statistic rarely mentioned in news reports: http://www.teenscreentruth.com/index. html

Massacres have drugs in common: Dr. Julian Whitaker, MD, "Prescription Drugs— The Reason Behind the Madness," *Health and Healing* (November 1999)

PSYCHIATRY AND EUGENICS

History of Psychiatry

Missing Central piece of the puzzle: Röder, Kubillus, and Burwell, p. 8

Kaufmann Therapy and psychiatrist as judge of illness: Ibid., pp. 26–28

Rüdin followed his convictions: Ibid., p. 95

Montagu Norman appoints John Rawlings Rees: Webster Griffin Tarpley and Anton Chaitkin, *George Bush: The Unauthorized Biography* (Washington, D.C.: Executive Intelligence Review, 1992), p. 69

Secret fifth columnists: Dr. John Rawlings Rees, "Strategic Planning for Mental Health," *Mental Health* 1, no. 4 (June 18, 1940): 103–104.

Beverly Eakman on cadre of experts: http://www.thenewamerican.com/index.php/culture/family/2074-the-new-face-of-psychiatry

Beverly Eakman on averting dissent under the pretext of preventing emotional disease: Ibid.

Eugenics

Oliver Wendell Holmes quote: Robert N. Proctor, *The Nazi War on Cancer* (Princeton, NJ: Princeton University Press, 1999), p. 21

Edwin Black on eugenics: http://hnn.us/articles/1796.html

Planned Parenthood figures and directors: http://www.plannedparenthood.org/files/AR08_vFinal.pdf

The Psychology of Conservatism

The psychology of conservatism: http://www.berkeley.edu/news/media/releases/2003/07/22_politics.shtml

Political conservatism as motivated social cognition: http://terpconnect.umd.edu/~hannahk/bulletin.pdf

Dr. José M. R. Delgado on electronic control of the brain: http://psychquotes.com/

Drug the Women and Children First

Evelyn Pringle on true goal of act: http://counterpunch.com/pringle04072009.html

Mike Adams on rise of autism and suspect vaccines: http://www.naturalnews.com/027179_John_Travolta_vaccines_autism.html

Dawbarns Law firm report on vaccines: http://www.hans.org/magazine/164/

Dr. Joseph Mercola on thimerosal and autism: http://articles.mercola.com/sites/articles/archive/2009/08/06/Proof-That-Thimerosal-Induces-AutismLike-Neurotoxicity.aspx

Mike Adams on bipolar disorder and handsome profits from drugs: http://www.naturalnews.com/019390.html

David Healy on "biobabble" and dangerous psychiatric drugs: http://www.psychologytoday.com/blog/side-effects/200904/bipolar-disorder-and-its-biomythology-interview-david-healy?page=2

FLU AND OTHER SWINISH IDEAS

Big Pharm Pays Off

Dr. Bruce Levine on Big Pharm: http://onlinejournal.com/artman/publish/article_4638.shtml

Emory's psychiatry department: http://brodyhooked.blogspot.com/search?q=emory+university+justify

Payments to Dr. Charles Nemeroff: http://s.wsj.net/public/resources/documents/SenateLetter081003.pdf

Advocacy groups funded by Big Pharm: http://www.cchrint.org/psycho-pharmaceutical-front-groups/

Dr. Julie Gerberding named president of Merck vaccine division: http://topnews.us/content/29115-dr-julie-gerberding-named-president-mercks-vaccine-division

Mike Adams on cross-contamination: http://www.naturalnews.com/027789_Dr_Julie_Gerberding_Merck.html

Adjuvants and Squalene

Dr. Wolfgang Wodarg on flu hoax: http://www.nspm.rs/nspm-in-english/swine-flu-they-organized-the-panic.html

Dr. Russell Blaylock on Novartis and adjuvants: http://socioecohistory.wordpress.com/2009/07/15/dr-russell-blaylock-vaccine-may-be-more-dangerous-than-swine-flu/

Squalene injected into U.S. soldiers: http://groups.google.la/group/misc.health.alternative/browse_thread/thread/fd4bce97971da453

The Military Vaccine Resource Directory: http://www.mvrd.org/showpage.cfm?ID=69

Robert F. Garry testimony: http://www.autoimmune.com/Subcommittee RFGarry24Jan02.html

Squalene in vaccines and Danish medical authorities: http://groups.google.la/group/misc.health.alternative/browse_thread/thread/fd4bce97971da453

Adverse reactions to swine flu vaccination: http://www.fluscam.com/

Vaccine_Package_Inserts_files/Novartis_A-H1N1_2009_Monvalent_Vaccine PackageInsert_BasedOn1980Approvalfor%20Fluvirin_UCM182242.pdf

Swine flu poll of parents: http://www.latimes.com/news/nationworld/nation/la-sci-parents-flu25-2009sep25,0,579663.story

Dr. Ethan Rubinstein and Canadian study showing vaccinations encourage swine flu: http://www.theglobeandmail.com/news/technology/science/study-prompts-provinces-to-rethink-flu-plan/article1303330/

Health-care workers dubious of swine flu vaccine: http://www.bmj.com/cgi/content/abstract/339/aug25_2/b3391

"Corrective action" and termination against workers who refuse flu shot: http://www.timesunion.com/AspStories/story.asp?storyID=836256&category=BUSINESS

Swine flu cases overestimated and Georgetown University: http://www.cbsnews.com/stories/2009/10/21/cbsnews_investigates/main5404829.shtml

Sanofi Pasteur recalls H1N1 vaccine: http://www.boston.com/news/local/connecticut/articles/2010/02/02/conn_says_h1n1_manufacturer_recalling_doses/

The Kansas City Pandemic of 1921

Dr. A. True Ott and the Kansas City smallpox fraud: http://www.open.salon.com/blog/gordon_wagner/2009/08/24/vaccine-induced_disease_epidemic_outbreaks

Polio vaccine adulterated with SV-40: http://www.thinktwice.com/Polio.pdf; also see Ed Haslam's book *Dr. Mary's Monkey* (Walterville, OR: TrineDay, 2007)

Kathleen Sebelius decree granting immunity to drugmakers: http://www.ktradionetwork.com/2009/07/24/makers-of-swine-flu-vaccine-cant-be-sued/

Lance Corporal Josef Lopez and the VA: http://www.kansascity.com/105/story/1415095.html

Military personnel must take vaccination: http://hamptonroads.com/2009/09/dod-service-members-must-get-swine-flu-vaccine

Making people government dependent: http://www.worldaffairsbrief.com

Schoolchildren vaccinated: http://www.kxan.com/dpp/health/health_centers/wwlp_ap_health_widespreadvaccinationsforschoolkids_200908171013_2772121

Flu Fears

Obama adds avian flu amendment: http://www.democrats.senate.gov/newsroom/record.cfm?id=248082&

Christopher Bona on Baxter's error: http://www.torontosun.com/news/canada/ 2009/02/27/8560781.html

Deadly mixture: http://preventdisease.com/news/09/031109_baxter.shtml

Baxter acts like biological terrorism: http://preventdisease.com/news/09/031109_ baxter.shtml

Suit against Baxter over HIV-contaminated blood: http://www.guardian.co.uk/ society/2007/sep/03/health

Malfunctioning dialysis machines: http://www.berkeleydailyplanet.com/issue/ 2001-10-19/article/7605

Baxter distributed Chinese contaminated heparin: http://www.washingtonpost. com/wp-dyn/content/article/2008/03/14/AR2008031403050.html

Kentucky settles Baxter over overcharging: http://www.kypost.com/content/news/ commonwealth/story/Conway-Announces-Multi-Million-Dollar-Settlement/ srxPJ5GaiU2gqFfhozY9-g.cspx

Baxter's patent application: http://www.theoneclickgroup.co.uk/documents/ vaccines/Baxter%20Vaccine%20Patent%20Application.pdf

Vaccine-contaminating viruses including HIV: http://socioecohistory.wordpress. com/2009/07/15/dr-russell-blaylock-vaccine-may-be-more-dangerous-than-swine-flu/

Novartis and I. G. Farben: http://www.britannica.com/EBchecked/topic/282192/ IG-Farben

Pot Busts Are High

Record high arrests: http://www.norml.org/index.cfm?Group_ID=7698

Mike Adams on failure of war on drugs: http://www.naturalnews.com/027257_ hemp_America_farmers.html

Jeffrey A. Tucker on the real horror of Prohibition: http://mises.org/story/3772

DUMBED-DOWN EDUCATION

Oklahoma School Study

Brandon Dutcher on school study: http://www.newson6.com/global/story. asp?s=11141949

Matthew Ladner and high school study: See complete study at http://www.

ocpathink.org/publications/perspective-archives/september-2009-volume-16-number-9/?module=perspective&id=2321

High school exit exams softened: http://www.nytimes.com/2010/01/12/education/12exit.html

Sad and ominous condition: Mark Bauerlein, *The Dumbest Generation: How the Digital Age Stupefies Young Americans and Jeopardizes Our Future* (New York: Jeremy P. Tarcher/Penguin, 2009), p. 30

Gothic churches can't compete: Bauerlein, p. xii

The Video Generation

Generation M study: http://www.kff.org/entmedia/entmedia030905nr.cfm

TV and reading consequences not the same: Bauerlein, p. 89

Screen intelligence does not transfer well: Bauerlein, p. 95

Reading is counterproductive: Bauerlein, pp. 41–43

Dangerous Teaching

Economy could not survive critical thinkers: John Taylor Gatto, *Dumbing Us Down: The Hidden Curriculum of Compulsory Schooling* (Philadelphia, PA: New Society Publishers, 1992), p. xiii

Workers Not Thinkers

Education aimed at destroying free will: http://newcitizenship.blogspot.com/2008/03/fichte.html

The Lincoln School: http://www.britannica.com/eb/article-9001067/New-Lincoln-School

Eustace Mullins on pernicious influence: http://www.mega.nu:8080/ampp/rockroth.html

Encyclopaedia Britannica and William Benton: http://www.nytimes.com/1995/05/16/business/slow-to-adapt-encyclopaedia-britannica-is-for-sale.html; http://en.wikipedia.org/wiki/William_Burnett_Benton

Paolo Lionni on high school graduate of 1900: http://www.sntp.net/education/leipzig_connection_6.htm

Rockefeller shaping a new industrial social order: Williams H. Watkins, *The White*

Architects of Black Education: Ideology and Power in America, 1865–1954 (New York: Teachers College Press, 2001), pp. 133–134

Gates and dreams of limitless resources: http://www.sntp.net/education/leipzig_connection_6.htm

Dr. Chester M. Pierce on insane children: http://psychquotes.com/

Lionni on tremendous control for one group: http://www.sntp.net/education/leipzig_connection_6.htm

Rockefeller-supported entities: http://archive.rockefeller.edu/publications/resrep/rose1.pdf

Norman Dodd on working to control education: A. Ralph Epperson, *The Unseen Hand: An Introduction to the Conspiratorial View of History* (Tucson, AZ: Publius Press, 1985), p. 209

Eakman on education as a means to change the student's fixed beliefs: http://www.thenewamerican.com/index.php/culture/family/2074-the-new-face-of-psychiatry

A tsunami of school shootings and mass murders: Eakman, ibid.

CIA on campuses: http://www.counterpunch.org/gibbs04072003.html

Games encourage skills sought by employers: http://www.fas.org/gamesummit/Resources/Summit%20on%20Educational%20Games.pdf

What students do in the classroom: Neil Postman and Charles Weingartner, *Teaching as a Subversive Activity*, (New York: Dell Publishing, 1969), pp. 19–20

Totally dependent on teacher authority: Ibid., p. 143

Twixters

Number living with parents doubled since 1970: http://www.time.com/time/magazine/article/0,9171,1018089-2,00.html

Crippling debt and Twixters: http://www.time.com/time/magazine/article/0,9171,1018089-3,00.html

PART III—HOW TO CONTROL ZOMBIES

MEDIA CONTROL AND FEARMONGERING

Ted Turner about five companies that control: http://www.brainyquote.com/quotes/keywords/control_2.html

Government-Dictated News

Leo Bogart about pressure on journalists: http://www.freedomforum.org/ publications/msj/courage.summer2000/t09.html

Robert McChesney on journalists don't ask questions: http://www.socialistproject. ca/bullet/246.php

Fearmongering

George W. Bush on threats to the nation: http://seattletimes.nwsource.com/html/ nationworld/2002795922_bush10.html

Bush and Los Angeles plot: Deb Riechmann, "Bush Says Cooperation Thwarted 2002 Attack," Associated Press (February 9, 2006)

Los Angeles mayor blindsided: Michael R. Blood, "L.A. Mayor Blindsided by Bush Announcement," Associated Press (February 9, 2006); http://www.sfgate.com/cgi-bin/article.cgi?f=/n/a/2006/02/10/national/a023740S10.DTL&type=printable

Doug Thompson's blog: http://www.capitolhillblue.com/artman/publish/article_ 8124.shtml

Powell's comments: Frank Bruni, "Bush Taps Cheney to Study Antiterrorism Steps," *New York Times* (May 8, 2001)

PATRIOT Act

Representative Ron Paul: http://www.federalobserver.com/archive.php?aid=847

Paul on sneak and peek: Ibid.

Kelly O'Meara on Fourth Amendment abrogated: Kelly Patricia O'Meara, "Police State," *Insight Magazine* (November 9, 2001)

The Enemy Belligerent Act and Glenn Greenwald: http://www.alternet.org/ rights/146081/mccain_and_lieberman%27s

Laser or Taser

David Banach: Wayne Perry, "Man Charged Under PATRIOT Act—Feds Admit Not a Terrorist," Associated Press (January 5, 2005); http://wcbstv.com/topstories/ David.Banach.Laser.2.233378.htm

Lasers in use by government: http://www.cnn.com/2005/TECH/04/15/laser. warn/

Taser death of Kevin Omas: http://www.amnestyusa.org/document.php? lang=e&id=ENGAMR510302006

Police Tactics and FEMA

Former state trooper Greg Evensen on roadblocks: http://www.newswithviews. com/Evensen/greg142.htm

Obama on intelligence estimates and more troops: http://www.globalresearch.ca/ index.php?context=viewArticle&code=BUR20090329&articleId=12943

Designated Terrorists

Paul Joseph Watson on potential violent domestic terrorist: http:// republicbroadcasting.org/?p=1719

Texas terrorist pamphlet: "Terrorism: What the Public Needs to Know," prepared and distributed by the Texas Department of Public Safety's Counterterrorism Intelligence Unit, July 2004. Copy in author's files.

Heavy police response in Oakland: http://www.post-gazette.com/pg/09271/ 1001494-100.stm?cmpid=latest.xml

No more America: http://www.wnd.com/index.php?fa=PAGE.view&pageId= 108307

Obama's School Talk

White House spokesman Tommy Vietor on changing language: http://blogs. abcnews.com/politicalpunch/2009/09/obamas-back-to-school-message— scribbled-with-some-controversy.htm

Parents object but Superintendent Tate says no opting out: http://www.foxnews. com/politics/2009/09/03/parents-object-obamas-national-address-students/

Obama song draws mild rebuke: http://www.msnbc.msn.com/id/33031485/ns/ us_news-education/

LEADER CONTROL

General Jones takes orders from Kissinger: http://www.cfr.org/publication/18515/ remarks_by_national_security_adviser_jones_at_45th_munich_conference_on_ security_policy.html

Nancy Gibbs on Obama's Nobel Prize: http://www.time.com/time/politics/ article/0,8599,1929395,00.html

Hillary Clinton used missile system as incentive: http://www.nepalnews.net/story/465015; http://www.usatoday.com/news/world/2009-03-03-missile_N.htm

Missile defense system scrapped: http://www.nytimes.com/2009/09/18/world/europe/18shield.html?_r=2&hp

A Council Cabinet

Council on Foreign Relations members of Obama's cabinet: Robert Gaylon Ross Sr., *Who's Who of the Elite* (Spicewood, TX: RIE, 1995), pp. 15–89

Trilateralists in Obama's administration and Patrick Wood: http://www.projectcensored.org/top-stories/articles/22-obamas-trilateral-commission-team/

Barry Goldwater on Trilateral Commission: http://buchanan.org/blog/a-chronological-history-of-the-new-world-order-604

A Comfortable Staff

Obama golf with CEO and Birkenfeld's punishment: http://www.indybay.org/newsitems/2009/08/27/18619904.php

Dr. Paul L. Williams lament: http://canadafreepress.com/index.php/article/12652

Michelle Obama's quote: http://www.afro.com/DesktopModules/EngagePublish/printerfriendly.aspx?itemId=1457&PortalId=1&TabId=456

White House staffers and salary: http://www.whitehouse.gov/assets/documents/July1Report-Draft12.pdf

2008 White House staff list: http://www.snopes.com/politics/obama/firstlady.asp; http://www.washingtonpost.com/wp-srv/opinions/graphics/2008stafflistsalary_title.html

Helping Hamas Terrorists

Presidential Determination: http://edocket.access.gpo.gov/2009/E9-2488.htm

Mu'ammar Gadhafi: http://www.politicsforum.org/forum/viewtopic.php?t=96914

Obama's "Civilian Army"

Civilian security force and word softening: http://www.wnd.com/index.php?fa=PAGE.view&pageId=80539

ACORN and SEIU

David Brown describes attack on Kenneth Gladney: http://www.americanthinker.com/blog/2009/08/will_obama_condemn_racist_unio.html

ACORN videos and backlash: http://online.wsj.com/article/SB125271412822705239.html?mod=googlenews_wsj

ACORN blocked for HUD grants: http://www.breitbart.com/article.php?id=D9ANCH580&show_article=1

This corrupt organization: http://news.yahoo.com/s/ap/20090917/ap_on_go_co/us_congress_acorn

TIPS and Other Snoops

Cuba's Committees for the Defense of the Revolution: Isabel Garcia-Zarza, "Big Brother at 40: Cuba's revolutionary neighborhood watch system," Reuters (October 12, 2000)

ACLU opposition to TIPS: Randolph E. Schmidt, "Postal Service Won't Join TIPS Program," Associated Press (July 17, 2002)

John Whitehead on government snoops: http://www.issues-views.com/index.php?print=1&article=23040

Tom Ridge did not want Americans spying on Americans: http://www.foxnews.com/story/0,2933,57874,00.html

TIPS website changes: http://www.thememoryhole.org/policestate/tips-changes.htm

Police chiefs endorse iWATCH program: http://www.usatoday.com/news/topstories/2009-10-03-197785316_x.htm

Asset Forfeiture Fund

Informants pay, regulation, and AFF: Dennis G. Fitzgerald, *Informants and Undercover Investigations: A Practical Guide to Law, Policy, and Procedure* (Boca Raton, FL: CRC Press, 2007), p. 64

Representatives Henry Hyde and Bob Barr on property seizure: http://www.law.cornell.edu/background/forfeiture/

Chemtrails

John Holdren on shooting pollutants into the atmosphere: http://www.nypost.com/p/news/politics/bam_man_cool_idea_block_sun_2Opipflho393Yi7gYoJLXP

Paul Crutzen on sending 747s: http://www.prisonplanet.com/secret-geo-engineering-projects-threaten-unknown-environmental-dangers.html; also see http://www.omega432.com/scalar.html

Global Swarming

Al Gore's energy investments: http://www.capitalresearch.org/pubs/pdf/v1185 475433.pdf

TCPR on Gore home's electricity use: http://www.chattanoogan.com/articles/article_114979.asp

Al Gore equates warming critics to con man Bernie Madoff: http://congress.blogs.foxnews.com/tag/al-gore/

Sam Kazman and destruction of CRU's temperature data: http://cei.org/news-release/2009/10/05/govt-funded-research-unit-destroyed-original-climate-data

Alfred Lambremont Webre on dire news: http://www.examiner.com/examiner/x-2912-Seattle-Exopolitics-Examiner~y2009m5d23-Solar-cycle-24-solar-flares—social-collapse-or-crushing-cold-temperatures-and-global-famine

Human expiration declared health hazard: http://www.cleveland.com/business/index.ssf/2009/12/epa_declares_carbon_dioxide_a.html

Richard S. Lindzen said CO_2 is not a pollutant: http://www.populartechnology.net/2008/11/carbon-dioxide-co2-is-not-pollution.html

Nancy Pelosi on transformational legislation: http://www.politico.com/news/stories/0609/24232.html

Obama talks to NH students: http://www.barackobama.com/2007/04/20/barack_obama_unveils_initiativ.php

A POLICE STATE

Model State Emergency Health Powers Act

Model State Emergency Health Powers Act: http://www.usatoday.com/news/healthscience/2002-07-22-states-healthlaw_x.htm

Military-grade anthrax: http://www.nytimes.com/2001/12/03/national/03POWD.html

Government Camps

History of repression in emergencies: Pamela Sebastian Ridge and Milo Geyelin,

"Civil Liberties of Ordinary Americans May Erode—Legally—Because of Attacks," *New York Times* (September 17, 2001)

Janet Reno's remarks: Jim Burns, "William Bennett Hopes to Shape Public Opinion of War on Terrorism," Cybercast News Service (March 12, 2002)

Durbin's comments on torture: http://www.freerepublic.com/focus/f-news/1425102/posts

Durbin apologizes: http://www.washingtonpost.com/wp-dyn/content/article/2005/06/21/AR2005062101654.html

Ashcroft a threat to liberties: http://articles.latimes.com/2002/aug/14/opinion/oe-turley14

KBR contract to build facilities: http://www.marketwatch.com story/kbr-awarded-homeland-security-contract-worth-up-to-385m?dateid=38741.5136277662-858254656

Camp FEMA

FEMA plans tent cities: http://archive.newsmax.com/archives/articles/2002/7/14/214727.shtml

Detention camps: author's interview with retired Lieutenant Colonel Craig Roberts, September 22, 2009.

Emergency centers on military installations: http://www.opencongress.org/bill/111-h645/show

Virginia Fair grounds as emergency holding area: http://www2.timesdispatch.com/rtd/news/local/article/MEAD15_20090914-215004/292878/

Army internment resettlement specialist: http://www.goarmy.com/JobDetail.do?id=292

American Police Force

The American Police Force in Hardin, Montana: http://www.kulr8.com/news/local/62465902.html; also see http://www.youtube.com/watch?v=8Y5qL3Vi9H0

American Police Force's "virtual office": http://www.guardian.co.uk/world/feedarticle/8705379

The American Police Force with Serbian logo: http://www.americanpolicegroup.com/index.html

Anonymous Hardin posting: http://disc.yourwebapps.com/discussion.cgi?id=149495;article=126560

Paul Joseph Watson on APF connected to Blackwater: http://www.prisonplanet. com/exposed-american-police-force-is-a-blackwater-front-group.html; http://www. prisonplanet.com/investigation-could-sink-american-police-force.html

Blogger William N. Grigg on paramilitary organization: http://freedominourtime. blogspot.com/2009/09/martial-law-is-their-business-and.html

The Posse Comitatus Act

The Phoenix Program: Douglas Valentine, "US Terrorist Attacks: Homeland Insecurity," *Disinformation* (October 9, 2001); http://old.disinfo.com/archive/ pages/article/id1631/pg1/index.html

The Third Infantry Division's First Brigade Combat Team deployed in USA: http:// www.armytimes.com/news/2008/09/army_homeland_090708w/

Military as panacea for domestic problems: Matthew Carlton Hammond, "The Posse Comitatus Act: A Principle in Need of Renewal," *Washington University Law Quarterly* (Summer 1997)

Attack on Kingsville: http://www.wnd.com/news/article.asp?ARTICLE_ID=16957

101st Airborne in Troy, Tennessee: http://www.clarksvilleonline.com/2009/ 09/23/101st-airborne-soldiers-to-conduct-air-assault-training-into-troy-tn/

Britt Snider and Garden Plot: Ron Ridenhour with Arthur Lubow, "Bringing the War Home," *New Times* (November 28, 1975); http://www.namebase.org/ppost14.html

Cable Splicer and reactions: http://www.namebase.org/ppost14.html

Rex-84 testing military use against the civilian population: http://www.publiceye. org/liberty/fema/Fema_3.html

Diana Reynolds and William French Smith: Ibid.

Strategic Support Branch: http://www.cnn.com/2005/ALLPOLITICS/01/23/ pentagon.intel/

Military recruiters denied, Representative Vitter and Jill Wynns: David Goodman, "No Child Unrecruited," *Mother Jones* (November–December 2002)

Kindergartners disciplined: Editors, "'Gun-Toting' Tot Loses Suspension Suit," Associated Press (May 1, 2002)

Ellen Schrecker and Nadine Strossen: http://www.progressive.org/0901/roth0102. html

Paul Proctor on snooping as un-American: http://www.newswithviews.com/war_ on_terror/war_on_terrorism1.htm

Deficient Border Patrol

Senator John Warner: http://jurist.law.pitt.edu/forum/forumnew62.php

Representative Tom Tancredo and irrelevance of the PCA: James P. Tucker Jr., "Defend US Borders with US Army Troops," *American Free Press* (October 21, 2002)

John Brinkerhoff on irrelevance of PCA: http://www.homelandsecurity.org/journal/Articles/brinkerhoffpossecomitatus.htm

Only 815 of 8,607 miles of U.S. border effectively controlled: http://www.cnsnews.com/news/article/54514

Office of Strategic Influence

The *New York Times* on Pentagon's credibility: http://www.nytimes.com/2002/02/19/international/19PENT.html?pagewanted=all

Office of Strategic Influence closed: Editors, "US Closes 'Disinformation' Unit," BBC News (February 26, 2002)

Henny Penny, the sky is going to fall: http://www.fair.org/index.php?page=1859&printer_friendly=1

Oath Keepers

Oath Keepers: http://oathkeepers.org/oath/

Bob Hanafin on illegality of disobeying orders: http://www.veteranstoday.com/modules.php?name=News&file=article&sid=8752&mode=thread&order=0&thold=0

Patrick M. Fahey and the UCMJ: http://republicdefenders.blogspot.com/2009/10/veterans-today-piece-on-oath-keepers.html

Oath Keepers will not obey these orders: http://oathkeepers.org/oath/2009/10/06/veterans-today-hit-piece-and-an-unofficial-response/

Mark Potok and the SPLC: http://www.lvrj.com/news/oath-keepers-pledges-to-prevent-dictatorship-in-united-states-64690232.html

Hate Crimes

SPLC one of most profitable charities: http://www.americanpatrol.com/SPLC/ChurchofMorrisDees001100.html

Minutemen as resurgent antigovernment patriots: http://www.splcenter.org/blog/index.php?s=minutemen&submit=

Judy Andreas on wildly flailing SPLC: http://www.borderguardians.org/03.html

Joe Solmonese and Jim DeMint on hate crimes prevention act: http://www.southbendtribune.com/article/20091023/News01/910230351/-1/XML

Jeff Sessions on troubling language: Ibid.

DARPA

Christopher H. Pyle on army plan to spy on Americans: http://www.mtholyoke.edu/offices/comm/oped/spying2.shtml

James Bamford on new databases for the NSA: http://www.nybooks.com/articles/23231

Electronic Surveillance

Senator Frank Church: James Bamford, "The Agency That Could Be Big Brother," *New York Times* (December 25, 2005)

Representative Paul on "Know Your Customer": Ron Paul, "Privacy Busters: Big Bank Is Watching," *Ron Paul Newsletter* (December 1998)

Walter Soehnge: Bob Kerr, "Pay Too Much and You Could Raise the Alarm," *Providence Journal* (February 28, 2006)

Representative Butch Otter: http://www.federalobserver.com/archive.php?aid=847

Editorial comment: Editors, "Another Cave-In on the Patriot Act," *New York Times* (February 11, 2006)

Bruce Fein on unchecked abuse: Liz Halloran, "Everyone's Spinning the Spying," *US News & World Report* (February 13, 2006)

ACLU's Nadine Strossen: http://www.ratical.org/ratville/CAH/policeState.html

Act used hundreds of times: Mark Mueller, "To Catch a Monster, Using Anti-Terror Law," *Star-Ledger* (August 14, 2005)

Tucker Bounds comment: http://www.sfgate.com/cgi-bin/article.cgi?file=/c/a/2008/07/10/MN3H11ME7C.DTL

Obama administration supporting the PATRIOT Act: http://www.huffingtonpost.com/2009/09/22/obama-patriot-act-surveil_n_295194.html

Kevin Bankston on Obama administration sounding like Bush administration: http://www.eff.org/press/archives/2009/04/05

Ron Paul quote: http://www.federalobserver.com/archive.php?aid=847

Magic Lantern, Fluent, dTective, and Encase

Key logger used by FBI: http://www.nytimes.com/2001/12/31/technology/ebusiness/31TECH.9.html; Ted Bridis, "Anti-Terror Tools Include High-Tech," Associated Press (October 28, 2001); http://multimedia.belointeractive.com/attack/response/1028tech.html

William Newman: Nat Hentoff, "The Sons and Daughters of Liberty," *Village Voice* (June 21, 2002)

Appeals court approves no warrant surveillance: http://www.wired.com/politics/law/news/2007/07/fbi_spyware?currentPage=all

Fluent and Oasis programs: http://www.theregister.co.uk/2001/03/06/cia_patching_echelon_shortcomings/

Video improvement with dTective: http://www.oceansystems.com/dtective/

Encase to recover computer disks: http://www.encaseenterprise.com/support/articles/restore.aspx

Russ Kick's worst-case scenario: http://www.villagevoice.com/2001-02-20/news/gotcha/

National ID Act

National ID system: http://www.federalobserver.com/archive.php?aid=4309

Janet Napolitano on repeal of the REAL ID Act: http://www.cnn.com/2009/POLITICS/04/22/real.ID.debate/

A Chip in Your Shoulder

NYPD chips: http://policechiefmagazine.org/magazine/index.cfm?fuseaction=display_arch&article_id=127&issue_id=102003

National ID system: http://www.time.com/time/nation/article/0,8599,191857,00.html

Comprehensive national ID system: http://www.lewrockwell.com/yates/yates64.html

ChipMobile: http://www.hoise.com/vmw/02/articles/vmw/LV-VM-06-02-6.html

CityWatcher.com and Tommy Thompson: http://www.wnd.com/news/article.asp?ARTICLE_ID=48760

VeriChip stock triple with implantable microchips for swine flu: http://www.reuters.com/article/hotStocksNews/idUSTRE58K4BZ20090921

Novartis and microchipped pill: http://www.dailymail.co.uk/health/article-1215200/Forgetful-patients-fitted-microchips-remind-pills.html#ixzz0V3lb7dOd

Sheriff Don Eslinger and GPS tracking: http://www.wired.com/politics/security/news/2002/10/55740

Texas representative Larry Phillips: http://www.engadget.com/2005/04/07/texas-state-representative-wants-transponders-in-all-cars/

Siemens tracking device and teens: http://www.nytimes.com/2001/05/24/living/24QUEE.html?pagewanted=1; also see http://www.digitalangel.com/

Court rules GPS tracking legal: http://www.boston.com/news/local/massachusetts/articles/2009/09/18/sjc_oks_secret_use_of_gps_devices/

Echelon and TEMPEST

Jim Wilson on Echelon and TEMPEST: http://www.popularmechanics.com/science/defense/1281281.html

New Cyberspace security plan: Ted Bridis, "US Considers Cybersecurity Plan," Associated Press (September 7, 2002)

The Cybersecurity Act of 2009

Leslie Harris on cybersecurity threat: http://www.cdt.org/headlines/1196

Larry Selzer on direct control of the president: http://www.wnd.com/index.php?fa=PAGE.view&pageId=93966

Jennifer Granick on loss of privacy: http://www.motherjones.com/politics/2009/04/should-obama-control-internet

Homeland Security

Bush's urgent need: http://news.bbc.co.uk/2/hi/americas/2031255.stm

Bush quote on Homeland Security: Editors, "Bush Signs Homeland Security Bill," CNN News (November 25, 2002)

James Joyner on dispersion of Homeland Security: http://www.outsidethebeltway.com/archives/dhs_new_hq_in_lunatic_asylum/

Obama releases some Reagan records: http://www.allgov.com/ViewNews/Obama_Opens_Some_Reagan_Records_Kept_Secret_by_Bush_90413

Obama blocks release of White House visitor list: http://www.msnbc.msn.com/id/31373407/ns/politics-white_house/

Oliver North's martial law plan: Alfonso Chardy, "Plan Called for Martial Law in US," Knight-Ridder News Service (July 5, 1987); also see http://www.bcrevolution.ca/fema_secrets.htm

John Dean's concern: http://www.bcrevolution.ca/fema_secrets.htm

Timothy H. Edgar and ACLU objections to Homeland Security: http://www.aclu.org/natsec/emergpowers/14418leg20020625.html

No-Fly List

Timothy Sparapani and Alberto Gonzales: Walter Pincus and Dan Eggen, "325,000 Names on Terrorism List," *Washington Post* (February 15, 2006)

Senator Kennedy and ACLU counsel Shuford: http://www.washingtonpost.com/ac2/wp-dyn/A17073-2004Aug19?language=printer

Arizona treasurer Dean Martin: http://www.theregister.co.uk/2009/07/17/arizona_treasurer_dean_martin_nofly_list/

Dr. Robert Johnson: www.thenation.com/blogs/thebeat?pid=63406

Michael Chertoff working for body scanner manufacturer: http://www.pbs.org/ombudsman/2010/01/scanning_the_source_1.html

Kate Hanni critizes Chertoff: http://www.boston.com/news/nation/washington/articles/2010/01/02/group_slams_chertoff_on_scanner_promotion/

Ben Wallace on scanners not picking up plastic, chemicals, and liquids: http://www.independent.co.uk/news/uk/home-news/are-planned-airport-scanners-just-a-scam-1856175.html

Peeping TSA

Boian Alexandrov and team on DNA damage: http://arxiv.org/abs/0910.5294

Evidence of damage hard to find: http://republicbroadcasting.org/?p=6086

Office of Transport Security manager Cheryl Johnson: http://www.prisonplanet.com/admitted-airport-body-scanners-provide-crisp-image-of-your-genitals.html

Virtual strip-searching: http://www.guardian.co.uk/uk/2010/jan/07/full-body-scan-uk-airport

Paul Joseph Watson on high definition and inverted images: http://www.infowars.com/inverted-body-scanner-image-shows-naked-body-in-full-living-color/

Marc Rotenberg on potential for misuse of scanning machines: http://cnn.org/2010/TRAVEL/01/11/body.scanners/index.html

Security Abuses

Robert Lee Lewis: author's interviews, summer 1999

Library incident: http://www.washingtonpost.com/wp-dyn/content/article/2006/02/16/AR2006021602066.html

Rebecca Soloman and bag of white powder: Editors, "Student Falls Victim to TSA Worker's Prank," *Austin American-Statesman* (January 24, 2010)

Photographers Under Fire

Mike Maginnis arrested: http://vigilant.tv/article/2528/photographer-arrested-camera-confiscated-for-taking-snaps-of-hotel

Photography under attack: http://photographernotaterrorist.org/events/

David Proeber and memory card: http://carlosmiller.com/2009/02/26/illinois-police-had-confiscated-dramatic-photos-of-gun-wielding-man/

Carlos Miller: http://carlosmiller.com/

SPJ president Clint Brewer: http://www.spj.org/news.asp?REF=812#812

Laura Sennett and DOJ raid: http://www.courthousenews.com/2009/09/25/Photog_Sues_Feds_for_Heavy-Handed_Raid.htm

You're on Camera

Vehicle surveillance in Medina, Washington: http://seattletimes.nwsource.com/html/localnews/2009873854_medina16m.html

D.C. mayor Anthony A. Williams: http://goliath.ecnext.com/coms2/gi_0199-1508565/Mayor-cites-need-for-surveillance.html

Internet Eyes instant event notification system: http://interneteyes.co.uk/

PART IV—HOW TO FREE ZOMBIES:
THE THREE BOXES OF FREEDOM

Andrew Gavin Marshall on listening to people who have been nothing but wrong: http://www.globalresearch.ca/index.php?context=va&aid=15501

THE SOAP BOX

An Unfettered News Media

Research on mass media including the Internet: http://usa.usembassy.de/media.htm

Iraq as unmitigated disaster: http://legacy.signonsandiego.com/uniontrib/20060303/news_lz1e3madsen.html

Firms with Pentagon contracts: http://www.usatoday.com/news/washington/2005-12-13-propaganda-inside-usat_x.htm

Wayne Madsen on independent reporting: http://www.signonsandiego.com/uniontrib/20060303/news_lz1e3madsen.html

Larry Siems and PEN court challenge to FISA Amendments Act: http://www.huffingtonpost.com/larry-siems/why-were-challenging-the_b_242843.html

Principles for an unfettered and free press: http://www.newswatch.in/newspaedia/395

Phil Donahue on support for journalists: http://mediachannel.org/wordpress/2007/01/26/no-substitute-for-free-and-unfettered-news-gathering/

StopBigMedia.com Coalition: http://www.stopbigmedia.com/=about

Robert McChesney on I. F. Stone and new Internet journalists: http://www.socialistproject.ca/bullet/246.php

Back-to-Basics Education

Mark Taylor on the essence of education: Mark Taylor e-mail to author, October 7, 2009

Homeschooling

Jon Reider and the NCHE: http://www.hslda.org/docs/nche/000002/00000234.asp

Isabel Shaw and Kelley Hayden: http://school.familyeducation.com/home-schooling/college-prep/41108.html

Hal Young on colleges pursing homeschoolers: http://davidnbass.com/2007/04/03/colleges-courting-homeschoolers-self-discipline-work-ethic-and-morals-catching-eye-of-recruiters/

Chris Klicka on missing out on crime: http://www.hslda.org/docs/nche/000000/00000068.asp

Online education grows with homeschooling: http://online.wsj.com/article/SB125374569191035579.html

Matthew Ladner on alternatives to traditional public education: http://www.ocpathink.org/publications/perspective-archives/september-2009-volume-16-number-9/?module=perspective&id=2321

Caring for Health

Dr. Len Saputo on disease care: Len Saputo, MD, with Byron Belitsos, *A Return to Healing: Radical Health Care Reform and the Future of Medicine* (San Rafael, CA: Origin Press, 2009), p. xxv

Saputo on allowing natural processes to take charge: Ibid.

Dr. Saputo on the integral-health medicine of the future: Ibid., p. 242

National Health Interview Survey: http://www.naturalnews.com/027291_health_medicine_biofeedback.html

Dr. Oliver Fein and doctors' letter to candidates: http://www.pnhp.org/news/2008/october/doctors_to_candidate.php

U.S. life expectancy: https://www.cia.gov/library/publications/the-world-factbook/rankorder/2102rank.html

U.S. child mortality: https://www.cia.gov/library/publications/the-world-factbook/rankorder/2091rank.html

Single-payer financing as the only way to recapture money: http://www.pnhp.org/facts/single_payer_resources.php

John C. Goodman on patient control over their money: http://www.hillsdale.edu/news/imprimis/archive/issue.asp?year=2009&month=03

Dr. John Geyman on increasing health insurance premiums: http://www.pnhp.org/news/2008/june/doctors_to_join_nati.php

THE BALLOT BOX

Michel Chossudovsky on overhaul of monetary system: http://www.globalresearch.ca/index.php?context=va&aid=12517

William K. Black remedies: http://www.pbs.org/moyers/journal/04032009/watch.html

Dr. Charles K. Rowley on prosperity and full employment: http://www.telegraph.co.uk/finance/comment/6146873/Adam-Smith-would-not-be-optimistic-in-todays-economic-world.html

Audit the Fed

The ambiguity must cease: http://www.monetary.org/federalreserveprivate.htm

Three courses for monetary reform: http://www.monetary.org/need_for_monetary_reform.html

Walter Burien on CAFRs: http://cafr1.com/

Hens, foxes, and the critical element for success: http://www.newsmakingnews.com/contents7,5,00.htm

Corrective measures and cooperation: http://cafr1.com/Revolution.html

Bruce Wiseman on congressional control over FSB: http://canadafreepress.com/index.php/article/10954

Toby Birch and the Guernsey experience: http://goldnews.bullionvault.com/guersney_experiment_credit_creation_gold_standard_051920083

BerkShares as local money: http://www.berkshares.org/whatareberkshares.htm

Fire Congress

Wright Patman on congressional blame: http://newswithviews.com/Devvy/kidd122.htm

Rasmussen Reports on low public opinion of Congress: http://www.rasmussenreports.com/public_content/business/general_business/september_2009/americans_now_view_congress_as_least_respected_job

Representatives have sold us out: http://www.kickthemallout.com/

Uncle Sam posters: http://www.kickthemallout.com/article.php/Story-Free_Uncle_Sam_Poster

Get Out of Our House: http://abclocal.go.com/ktrk/video?id=7147953

Three million pink slips to Congress: http://www.wnd.com/index.php?fa=PAGE.view&pageId=112847

Fire Congress website: http://www.firecongress.org/

De facto term limits on Congress: http://firecongress.meetup.com/

U.S. term limits: http://www.termlimits.org/content.asp?pl=2&contentid=2

Billboards on firing Congress: http://www.weshouldfirecongress.com/#/about-us/4534589659

Paul Volcker on controlling both parties: http://www.government-propaganda.com/george-green.html

Harvey Wasserman on turning from both Republican and Democratic parties: http://www.opednews.com/articles/Part-Two-Talking-with-Har-by-Joan-Brunwasser-091006-494.html

Poll Watchers and Paper Ballots

Average voter turnout in other nations: http://www.nonprofitvote.org/voterturnout2008

Ralph Nader on serious undermining effects: http://archives.cnn.com/2000/ALLPOLITICS/stories/10/31/lkl.nader/index.html

Bev Harris on vote-counting secrecy: http://www.scoop.co.nz/stories/HL0309/S00150.htm

Diebold abused copyrights: http://www.usatoday.com/tech/techinvestor/earnings/2004-10-11-diebold_x.htm

Miami Herald editorial: http://www.miamiherald.com/opinion/editorials/story/1258667.html

Harvey Wasserman on need for paper ballots: http://www.opednews.com/articles/Part-Two-Talking-with-Har-by-Joan-Brunwasser-091006-494.html

Diebold software decertified in 2009: http://www.sos.ca.gov/elections/voting_systems/premier/premier-11819-withdrawal-approval033009.pdf; also see http://www.sos.ca.gov/elections/elections_vsr.htm

Kim Alexander on worst fears confirmed: http://www.cfvi.us/?q=node/65

Paper ballot plans fails in the House: http://fcw.com/articles/2008/07/21/house-defeats-paper-ballot-funding.aspx

Enforce the Tenth Amendment

Thomas J. DiLorenzo and on federal powers: http://www.perspectives.com/forums/view_topic.php?id=214002&forum_id=4&jump_to=4403721

Charles Key and a definite follow-up: http://www.tenthamendmentcenter.com/2009/09/29/ohio-senate-affirms-state-sovereignty/

San Francisco Examiner editorial: http://www.sfexaminer.com/opinion/Examiner-Editorial-States-reassert-sovereignty-with-legislation-59954857.html

Virginia prohibits mandatory federal health care and Senator Frederick M. Quayle: http://www.businessweek.com/ap/financialnews/D9DJJV500.htm

Michigan solar manufacturing plant: http://www.connectmidmichigan.com/news/story.aspx?id=359746

Dow Chemical solar shingles: http://news.dow.com/dow_news/corporate/2009/20091005b.htm

Thirty-six Remedies for a Broken Society

Ron Paul on WTO: http://www.house.gov/paul/tst/tst2002/tst012102.htm

National Commission on Marihuana: http://proxy.baremetal.com/csdp.org/research/shafernixon.pdf

FEAR on asset forfeiture: http://www.fear.org/

Defeat Fascism

Plan to defeat fascism: http://freepeopleontheland.wordpress.com/about-fascism/how-to-defeat-fascism/

Chris Martenson's lifestyle change: http://www.chrismartenson.com/about

Officer challenges Obama's qualifications: http://www.military-money-matters.com/soldier-challenges-barack-obama.html

Others challenge Obama: http://www.wnd.com/index.php?fa=PAGE.view &page Id=90574

THE AMMO BOX

One firearm for every adult American: http://injuryprevention.bmj.com/cgi/content/full/13/1/15

IRS purchasing 12-gauge shotguns: https://www.fbo.gov/index?s=opportuni ty&mode=form&id=8d3b076bd4de14bbda5aba699e80621d&tab=core&_ cview=1&cck=1&au=&ck=

Aleksandr Solzhenitsyn on halting the cursed machine: http://www.lewrockwell.com/gaddy/gaddy53.html

Enough weaponry to arm both the Chinese and Indian armies: http://www.ammoland.com/2009/04/27/update-usa-buys-enough-guns-in-3-months-to-outfit-the-entire-chinese-and-indian-army/

Gun and Ammo Sales Booming

Obama on belief in Second Amendment: http://www.youtube.com/watch?v =kBHkMADXnOw

Brad DeSaye and his unprecedented business: http://www.amconmag.com/article/2009/may/18/00024/

Roy Eicher on fear of no ammo: http://www.local12.com/news/local/story/Gun-

Ammo-Sales-Remain-Strong-Despite-Economy/LpuDiNz5XUi-njCfQpZn1Q.
cspx?rss=30

John Woniewski on whether gun and ammo demand would continue: http://www.
appeal-democrat.com/articles/ammo-76288-detroit-gun.html

Sheriff Darren White on his ammo back order: http://www.newwest.net/topic/
article/please_save_some_bullets_for_the_cops/C530/L37/

Curtis Shipley and rescinded Pentagon directive: http://www.cnsnews.com/public/
content/article.aspx?RsrcID=47112

INDEX

drug advertising, 137–43, 376
drug corporations. *See* Big Pharm
Drug Enforcement Agency (DEA), 151, 188, 244–45, 362
drugging the population, 132–52
 aspartame, 143–48
 Big Pharm, 133–38
 children and women, 148–52, 162–66
 DTC advertising, 137–43
drug patents, 135–37
drug vaccines, 163–65
dTective, 289
Dulles, Allen, 15, 107–8
Dulles, John Foster, 15, 108
dumbed-down education. *See* education
Dumbest Generation, The (Bauerlein), 193, 194–95
Dumbing Us Down (Gatto), 195, 196
Durbin, Richard J., 254–55
Dutcher, Brandon, 191

Eakman, Beverly K., 156–57, 202, 203, 205
Easterling, Lloyd, 272
Easterling, Scott, 380
Echelon, 297–99
E. coli (Escherichia coli), 85
economic crisis of 2008. *See* financial crisis of 2008
economic decline of America, 9–11
economic stimulus package of 2009, 35–37
Ecoscience (Ehrlich), 120
Edgar, Timothy H., 305–6, 307
education, 190–207
 back-to-basics, 329–30, 375–76
 dangerous teaching, 195–96
 history in America, 197–206
 homeschooling, 331–34
 Oklahoma school study, 191–93
 remedies for, 329–30, 375–76
 Twixters, 206–7
 video generation, 193–95
Eglin Air Force Base, 257
Ehrlich, Paul and Anne, 120
Eicher, Roy, 383–84
Election Systems & Software (ES&S), 357–58
Electronic Frontier Foundation (EFF), 287–88, 300

Electronic Privacy Information Center (EPIC), 311
electronic surveillance, 282–90
electronic voting machines, 356–59
electroshock therapy, 154
Eli Lilly and Company, 168
Elliott, Douglas, 41
Elmendorf Air Force Base, 257–58
Elsas, Louis J., 144
Emanuel, Rahm, 238–39
Emory University, 168
Enabling Act of 1933, 215–16
Encase, 289
Encyclopaedia Britannica, Inc., 198–99
Enemy Belligerent Interrogation, Detention, and Prosecution Act of 2010, 218–19
"enemy combatants," 218–19, 255
Energy Index Point Score, 118
Engdahl, F. William, 39–40, 41
English language, 375
Enron, 27, 28
Environmental Protection Agency (EPA), 89, 249, 250
Epstein-Barr virus, 117–18
Ervin, Sam, 267–68, 280
Eslinger, Don, 295
Esquivel, Phil, 266
eugenics, 119, 120, 157–61
European Central Bank, 77
Evensen, Greg, 220–21
Ewen, Stanley, 94
Ewing, Oscar, 102–3
eWorldtrack, 296
executive salaries and bonuses, 30, 36–37, 71

Fahey, Patrick M., 275
faith, money for debt and, 56–57
Fannie Mae, 36, 37–39, 51, 366
Farah, Joseph, 352
Farmer, John, 370
farm subsidies, 374
fascism, 377–81
 defined, 4
fearmongering
 media control and, 213–15
 police tactics and FEMA, 220–22

Northwest Flight 253, 308–9
Norton, Charles D., 64
Novartis, 133, 143, 170, 173, 184–86, 294
NutraSweet, 145–46

Oath Keepers, 274–76
Obama, Barack
  avian flu pandemic and, 182
  bank cover-up and, 32
  Big Pharm and, 147
  birth certificate of, 304, 380–81
  campaign promises of, 147, 229–31, 304,
    350–51
  "civilian army" of, 238–39
  corporate buddies of, 233–34
  economic stimulus package and, 17, 35–37
  electronic surveillance and, 286–88
  fearmongering of, 221–22
  FSB approval and, 70–71, 345–46
  global warming and, 251
  Hamas terrorists and, 237
  health-care plan of, 105–6, 120, 180, 239,
    336
  Marxist Socialism of, 3, 18
  right to bear arms and, 383
  school talk of, 225–28
  swine flu emergency and, 172, 175
  tea parties and, 13–14, 225
  Trilateralists in administration, 231–33
  U.S. debt and Chinese, 23–24
Obama, Michelle, staff of, 234–37
"Obama: Trilateral Commission Endgame"
  (Wood), 232
Office of Information Awareness (OIA),
  279–80
Office of Management and Budget (OMB),
  34
Office of Strategic Influence (OSI), 273–74
Office of Strategic Services (OSS), 108
O'Keefe, James, 240–41
Oklahoma, school study, 191–93
O'Meara, Kelly Patricia, 151, 218, 286
160th Special Operations Aviation
  Regiment (Night Stalkers), 266–67
100,000,000 Guinea Pigs (Kallet and
  Schlink), 88
online schooling, 333–34

Operation Cable Splicer, 267–68
Operation Cure All, 130
Operation Diomedes, 267
Operation Garden Plot, 267–68
Operation Last Dance, 266–67
Operation Paperclip, 100, 108–11, 113–14,
  149
Operation Shamrock, 298
Operation TIPS, 242–44
Orwell, George, 243, 301
osteosarcoma, 102
Oswald, Lee Harvey, 370
Ott, A. True, 177–78
Otter, C. L. "Butch," 284–85
outsourcing policies, 372
overpopulation, 119–20, 124–26
over-the-counter (OTC) derivatives. *See*
  derivatives
Overtreated (Brownlee), 104
Owens, Wilbur, 12
Oz, Mehmet, 167

Pakistan, 222
Palestinians, 237
Pan Am Flight 103, 312
Pandremix, 172
paper ballots, 355–59
Parks, Lawrence, 52–53
Parry, Nigel, 225
Patman, Wright, 350
PATRIOT Act, 215–19, 282–86, 370
Patterson, Robert P., 113
Patton State Hospital, 159
Paul, Ron, 76–79, 216, 217–18, 284, 288,
  371–72
Paulson, Henry "Hank," 29–33, 70
Pavlov, Ivan, 156
Paxil, 139, 150, 152, 168
Peace Corps, 238
Peerwani, Nizam, 220
Pellegrini, Frank, 290–91, 292
Pelosi, Nancy, 250
PEN American Center, 325
Pennock, Paul, 179
Pension Benefit Guaranty Corp. (PBGC), 37
pentaerythritol tetranitrate (PETN),
  309–10